Agricultural Animal Physiology and Morphology

Agricultural Animal Physiology and Morphology

Edited by **Mia Steers**

SYRAWOOD
PUBLISHING HOUSE

New York

Published by Syrawood Publishing House,
750 Third Avenue, 9th Floor,
New York, NY 10017, USA
www.syrawoodpublishinghouse.com

Agricultural Animal Physiology and Morphology
Edited by Mia Steers

© 2016 Syrawood Publishing House

International Standard Book Number: 978-1-68286-162-2 (Hardback)

Printed in the United States of America.

Contents

Permissions

List of Contributors

Preface

This book aims to highlight the current researches and provides a platform to further the scope of innovations in this area. This book is a product of the combined efforts of many researchers and scientists from different parts of the world. The objective of this book is to provide the readers with the latest information in the field.

Agricultural animal physiology focuses on the study of economically significant animals such as cows, camels, chickens, etc. The topics included in this book on animal physiology are of utmost significance and are bound to provide incredible insights to readers. Reproductive performance, immunology, morphology, growth, genetics, etc. are some of the topics that have been covered in this book. It includes contributions of experts and scientists which will provide advanced knowledge into this field. This text is meant for all those who are looking for an elaborate reference on agricultural animal physiology.

I would like to express my sincere thanks to the authors for their dedicated efforts in the completion of this book. I acknowledge the efforts of the publisher for providing constant support. Lastly, I would like to thank my family for their support in all academic endeavors.

Editor

Efficacy of immunization with the recombinant collagen adhesin A region against *Staphylococcus aureus*–induced mouse mastitis

Changmin Hu and Aizhen Guo*

Faculty of Animal Science and Veterinary Medicine, Huazhong Agricultural University, Wuhan 430070, China.

Staphylococcus aureus (*S. aureus*) is a major cause of bovine mastitis. In this present study, we assessed the efficacy of the recombinant A region of collagen adhesin (rCna-A) as a mastitis vaccine in a mouse model. Three groups of mice were immunized with either rCna-A, inactivated *S. aureus*, or phosphate-buffered saline (PBS) as a control. IgG and IgG subtype titers of mice in the rCna-A vaccine group were significantly higher than those in the killed vaccine group (P<0.01). Immunized lactating mice were challenged with *S. aureus* via the intramammary route. Significantly, fewer bacteria were recovered from mice in the rCna-A group than from those in the killed vaccine group (P<0.001). Histopathology indicated that the mammary structure showed greater integrity and a milder inflammatory response in the rCna-A group as compared with both the inactivated vaccine and PBS control groups. These results suggested that the rCna-A protein may be an attractive target for a vaccine against *S. aureus*-induced mastitis.

Key words: *Staphylococcus aureus,* collagen adhesion, inactivated vaccine, mastitis.

INTRODUCTION

Mastitis is recognized as one of the most costly diseases affecting the dairy industry (Miller et al., 1993; Xu et al., 2011). *Staphylococcus aureus* (*S. aureus*) is a major bovine mastitis pathogen, and causes nearly 50% of bovine mastitis (Hynes, 1992; Haas and Plow, 1994). Because of the dramatic increase in antibiotic multi-resistant staphylococci and the slow development of new antibacterial agents, resolution of *S. aureus*-induced mastitis remains as a major problem to control. Vacci-nation is a logical approach to the control of infectious diseases in food production animals (Nilsson et al., 1998; Leitner et al., 2003). Over the past several decades live, heat-killed, and formalin-inactivated preparations of *S. aureus* cells or toxins have been tested as vaccines to prevent staphylococcal infections (Gong et al., 2010). However, these attempts to vaccinate against *S. aureus* infections have most often been unsuccessful (Flock, 1999; Hu et al., 2010).

Microbial adhesion to host tissue is the initial critical event in the pathogenesis of *S. aureus* infections (Patti et al., 1994a), and immunotherapies designed to inhibit the adhesion of pathogenic microorganisms appear to be an effective way to prevent or minimize the severity of an infection (Talbot and Lacasse, 2005). Collagen adhesin (Cna) has been reported to be the major adhesin responsible for high affinity collagen binding and is a major virulence factor in *S. aureus* infection (Holderbaum et al., 1986; Gillaspy et al., 1998). Cna was shown to be involved in the pathogenesis of *S. aureus* infection of the cornea in rabbit (Rhem et al., 2000), osteomyelitis (Elasri et al., 2002) and septic arthritis in mice, respectively (Patti et al., 1994b). Moreover, passive immunization with antibodies against Cna also showed promising protection

*Corresponding author. E-mail: azg305@yahoo.cn.

in a mouse model of sepsis (Flock, 1999).

S. aureus Cna consists of an N-terminal signal peptide, a non-repetitive A region, and one to four repeating units, followed by a cell wall anchor region, a transmembrane segment, and a short, positively-charged cytoplasmic tail. The A region of Cna was found to be fully responsible for the collagen binding activity of S. aureus (Rich et al., 1998). Nilsson et al. (1998) reported that mice vaccinated with the collagen binding portion of Cna were partially protected from septic death. Molecular characterization of the Cna gene has shown that the A region is conserved among clinical S. aureus isolates. It is thus an attractive target for development of Cna-A vaccines against S. aureus mastitis. However, there is at present no reports on the efficacy of Cna-A as a vaccine against S. aureus mastitis. In this present study, rCna-A was expressed and subunit vaccines were prepared and evaluated in a mouse mastitis model. The mechanism of protection against S. aureus was further investigated.

MATERIALS AND METHODS

Bacterial strains, plasmids and culture conditions

Escherichia coli strain DH5α, grown in Luria-Bertani broth or agar (Difco), was used for cloning plasmids, and pET-28a (+) was used as an expression vector. E. coli BL21-CodonPlus (DE3) was used as a host for protein production. S. aureus strain J9 was isolated from bovine mastitis in China, and grown in either Tryptic Soy Broth or Agar (BD Difco, Sparks, MD, USA).

Cloning of Cna-A

Genomic DNA was extracted from S. aureus strain J9 using an extraction kit (Clontech, Palo Alto, CA, USA). Extracted DNA was then used as the template for polymerase chain reaction (PCR). The gene encoding Cna-A was amplified using the forward primer Can-A-Fwd (5'- TATGGATCCGTAGCTGCAGATGCACC-3'), which includes a BamHI restriction site (underlined), and the reverse primer Cna-A-Rev (5'-CGCCTCGAGCTCTGGAATTGGTTCAATTTC-3'), which includes a XhoI restriction site. PCR was conducted under the following conditions: an initial 5 min denaturation at 94°C followed by 30 cycles of 30 s at 94°C, 45 s at 64°C, and 2 min at 72°C with a final extension at 72°C for 10 min. PCR product was purified with a Qiagen PCR Product Purification kit (Qiagen, Chatsworth, CA) as described by the manufacturer, and the rCna-A was subsequently cloned into pET-28a (+) digested with BamHI and XhoI to generate the recombinant plasmid pET-28a-Cna-A. Restriction digestion, ligation, and transformation of DNA into competent E. coli cells were performed using standard methods (Sambrook and Rusell, 2001) or following the manufacturer's instructions.

Expression and purification of rCna-A

Recombinant Can-A protein expression was induced by the addition of 1 mM isopropy-β-D-thiogalactoside (IPTG), and the cells were harvested 4 h post-induction. The cells were then collected by centrifugation at $10,800 \times g$ for 10 min at 4°C, washed in 0.1 M PBS (pH 7.2), and disrupted by sonication. rCna-A was purified by affinity chromatography using nickel-nitrilo triacetic acid resin (Ni-NTA, Qiagen) according to the manufacturer's specifications.

Western blot analysis

The protein samples were separated by sodium dodecyl sulfate polyacrylamide gel electrophoresis (SDS-PAGE), transferred onto Immobilon-P polyvinylidene difluoride membranes and blocked with 3% (w/v) bovine serum albumin (BSA) in phosphate-buffered saline (PBS) for 1 h at 25°C. The membranes were washed with PBS–Tween 20 (PBST) three times and further incubated for 4 h with rabbit anti-S. aureus antibody prepared in our laboratory. After extensive washing, bound antibody was detected using horseradish peroxidase (HRP)-conjugated goat anti-rabbit IgG antibody (Southern Biotech, Birmingham, AL, USA). The membranes were washed with PBST, followed by washing with 0.1 M sodium acetate for 5 min. Finally, the reaction on the membranes was developed with 3, 3'-diaminobenzidine in 0.1 M sodium acetate buffer containing hydrogen peroxide.

Vaccine preparation

In order to prepare the protein vaccines, 40 µg of purified rCna-A was dissolved with 100 µl of PBS and emulsified with an oil adjuvant. S. aureus strain J9 was cultured and adjusted to 1×10^6 CFU/ml, and then inactivated at 37°C for 24 h by the addition of formaldehyde. After confirmation that no bacteria had survived by culture assay, the bacteria were emulsified at a 1:1 ratio with oil adjuvant to prepare the bacterial vaccine.

Mouse immunization

Forty-eight specific pathogen free (SPF) BALB/c mice (12 male, 36 female, age 7 to 9 weeks) were maintained and handled according to the Regulations for the Administration of Affairs Concerning Experimental Animals. The female mice were randomly divided into three groups of twelve, one of which was injected intraperitoneally with rCna-A protein vaccine. Another group was immunized with inactivated S. aureus. Aliquots of 100 µl of each vaccine were injected intraperitoneally into mice. Booster injections were performed two weeks later. The control group was injected with PBS following the same protocol.

IgG detection

Enzyme-linked immunosorbent assays (ELISAs) were used to determine the IgG titers against rCna-A according to the method of Gong et al. (2010). Polystyrene Maxisorp 96-well plates (Nalgene Nunc International Corp., Rochester, NY, USA) were coated overnight at 4°C with 0.1 µg rCna-A per well. Following saturation of the plates with a skim milk solution (5% w/v) overnight at 4°C, the serum samples were added to the plate and incubated for 45 min at 37°C. HRP-conjugated goat anti-mouse IgG antibodies were then added and incubated for 30 min at 37°C. Chromogenic substrate solution (100 µl; 42 mM tetramethylbenzidine and 0.01% hydrogen peroxide) was added for 10 min and the enzymatic reaction was stopped by the addition of 50 µl hydrofluoric acid (HF). Three washes with PBS-0.05% Tween-20 were performed between each step. The optical density (OD) was read on a plate reader (BioTek Instruments, Winooski, VT, USA) at 630 nm. Endpoint titers were calculated as the reciprocal of the last serum dilution that gave a value twofold higher than PBS control group.

The method used to determine antibodies titers against S. aureus was the same as that described above for rCna-A except that the

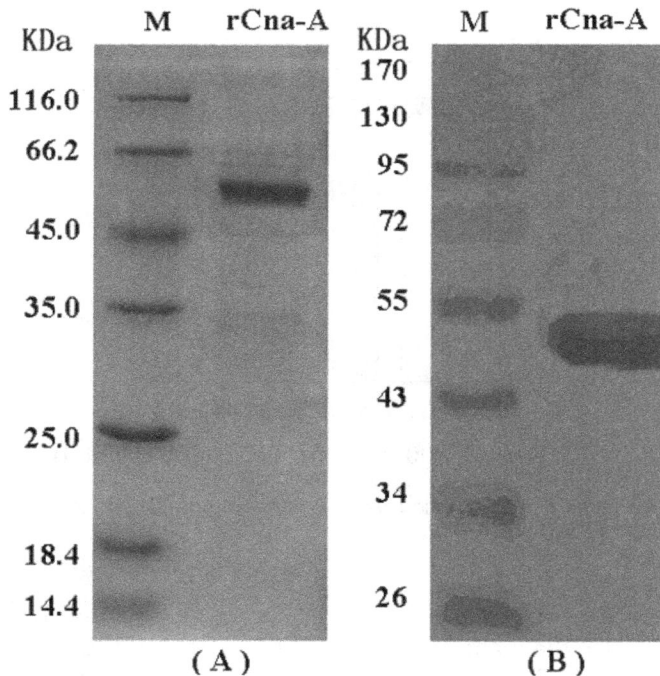

Figure 1. SDS–PAGE analysis of rCna-A expressed in *E. coli* (A). Lanes rCna-A, rCna-A purified by His affinity chromatography; Lanes M, reference proteins with the molecular weight labeled on the left. Western blot analysis of rCna-A (B). Lanes rCna-A, rCna-A protein; Lanes M, pre-stained reference proteins with the molecular weight labeled on the left.

coated antigen was replaced with killed *S. aureus*.

Detection of IgG1 and IgG2a subtypes

ELISAs were performed as described above except that the secondary antibody was either mouse anti-IgG1-HRP (Southern) or mouse anti-IgG2a-HRP (Southern). The dilution was 1:100 in the inactivated *S. aureus* group and 1:102,400 in the rCna-A group.

Challenge of lactating mice and enumeration of bacteria

Female and male mice were placed in a 3:1 ratio in the same cage so that the female mice could become pregnant. The mice were anaesthetized and mammary glands were inoculated under a binocular stereomicroscope. According to the methods of Hu et al. (2010), mice were infected in the R4 (fourth on the right) and L4 (fourth on the left) mammary abdominal glands by injecting 50 μl of a bacterial suspension containing approximately 5×10^6 CFU in PBS into the mammary ducts of each mouse. After 24 h, the mice were killed and their mammary glands were aseptically removed. The glands were then placed in sterile PBS (1:10, w/v) and homogenized. Appropriate dilutions were seeded in tryptic soy agar (TSA) to determine the numbers of CFU per gland.

Histological examination of mammary glands

Mammary glands at 24 h after inoculation were fixed in 4% formaldehyde in PBS for 24 h then embedded in paraffin wax. Fixed mammary glands were dehydrated through a graded alcohol series, immersed in xylene, and embedded in paraffin. The slices were then cut and stained with hematoxylin and eosin (H&E). Light microscopy was performed at a magnification ×400 and histological micrographs were obtained. Morphological characteristics of the test and control groups were compared.

Statistical analysis

The arithmetic mean and standard error of the mean were calculated for each treatment group. The data were analyzed using an analysis of variance (ANOVA) with the SAS software package (SAS Institute, Cary, NC, USA). P values < 0.05 were considered statistically significant.

RESULTS

Expression of the recombinant plasmids

Recombinant plasmids containing the Cna-A protein were constructed in the backbone of a pET-28a (+) vector for expression of the recombinant protein. The rCna-A protein before and after purification was analyzed by SDS–PAGE, and the bands with the expected size (50 kDa) were observed (Figure 1A). Western blot assay confirmed that rCna-A was immunogenic and reacted with rabbit antiserum to *S. aureus*, as shown by the band present at the expected size on the blotted membrane (Figure 1B).

Antibody response detected by ELISA

Both inactivated *S. aureus* and rCna-A induced an immune response in mice. Two weeks after the first immunization, antibody titers reached over 1:200 in both groups. Furthermore, mice primed with rCna-A produced significantly higher antibody titers as compared to mice vaccinated with inactivated bacteria. The antibody levels continued to rise and reached a peak after the third immunization (1:526,628 in the rCna-A group and 1:3,200 in the inactivated *S. aureus* group, P < 0.001), when sample collection was stopped (Figure 2A). However, antibody production rates in the subunit vaccine group were faster than in the inactivated vaccine group for each immunization. IgG1 and IgG2a isotype levels were evaluated in each group (Figure 2B). Only the groups vaccinated with rCna-A demonstrated a higher proportion of IgG1 than IgG2a, with 2000-fold higher levels induced by the inactivated bacterial vaccine.

Bacterial loads of infected mammary glands

Immunized mice were then challenged with an injection of *S. aureus* (50 μl of a bacterial suspension containing approximately 5×10^6 CFU in PBS) in the R4 and L4 glands. Twenty-four hours after infection, the numbers of

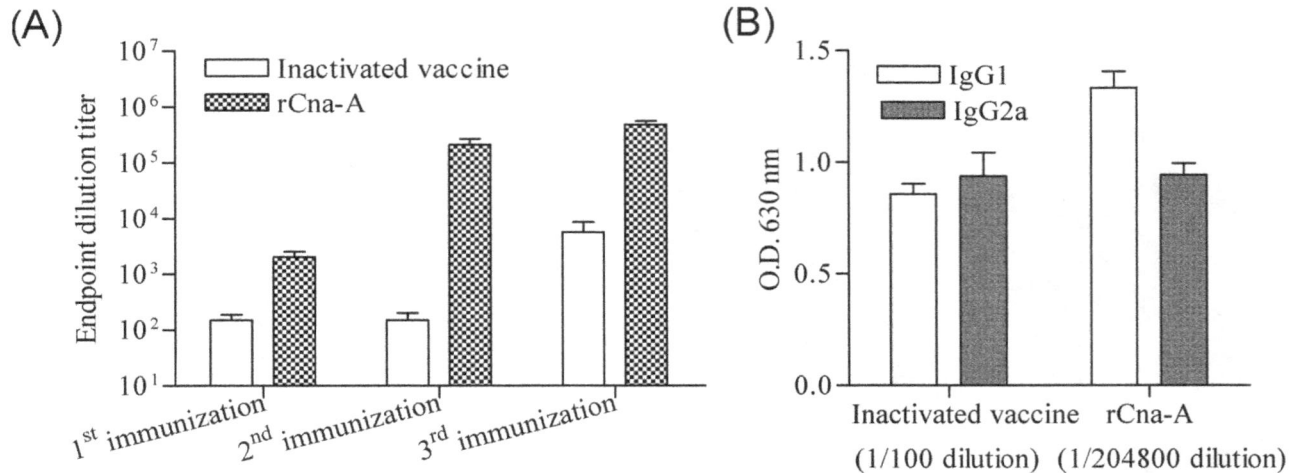

Figure 2. Serum total IgG (**A**) and IgG1 and IgG2a (**B**) titers as determined by ELISA. Mice were immunized three times, and sera were obtained after each vaccination. Results are shown as mean ± SD.

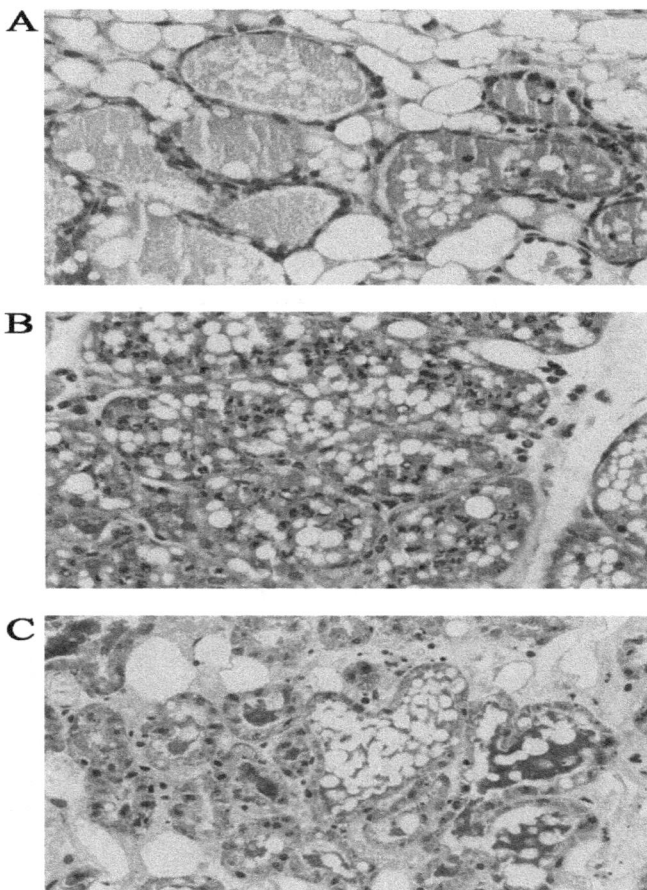

Figure 3. Microscopic images of mammary glands in the rCna-A group (A), inactivated S. aureus group (B), and PBS control group (C). A = The structure of breast tissue intact with only few PMNs visible. B = The mammary acinar structure partially destroyed with greater PMN infiltration. C = Most mammary glands damaged and necrotic. Magnification: 400.

CFU in the mammary glands of the rCna-A group ($6.40 \times 10^2 \pm 4.73 \times 10^2$ CFU/gland) were significant lower than those in both the inactivated vaccine ($3.24 \times 10^8 \pm 8.73 \times 10^7$ CFU/gland, $P < 0.001$) and PBS control ($5.16 \times 10^9 \pm 1.23 \times 10^8$ CFU/gland, $P < 0.001$) groups.

Histology

Mammary gland sections were stained with H&E and examined (Figure 3). In the rCna-A group, breast tissue structures were intact and some PMN infiltration had occurred in a portion of the mammary gland. In the inactivated vaccine group, many mammary glands inflammatory showed cell infiltration, and some proliferation of mammary stroma were observed. In the PBS control group, most mammary glands were damaged, mammary epithelial cells were swollen, milk was coagulated, and a large number of epithelial cells were necrotic and lost.

DISCUSSION

Although infection by *S. aureus* is a complex process, adhesins play the most important role in *S. aureus* infection and invasion (Joh et al., 1999; Miller and Cho, 2011). Thus, blocking bacterial adhesion to cells and colonization of the mucosal surface may be the most effective strategy for preventing *S. aureus* infection (Nour El-Din et al., 2006), and vaccines that target adhesins have shown promise in terms of preventing *S. aureus*–induced mastitis (Gong et al., 2010). Many adhesins have proved to be attractive for inclusion in vaccines against *S. aureus* infection. Cna, one of the most important adhesins, proved to be an ideal candidate component for inclusion in a vaccine against *S. aureus* infection (Flock,

1999). In this present study, the gene Can-A encoding the A region of Cna was cloned and rCna-A was successfully expressed. Western blot analysis showed that rCna-A possessed good immunogenicity and reactogenicity, which provided a basis for further evaluation of the efficacy of the mastitis vaccine against *S. aureus* infection.

In this study, mice were immunized with rCna-A, inactivated *S. aureus*, and PBS, respectively. The rCna-A group was superior to the inactivated vaccine group in terms of both the antibody production rate and the final titer. Higher antibody titers are thought to induce better protection against intramammary infection (Otto, 2008). In the rCna-A immunized group levels of the IgG1 subtype were 2000 times higher than in the inactivated vaccine group, showing that IgG1 is the dominant subclass. Further, the IgG1 subtype is the most effective inducer of opsonophagocytosis (Schlageter and Kozel, 1990; Gong et al., 2010).

In order to evaluate the potential effects of these vaccines on *S. aureus* mastitis, a mouse mastitis model was used as described by Gong et al. (2010). Only a small number of *S. aureus* (500 to 5000 CFU) are required for induction of mouse mastitis symptoms, and both the appearance and histology of infected mammary glands in mice are similar to those in naturally-infected bovine mammary glands (Kim et al., 1999; Akers and Nickerson, 2011). In this study, the numbers of bacteria in the mammary glands of the rCna-A immunized group were significantly lower than in either the inactivated *S. aureus* or PBS groups. These results indicate that rCna-A could prevent *S. aureus* invasion more effectively than inactivated bacteria. Histological examination lent support to this view.

The mammary glands of mice in the rCna-A group showed greater structural integrity and a milder inflammatory response as compared to the inactivated vaccine and PBS control groups. These observations indicate that vaccination with rCna-A protected mice against *S. aureus* infection.

In summary, the rCna-A subunit vaccine developed in this study provided significantly better protection against *S. aureus*–induced mastitis compared to that produced either by the inactivated *S. aureus* vaccine or the mock immunization in mouse model. The rCna-A protein may thus be a potential candidate of a vaccine against *S. aureus*-induced mastitis. The precise mechanism by which rCna-A provides protection against *S. aureus* requires further elucidated.

ACKNOWLEDGEMENTS

This work was supported by the national natural science foundation of China (31101874), and the Fundamental Research Funds for the Central Universities (grant no. 2010QC002).

REFFRENCES

Akers RM, Nickerson SC (2011). Mastitis and its impact on structure and function in the ruminant mammary gland. J Mammary Gland Biol. Neoplasia 16:275-89. .

Elasri MO, Thomas JR, Skinner RA, Blevins JS, Beenken KE, Nelson CL, Smeltzer MS (2002). *Staphylococcus aureus* collagen adhesin contributes to the pathogenesis of osteomyelitis. Bone 30:275-280.

Flock JI (1999). Extracellular-matrix-binding proteins as targets for the prevention of *Staphylococcus aureus* infections. Mole. Med. Today 5:532-537.

Gillaspy AF, Lee CY, Sau S, Cheung AL, Smeltzer MS (1998). Factors affecting the collagen binding capacity of *Staphylococcus aureus*. Infect. Immunity 3170-3718.

Gong R, Hu C, Xu H, Guo A, Chen H, Zhang G, Shi L (2010). Evaluation of clumping factor A binding region A in a subunit vaccine against *Staphylococcus aureus*-induced mastitis in mice. Clin. Vaccine Immunol. 17:1746-1752.

Haas TA, Plow EF (1994). Intergrin-ligand interactions: a year in a review. Curr. Opin. Cell Biol. 6:656-602.

Holderbaum D, Hall GS, Ehrhart LA (1986). Collagen binding to *Staphylococcus aureus*. Infect. Immunity 54:359-364.

Hu C, Gong R, Guo A, Chen H (2010). Protective effect of ligand-binding domain of fibronectin-binding protein on mastitis induced by *Staphylococcus aureus* in mice. Vaccine 28:4038-4344.

Hynes RO (1992). Integrins: versatility, modulation, and signaling in cell adhesion. Cell 69:11-25.

Joh D, Wann ER, Kreikemeyer B, Speziale P, Höök M (1999). Role of fibronectin-binding MSCRAMMs in bacterial adherence and entry into mammalian cells. Matrix Biol. 18:211-223.

Kim SK, Ragupathi G, Musselli C, Choi SJ, Park YS, Livingston PO (1999). Comparison of the effect of different immunological adjuvants on the antibody and T-cell response to immunization with MUC1-KLH and GD3-KLH conjugate cancer vaccines. Vaccine 18:597-603.

Leitner G, Yadlin N, Lubashevsy E, Ezra E, Glickman A, Chaffer M, Winkler M, Saran A, Trainin Z (2003). Development of a *Staphylococcus aureus* vaccine against mastitis in dairy cows. II. Field trial. Veter. Immunol. Immunopathol. 93:153-158.

Miller GY, Bartlett PC, Lance SE, Anderson J, Heider LE (1993). Costs of clinical mastitis and mastitis prevention in dairy herds. J. Am. Veter. Medicine Assoc. 202: 1230-1236.

Miller LS, Cho JS (2011). Immunity against *Staphylococcus aureus* cutaneous infections. Nat Rev Immunol 11: 505-18. doi: 10.1038/nri3010.

Nilsson IM, Patti JM, Bremell T, Höök M, Tarkowski A (1998). Vaccination with a recombinant fragment of collagen adhesin provides protection against *Staphylococcus aureus*-mediated septic death. J. Clin. Investigat. 101:2640-2649.

Nour El-Din AN, Shkreta L, Talbot BG, Diarra MS, Lacasse P (2006). DNA immunization of dairy cows with the clumping factor A of *Staphylococcus aureus*. Vaccine 24:1997-2006.

Otto M (2008). Targeted immunotherapy for staphylococcal infections: focus on anti-MSCRAMM antibodies. BioDrugs. 22:27-36.

Patti JM, Allen BL, McGavin MJ, Höök M (1994a). MSCRAMM-mediated adherence of microorganisms to host tissues. Annual. Rev. Microbiol. 48:585-617.

Patti JM, Bremell T, Krajewska-Pietrasik D, Abdelnour A, Tarkowski A, Rydén C, Höök M (1994b). The *Staphylococcus aureus* collagen adhesin is a virulence determinant in experimental septic arthritis. Infect. Immunity 62:152-161.

Rhem MN, Lech EM, Patti JM, McDevitt D, Höök M, Jones DB, Wilhelmus KR (2000). The collagen-binding adhesin is a virulence factor in *Staphylococcus aureus* keratitis. Infect. Immunity 68:3776-3779.

Rich RL, Demeler B, Ashby K, Deivanayagam CC, Petrich JW, Patti JM, Narayana SV, Höök M (1998). Domain structure of the *Staphylococcus aureus* collagen adhesin. Biochemistry 37:15423-15433.

Sambrook J, Rusell DW (2001). Molecular Cloning: A laboratory manual. 3 ed. Cold Spring Harbor Laboratory Press. p. 323.

Schlageter AM, Kozel TR (1990). Opsonization of *Cryptococcus neoformans* by a family of isotype-switch variant antibodies specific for the capsular polysaccharide. Infect. Immun. 58:1914-1918.

Talbot BG, Lacasse P (2005). Progress in the development of mastitis vaccines. Livestock. Prod. Sci. 98:101-113.

Xu H, Hu C, Gong R, Chen Y, Ren N, Xiao G, Xie Q, Zhang M, Liu Q, Guo A, Chen H (2011). Evaluation of a novel chimeric B cell epitope-based vaccine against mastitis induced by either *Streptococcus agalactiae* or *Staphylococcus aureus* in mice. Clin Vaccine Immunol. 18:893-900.

Mineral nutrition in the control of nematodes

Simone de Melo Santana-Gomes[1], **Claudia Regina Dias-Arieira**[1], **Miria Roldi**[2], **Tais Santo Dadazio**[2], **Patricia Meiriele Marini**[2] **and Davi Antonio de Oliveira Barizão**[1]

[1]Universidade Estadual de Maringa, Pós Graduação em Agronomia, Avenida Colombo, no 5790 - CEP: 87020-900, Maringa, PR, Brazil.
[2]Departamento de Ciências Agronômicas, Universidade Estadual de Maringa, Avenida Colombo, no 5790 - CEP: 87020-900, Maringa, PR, Brazil.

Although plant tolerance and resistance to pathogens are genetically controlled, they are significantly influenced by environmental nutrition factors. Mineral nutrients can increase or decrease resistance or tolerance to pathogens and pests. Plants receiving ample nutrition have higher resistance to diseases and higher tolerance. This is based on the fact that more vigorous plants have greater capacity to offset a loss of photosynthates, for instance, or reduction in root and leaf surface area caused by infection by pathogens or poor nutrition. In this context, plant mineral nutrition can be considered an environmental nutrition factor that can be manipulated relatively easily, and is an important component in disease control, as well as a tool for managing phytonematodes, reported to be causing losses in countless important agricultural crops. Thus, the aim of this review was to discuss the importance of plant mineral nutrition in the control of phytoparasitic nematodes.

Key words: Management, plant nutrition, phytonematodes, resistance.

INTRODUCTION

Plant nutrition is a uniquely important factor in fostering life on earth, which in all its variety is still made up of atoms of chemical elements. The basic reservoirs of these elements are rocks, oceans and the atmosphere. If one of the chemical elements essential to plant life is not available in sufficient quantity or is present in chemical combinations that are difficult to absorb, the deficiency of this element will cause disruptions in plant metabolic processes. The occurrence of these disruptions and identification of nutrient deficiencies and their causes, as well as the toxic effects of excessive concentrations of certain elements, highlights the functional roles of these elements in plant metabolism (Epstein and Bloom, 2004).

Generally, speaking, under field conditions, the application of fertilizers affects the incidence of diseases and pests induced by the plants' nutritional status and indirectly produce dense stands and alterations in the interception of light and moisture within the crop (Agrios, 2005). The numerous effects of plant nutritional status and fertilizer use on diseases and pests are directly relevant to the control of these organisms by chemical means. The application of mineral fertilizer replaces or reduces the need for chemical control in some cases, whereas in others it can increase this need. For instance, chemical control is essential in the presence of high levels of nitrogen (Marschner, 1997).

In general, nutrients can directly or indirectly predispose plants to pathogen attack. They can reduce or increase disease severity, affect the environment to attract or deter pathogens and also induce resistance or tolerance in the host plant (Zambolim et al., 2001; Agrios, 2005). Applying fertilizer can, partially, offset nematode-induced damage by stimulating plant development (Ferraz et al., 2010). The general rule is that, if a nutrient is essential to a plant species, it should be supplied in balanced proportion to other essential nutrients, since deficiency can aggravate disease, especially in short-cycle crops (Zambolim et al., 2001).

Nematodes are among the pathogens that can be affected by plant nutrition. They are very small, almost transparent organisms that easily escape detection by the farmer. On penetrating the plant and feeding on its tissues, they draw off its nutrients and cause mechanical damage and physiological changes by injecting toxins (Cadet and Spaull, 2005). The symptoms of nematode attack vary according to the species, and mainly take the form of galls produced by *Meloidogyne* spp., cysts caused by *Heterodera* spp. and root lesions caused by the genus *Pratylenchus*. However, other symptoms can be presented, such as very poor root systems, peeling or tearing of the cortex and tuber necrosis (Agrios, 2005; Cadet and Spaull, 2005; Ferraz et al., 2010). These symptoms are often mistaken for abiotic problems or outbreaks of other pathogens (Agrios, 2005). In the aerial part, the main symptoms are a reduction in plant growth due to nutrient deficiency, wilting during the hottest times of day and a fall in crop yield, generally due to root damage that lowers the plant's capacity to absorb water and nutrients (Cadet and Spaull, 2005; Decraemer and Hunt, 2006; Ferraz et al., 2010).

Phytoparasitic nematodes upset the balance of agroecosystems and impose limits on the yields of numerous crops, such as cotton, maize, soybean, sugarcane and vegetables (Bond et al., 2000; Dinardo-Miranda et al., 2000; Barros et al., 2005; Barbosa et al., 2010; Obici et al., 2011; Dias-Arieira et al., 2012a). The nematode control is complex and the best results are obtained by integrated management.

In the soil, nematodes are attracted to their hosts by the concentration gradient formed by root exudates, which provide a recognition signal, but can also repel nematodes. However, it is not clear whether mineral nutrients play an important role in this process. Some studies show that nematodes cause a drop in root system activity and growth (Oteifa and Elgindi, 1976; Merwin and Stiles, 1989). For example, nematodes are cited as the main agents responsible for potassium deficiency in apples (Merwin and Stiles, 1989). In cotton, attack by *Rotylenchulus reniformis* Linford and Oliveira can cause significant losses, but does not affect cotton plant aerial part growth in the presence of high levels of available potassium (Oteifa and Elgindi, 1976). Similar responses have also been observed for micronutrients (Huber and Wilhelm, 1988).

Greenhouse studies have shown that applying macronutrients to sugarcane reduces the severity of the disease caused by *Meloidogyne* spp., allowing the plant

to develop normally (Asano and Moura, 1995). In contrast, the same study showed that there was no observed effect for micronutrients on plant development, or on nematode reproduction. Therefore, it is possible that interactions involving *Meloidogyne* spp. and macronutrient deficiencies are responsible for more severe meloidogynosis in sugarcane on the coastal plateaus of northeastern Brazil.

Among plant nutrients, nitrogen is essential for growth and yield. An abundance of nitrogen results in the production of new tissues and saps, and can extend the vegetative state and increase the number of feeding sites in the roots, encouraging nematode attack. On the other hand, a plant that is deficient in nitrogen can become debilitated, suffer slowed growth and become more susceptible (Zambolim et al., 2001; Ferraz et al., 2010). However, the form in which the nutrient is available, whether ammonium (NH_4^+) or nitrate (NO_3^-), has more effect on the severity of the nematode attack than the quantity of nitrogen available (Table 1) (Ferraz et al., 2010).

Nitrogen in ammonium form, present in fertilizers and organic matter, is more prejudicial to nematodes than in nitrate form due to the release of free ammonia (NH_3) into the soil during its decomposition (Rodríguez-Kábana et al., 1981; Rodríguez-Kábana, 1986). The nematicidal property of ammonia is mainly attributable to its plasmolytic effect around the point at which it is applied to the soil (Rodríguez-Kábana, 1986). Nevertheless, other indirect mechanisms, such as an increase in microbiota antagonistic to nematodes, may also have a significant effect. Many fungi in the soil prefer nitrogen in ammonium form (Rodríguez-Kábana et al., 1981), which means that ammonia-producing materials should encourage the proliferation of nematode parasitic fungi. Because of the nematicidal effect of ammonium nitrogen, urea has also been the subject of many studies, since it is rapidly converted into ammonia by the action of urease in the soil (Rodríguez-Kábana, 1986). Urea is a good nematicide when applied in at doses higher than 300 kg N ha^{-1}. However, doses at this level can be phytotoxic due to the low C : N ratio. Therefore, applying urea in conjunction with a supplementary source of carbon is a viable method of controlling nematodes as well as reducing the phytotoxic effect of the fertilizer (Huebner et al., 1983).

Working with mixtures of urea and molasses to control *Meloidogyne arenaria* (Neal) Chitwood in soil cropped with *Curcubita pepo* L., Rodríguez-Kábana and King (1980) found that it was possible to reduce the severity of nematode attacks using urea alone. The authors also reported that the results represent a specific case in which the NH_3 supplied prevents the urease in the soil from acting on the urea, resulting in the release of NH_3 and CO_2. Since the action of the urea in the soil is dependent on the urease (Kandeler and Gerber, 1988), any treatment that results in the formation of the enzyme

Table 1. Effect of different nutrients on some species of nematode.

Host	Nematode	NO_3^-	NH_4^+
Soybean	*Heterodera glycines* Ichinohe	Increases severity	Reduces severity
Tobacco	*Globodera* spp.	Increases severity	Reduces severity
Cucumber	*Meloidogyne* spp.	Increases severity	Reduces severity
Common bean	*M. incognita* (Kofoid and White) Chitwood	Increases severity	Reduces severity

Source: Adapted by Zambolim et al. (2001).

can increase the efficacy of chemicals in combating nematodes. Therefore, the addition of a source of carbon (molasses) stimulated the production of urease and improved the efficacy of the urea against *M. arenaria*, possibly by stimulating the activity of biological control agents (Rodríguez-Kábana and King, 1980). The authors also mention the possibility of using nitrification inhibitors to increase the efficacy of treatments, with the aim of reducing the quantity of urea necessary to produce nematicidal activity, significantly cutting treatment costs.

Phosphorus is essential to plant growth and can also influence diseases caused by nematodes (Ferraz et al., 2010). Plants with high levels of phosphorus, release fewer root exudates and are therefore less attractive to nematodes cutting decreasing the incidence of the diseases (Marschner, 1997). Furthermore, plants become more resistant when supplied with sufficient quantities of phosphorus (Zambolim et al., 2005), as a result of increases in protein synthesis, cell activity and production of polyphenols, peroxidase and ammonia (Wang and Bergeson, 1974). The effect of phosphorus in the control of nematodes can vary depending on the source used. Application of phosphorus in the form of triple superphosphate was more effective in controlling *Pratylenchus scribneri* Steiner than single superphosphate when used on soybean, maize and cotton (Collins and Rodríguez-Kábana, 1971).

Phosphite has been widely used in agriculture for the countless advantages that it has, mainly in increasing the absorption of phosphorus by the plant compared to phosphate-based products. Furthermore, it has other interesting features, such as boosting microbial activity (Cohen and Coffey, 1986) and activating plant defense mechanisms (Jackson, 2000).

In a study conducted by Dias-Arieira et al. (2012b), potassium phosphite was effective in cutting the population of *Pratylenchus brachyurus* and Filipjev and Schuurmans Steckhoven in maize. Similar results have been reported for other nematode species (Oka et al., 2007). The efficacy of phosphite in the control of microorganisms is attributed to 2 factors. The first is direct microbial activity against phytopathogens (Guest and Grant, 1991). However, in the study conducted by Dias-Arieira et al. (2012b), the phosphite was applied to the aerial part of the plant, that is, spatially separated from the nematode, and in this case its efficacy was related to a second factor, its capability of stimulating plant defense mechanisms, such as the production of phytoalexins (Derks and Creasy, 1989). This hypothesis finds further support in the study conducted by Salgado et al. (2007) in which potassium phosphite boosted the hatching of *Meloidogyne exigua* Goeldi, but did not kill off juveniles, that is, it had no direct action on the parasite. In addition, Oka et al. (2007) reported that potassium phosphite applied to the aerial part was effective in controlling *Heterodera avenae* Wollenweber and *Meloidogyne marylandi* Jepson and Golden in wheat and oats. This result is attributable to phosphite's ability to translocate along the plant's xylem and phloem (Quimette and Coffey, 1990).

Although, potassium (K) is the most abundant cationic mineral in plants, it is not an integral part of any metabolite that can be isolated in vegetal matter. However, it is present in high concentrations in the cytosol and vacuole as a free ion (K^+). In addition to playing an important role in enzymatic activation (over 60 enzymes), potassium also participates in membrane crossover, neutralization of anions and maintaining membrane potential (Epstein and Bloom, 2004).

It is thought that adequate plant nutrition with potassium helps reduce the incidence of disease due to increased resistance to the penetration and development of pathogens (Huber and Arny, 1985; Perrenoud, 1990), and is considered the nutrient that has the greatest influence on diseases (Perrenoud, 1990). The increase in plant resistance in the presence of plentiful supplies of potassium is mainly due to the increased thickness of the epidermal cell wall, boosting the structural rigidity of tissues and playing a fundamental role in many metabolic reactions in plants, regulating stomata functioning and promoting rapid recovery of injured tissue, due to the accumulation of phytoalexins and phenols around the infection site (Huber and Arny, 1985).

Barbosa et al. (2010) evaluated the use of potassium fertilizer (single and multiple doses) on populations of *H. glycines* in resistant and susceptible soybean cultivars and observed that increasing doses of potassium reduced the number of females in the root system and the nematode reproduction factor in the susceptible cultivar. Similarly, in an experiment developed by Pinheiro et al. (2009), doses of potassium significantly influenced the number of cysts pot[-1], eggs cyst[-1], females and cyst per root system and the reproduction factor of *H. glycines* in soybean. This reduction is thought to be due to the interference of the potassium in the reception of the signal by the cell membrane, reducing the number of syncytia (Barbosa et al., 2010). Potassium fertilizer also reduced nematode multiplication in the pathosystems of *M. exigua-Coffea arabica* L. (Santos, 1978). On the other hand, this kind of fertilizer boosted increased populations of *Tylenchulus semipenetrans* Cobb in *Citrus aurantium* L., *R. reniformis* in *Vigna sinensis* Endl. (Badra and Yousif, 1979), and *R. reniformis* in cotton (Pettigrew et al., 2005), a fact result that has been attributed to the stronger growth of the plant's root system.

Calcium (Ca) performs a number of functions in plants. In quantitative terms, it is the predominant nutrient in the apoplast, where it has at least 2 separate functions: a) interlinking pectin chains and therefore contributing to their stability, and b) affecting the mechanical properties (rheology) of the pectin gel. It is an essential element for the integrity of the plant cell's plasmatic membrane, and more specifically ion-transport selectivity (Epstein and Bloom, 2004). Like other nutrients, calcium must be present in sufficient quantity in the soil, since calcium-deficient plants are more susceptible to nematode attack (Hurchanik et al., 2003).

The calcium content of plant tissues can affect the incidence of diseases in two ways: a) when calcium levels are low, there is an increase in the efflux of compounds of low molecular weight (sugars) from the cytoplast to the apoplast, and b) calcium polygalacturonates are required for the middle lamella to stabilize the cell wall. Calcium plays a critical role in cell division and development, in the structure of the cell wall and in the formation of the middle lamella (Huber, 1991). Many phytopathogenic agents reach the plant tissue by producing extracellular pectolytic enzymes, such as galacturonase, that degrade the middle lamella (McGuire and Kelman, 1986), and the enzyme activity is dramatically inhibited by the presence of calcium (Marschner, 1997).

Application of calcium carbide (CaC_2) was effective in reducing the number of galls, egg masses and juveniles of *M. incognita* in zucchini (*Cucurbita pepo* var. Melopepo), regardless of the concentration used, increasing crop yield (Mohamed and Youssef, 2009). However, the form in which calcium is most widely used consists of liming, which is directly link to soil pH. Rocha et al. (2006) evaluated the effect of increasing doses of lime on the population of *H. glycines* in soybean roots and found that the number of females dropped as lime doses were increased, up to 3.039 t ha[-1]. The authors concluded that increasing the supply of calcium boosted the resistance of the cells, lowering root infection by the nematode. On the other hand, Anand et al. (1995) reported that higher pH values (6.5 and 7.5) resulted in higher populations of *H. glycines* in soil with a pH of 5.5. The authors explain that soils with pH values of 6.5 and

7.5 are more favorable to soybean growth, resulting in stronger increasing root development and consequently better conditions for the development of infection sites.

Soil pH seems to be important for nematode activity, although the probable effects are indirect, due to the alteration of the microbiota in the soil and the availability of micronutrients to plants (Rocha et al., 2006). High pH values and base saturation were favorable to *H. glycines* in soybean crops in the municipalities of Chapadão do Sul and Chapadão do Céu, Brazil, where pH values higher than 6.0 intensified the damage caused by the nematode and immobilized micronutrients in the soil, causing deficiencies in the plants and reducing natural parasitism on eggs and cysts (Silva et al., 1997).

Silicon (Si) is among the most abundant nutrients in the earth's crust, existing in quantities second only to oxygen. The majority of soils are predominantly silicates and aluminosilicates. Plants absorb silicon in variable quantities, and the usual amounts of silicon found in dry plant matter range from around 0.1 to 10.0% (Epstein and Bloom, 2004). Silicon has been researched especially as an inducer of plant resistance to diseases (Rodrigues and Datnoff, 2005). It can induce the plant's defense mechanisms by activating the synthesis of substances such as phenols, lignin, suberin and callose in the cell wall.

The mechanism by which silicon activates resistance has yet to be fully elucidated. The way in which silicon is deposited in the cell wall has given rise to the hypothesis of a possible physical barrier, by the ascending movement of the element from the roots to the leaves, undergoing polymerization in the extracellular spaces in the cell walls and the xylem vessels (Terry and Joyce, 2004). According to Epstein and Bloom (2004), plants absorb silicon, transporting it rapidly to the aerial part and when transpiration takes place, the dissolved silicon becomes supersaturated and can polymerize and form solids, known as opals or opal phytoliths. Incorporation of these phytoliths at the end of the respiratory flow of water in the cell walls makes them rigid. Although the majority of the silicon in plants is polymerized or solidified, the role of this element in resistance to diseases is mainly due to the silicon fraction in solution inside the plant, suggesting that defensive compounds are synthesized in the dicotyledons.

Silva et al. (2010) studied the silicon-mediated biochemical resistance responses of coffee to *M. exigua* and presented evidence indicating that the reproductive capability of the nematode in coffee roots supplied with silicon was adversely affected. The response was associated with the production of lignin and an increase in the activity of peroxidase (POX), polyphenol oxidase (PPO) and phenylalanine amonialiase (PAL), especially in the susceptible cultivar evaluated. According to Guimarães et al. (2008), potassium silicate was effective in inducing sugarcane resistance to *M. incognita*, since it reduced the number of pathogen eggs in the RB867515 and RB92579 varieties. However, it did not affect the aerial part biomass of the RB867515 and RB863129 varieties, nor the population density of *Pratylenchus zeae* Graham in the soil, 100 days after transplanting. Dutra et al. (2004) reported a greatest decrease in the number of galls and eggs of a number of species of *Meloidogyne* in common bean, tomato and coffee treated with calcium silicate.

In contrast to other heavy metal ions such as copper, iron and manganese, zinc (Zn) is a divalent cation (Zn^{2+}) that does not undergo valency exchange and, therefore, has no redox activity (Epstein and Bloom, 2004). Over 80 proteins containing zinc have been reported. Many enzymes require zinc at the active site, as occurs in carbonic anhydrase, superoxide dismutase (together with copper) and alcohol dehydrogenase. In other enzymes, zinc is an integral component of the protein but is not near the active site. Zinc ions regulate the conformation of the protein domain which connects to the DNA. This means that zinc deficiency seriously interferes with growth, which is dependent on protein synthesis, which in turn depends on transcription (Barker and Pilbeam, 2007). These considerations suggest that one of the inhibiting effects of zinc deficiency on growth could be due to inadequate protein synthesis (Epstein and Bloom, 2004).

Plants deficient in zinc contain low levels of superoxide dismutase and therefore high levels of superoxide radicals, promoting membrane lipid peroxidation and a loss of membrane integrity, increasing permeability (Barker and Pilbeam, 2007). Furthermore, the accumulation of free amino acids and amides occurs as a result of protein synthase inhibition due to zinc deficiency, boosting the quantity of these amino acids in root exudates (Cakmak and Marschner, 1988). Since nematodes are attracted by exudates, the higher root exudation in plants deficient in zinc can attract these parasites and, therefore, speed up the infection process (Streeter et al., 2001).

We still do not know the exact role played by zinc in relation to nematodes. However, supplying this mineral increased the activity of *Pseudomonas aeruginosa* IE-6S$^+$ and *P. fluorescens* CHA0, helping in the biological control of *Meloidogyne javanica* (Treub) Chitwood (Siddiqui and Shaukat, 2002). Shaukat and Siddiqui (2003) reported that even when zinc alone was applied, it caused a decrease in the numbers of *M. javanica* and boosted increased growth in tomatoes. The authors also confirmed an increase in the antagonistic activity of rhizobacteria in the rhizobio group associated with the application of zinc. In another study, Siddiqui et al. (2002) confirmed that alterations in the soil when zinc was applied at 0.8 or 1.6 mg per kg^{-1} soil, whether alone or combined with *Pseudomonas aeruginosa* IE-6S$^+$, significantly reduced the penetration of *M. javanica* into tomato roots. At all nematode population densities, zinc reinforced the efficacy of the bacteria in reducing parasite penetration and subsequent development of galls. Absence of the bacteria and/or zinc increased the nematode population in the soil, causing a significant decrease in plant height, fresh weight and root protein content.

The increase in concentrations of antibiotics after applications of zinc could provide protection against diseases with a rapid onset, such as damping-off (*Pythium* and *Rhizoctonia* root rot), exceeding the bacterium's capability to take hold and becoming established in the rhizosphere, producing antibiotics *in situ* (Duffy and Défago, 1995). Zinc is known to stimulate antimicrobial polypeptide biosynthesis by biocontrol bacteria, change the structure of microbial communities and alter the physiology of plants, and can be used advantageously in controlling nematodes (Behal and Hunter, 1995; Siddiqui and Shaukat, 2002; Shaukat and Siddiqui, 2003).

Manganese (Mn) activates a number of enzymes on the chemical acid and subsequent pathways, leading to the biosynthesis of aromatic amino acids such as tyrosine and a number of other secondary compounds such as lignin and flavenoids (Barker and Pilbeam, 2007). For example, Mn^{+2} affects the PAL enzyme and stimulates peroxidases necessary for lignin biosynthesis. The lower lignin content in plants deficient in manganese is an indication of the need for this element at a number of stages in lignin biosynthesis and the reduction in the amount of root material contributes to lower plant resistance to pathogen attack (Marschner, 1997). It has been reported that supplying manganese to barley did not alter the number of immature females of *Heterodera avenae* Wollenweber, but untreated plant growth was significantly impaired by comparison with the growth of plants treated with the nutrient (Wilhelm et al., 1985). The authors think that applying manganese may have offset a deficiency in the absorption of the nutrient, caused by nematode infection.

Copper (Cu) influences the formation and chemical composition of the cell walls. In deficient leaves, the proportion of α-cellulose increases and lignin content decreased by half in comparison to leaves adequately supplied with copper. This effect is even more marked in the sclerenchyma cells in the stem (Marschner, 1997). In plants with severe copper deficiency, the xylem vessels are also insufficiently lignified, but lignification responds rapidly to treatment with the nutrient (Barker and Pilbeam, 2007). Copper must be supplied via the leaves at the beginning of the rapid plant growth phase and at the onset of flowering in the majority of cultivated species (Fancelli, 2008). According to Marschner (1997), lignification is inhibited in tissue deficient in copper because two copper enzymes are involved in the lignin biosynthesis pathway: polyphenol oxidase, which catalyzes the oxidation of phenolic compound lignin precursors, and diamine peroxide, which supplies the H_2O_2 necessary for oxidation by peroxidases. Tissues deficient in copper therefore exhibit lower activity of these enzymes, accumulating phenolic compounds.

Fancelli (2008) points out that the most important micronutrients for preventing disease and that are usually ignored in production systems are Cu, B and Mn, since they play a fundamental role in the chemical acid pathway, the main plant defense pathway. It is also worth noting that, although Zn is the micronutrient most often taken into account in fertilizing programs, if Zn is supplied in high doses with no defined technical criterion, it can interfere with the use and metabolization of other nutrients, as well as encouraging the growth of fungi and fungal metabolite (micotoxin) production.

Boric acid is capable of forming complexes with diols and polyols, particularly with *cis*-diols. Polyhydroxyl compounds with a *cis*-diol configuration are necessary for forming these complexes. They include a number of sugars and their derivatives, and especially mannitol. These compounds serve as constituents of the hemicellulose fraction of the cell wall (Marschner, 1997). Some *o*-diphenols, such as caffeic acid and hydroxyferulic acid, are important precursors in the biosynthesis of lignin in the dicotyledons, because they have a *cis*-diol configuration and form stable borate complexes.

Boron deficiency is associated with morphological alterations and changes in tissue differentiation (Barker and Pilbeam, 2007). The formation of borate complexes with certain phenols is probably involved in the regulation of phenol levels and the synthesis rate for alcoholic phenols as precursors of lignin biosynthesis. Where there is a boron deficiency, phenols accumulate and polyphenol oxidase activity increases (Marschner, 1997). Depending on its dynamic in the soil and plant, boron should be supplied through the soil at the pre-sowing (or sowing) stage, using sources of medium solubility (Fancelli, 2008).

The balanced application of macro and micronutrients to the soil is the best way of ensuring that the crop is able to withstand the damage caused by nematodes. It is unlikely that fertilizing alone will control parasites of this kind. Therefore, fertilizers should be used initially to feed the plant, and be integrated with more effective methods of control (Ferraz et al., 2010).

It is worth pointing out that, in addition to studies aimed at evaluating the effects of plant nutrition on the nematode population, some researchers have been seeking answers using the opposite approach, i.e. the effect of nematodes on plant nutrition (Heffes et al., 1992; Hurchanik et al., 2004). Hurchanik et al. (2004) observed that coffee plantation infection by *Meloidogyne konaensis* Eisenback, Bernard and Schmitt significantly reduced the absorption of manganese and copper by the root system. Reductions in calcium and manganese were also observed, and attributed to root system damage caused by nematodes.

Heffes et al. (1992) evaluated the effect of *M. incognita* and *R. reniformis* on growth and nutrient contents in caupi beans and maize and observed that for each parasite-host combination there were major differences in root and aerial part nutrient concentrations when infected and uninfected plants were compared. Where the host was affected by the parasite, potassium concentrations were always lower and aluminum and vanadium concentrations always higher in the infected plants. In contrast, nematode infection resulted in a greatest decrease in iron concentration in the roots and aerial part. Although, nematode reproduction was higher in caupi beans than in maize, nutrient absorption through the roots was not affected, whereas in maize the absorption of nutrients was severely impaired. Nutrient translocation does not seem to be a limiting factor in host growth, since few differences were observed between the root/aerial part nutrient ratios of infected and uninfected plants. Other factors over and above nutrient absorption and translocation could be involved in the adverse effects of nematode parasitism in caupi beans, including water absorption, carbohydrate partitioning and phytoalexin toxicity.

In contrast, soybean plants parasitized by *H. glycines* exhibited yellowing and dwarfism as the main symptoms, which is why the disease is known as soybean yellow dwarf disease (Ferraz et al., 2010). This symptomatological situation is the result of nitrogen deficiency, stemming from a drastic drop in the nodulation of the symbiont bacterium *Bradyrhizobium japonicum* (Kirchner) Jordan (Andrade and Ponte, 1999). In many cases, the deficiency symptoms are common even if nutrients are available in the soil (Hurchanik et al., 2003), showing the dysfunction that occurs due to nematode infection of the root system.

Conclusions

There are still few studies on the influence of mineral nutrition on diseases caused by nematodes and some of the existing studies do not elucidate the mechanisms by which the nutrients reduce the phytoparasite population. In some nematode-plant interactions, the accumulation of cellulose, lignin and other elements brought about by plant nutrition confer nematode-resistance on the host. Some nutrients, such as silicon, seem to be directly related to the resistance induction process. This is, sometimes, accompanied by indirect responses, such increased activity of natural enemies. However, there are many contradictory results and numerous processes for which we still have no answers. In view of this, there is a need for conclusive research in order to elucidate the ways in which nutrients combat phytonematode infection.

REFERENCES

Agrios GN (2005). Plant Pathology. 5ª ed. London: Elsevier Academic

Press. P. 922.

Anand SC, Matson DW, Sharma SB (1995). Effect of soil temperature and pH on resistance of soybean to *Heterodera glycines*. J. Nematol. 27:478-482.

Andrade NC, Ponte JJ (1999). Efeito do sistema plantio direto em camalhão e do consórcio com *Crotalaria spectabilis* no controle de Meloidogyne incognita em quiabeiro. Nematol. Bras. 23:111-116.

Asano S, Moura RM (1995). Efeitos dos macro e micronutrientes na severidade da meloidoginose da cana-de-açúcar. Nematol. Bras. 19:15-20.

Badra T, Yousif GM (1979). Comparative effects of potassium levels on growth and mineral composition of intact and nematized cowpea and sour orange seedlings. Nematol. Mediterr. 7:21-27.

Barbosa KAG, Garcia RA, Santos LC, Teixeira RA, Araújo FG, Rocha MR, Lima FSO (2010). Avaliação da adubação potássica sobre populações de *Heterodera glycines* em cultivares de soja resistente e suscetível. Nematol. Bras. 34:150-157.

Barker AV, Pilbeam DJ (2007). Handbook of Plant Nutrition. London: Taylor & Francis Group. P. 613.

Barros ACB, Moura RM, Pedrosa EMR (2005). Estudo de interação variedade-nematicida em cana-de-açúcar, em solo naturalmente infestado por *Meloidogyne incognita*, M. javanica e Pratylenchus zeae. Nematol. Bras. 29:39-46.

Behal V, Hunter IS (1995). Tetracyclines. In: Vining LC, Stuttard C (eds) Genetics and Biochemistry of Antibiotic Production. Boston: Butterworth-Heinemann, pp. 359-384.

Bond JP, McGawley EC, Hoys JW (2000). Distribution of plant-parasitic nematodes on sugarcane in Louisiana and efficacy of nematicides. J. Nematol. 32:493-501.

Cadet P, Spaull V (2005). Nematode parasites of sugarcane. In: Luc M, Sikora RA, Bridge J (eds) Plant parasitic nematodes in subtropical and tropical agriculture. Wallingford: CAB International, pp. 645-674.

Cakmak I, Marschner H (1988). Enhanced superoxide radical production in roots of zinc-deficient plants. J. Exp. Bot. 39:1449-1460.

Cohen MD, Coffey MD (1986). Systemic fungicides and the control of Oomycetes. Ann. Rev. Phytopath. 24:311-338.

Collins RJ, Rodríguez-Kábana R (1971). Relationships of fertilizer treatments and crop sequence to populations of lesion nematode. J. Nematol. 3:306-307.

Decraemer W, Hunt D (2006). Structure and classification. In: Perry R, Moens M (eds) Plant Nematology. Wallingford: CAB International. pp. 3-32.

Derks W, Creasy LL (1989). Influence of fosetyl-Al on phytoalexin accumulation in the *Plasmopara viticola* grapevine interaction. Physiol. Mol. Plant Path. 3:203-213.

Dias-Arieira CR, Cunha TPL, Chiamolera FM, Puerari HH, Biela F, Santana SM (2012a). Reaction of vegetables and aromatic plants to Meloidogyne javanica and M. incognita. Hortic. Bras. 30:322-326.

Dias-Arieira CR, Marini PM, Fontana LF, Roldi M, Silva TRB (2012b). Effect of *Azospirillum brasilense*, Stimulate® and potassium phosphite to control *Pratylenchus brachyurus* in soybean and maize. Nematropica 42:170-175.

Dinardo-Miranda LL, Garcia V, Menegatti C (2000). Controle químico de nematoides em soqueiras de cana-de-açúcar. Nematol. Bras. 24:55-58.

Duffy BK, Défago G (1995). Influence of cultural conditions on spontaneous mutations in *Pseudomonas fluorescens* CHA0. Phytopathology 85:1146.

Dutra MR, Garcia ALA, Paiva BRTL, Rocha FS, Campos VP (2004). Efeito do Silício aplicado na semeadura do feijoeiro no controle de nematoide de galha. Fitopatol. Bras. 29:172.

Epstein E, Bloom A (2004). Mineral Nutrition of Plants. Sunderland: Sinauer Associates. P. 380.

Fancelli AL (2008). Influência da nutrição na ocorrência de doenças de plantas. Inform. Agron. 122:23-24.

Ferraz S, Freitas LG, Lopes EA, Dias-Arieira CR (2010). Manejo sustentável de fitonematoides. Viçosa: Editora UFV. P. 306.

Guest D, Grant B (1991). The complex action of phosphonates as antifungal agents. Biol. Rev. 66:159-187.

Guimarães LMP, Pedrosa EMR, Coelho RSB, Chaves A, Maranhão SRVL, Miranda TL (2008). Efeito de metil jasmonato e silicato de potássio no parasitismo de Meloidogyne incognita e Pratylenchus zeae

em cana-de-açúcar. Nematol. Bras. 32:50-55.

Heffes PT, Coates-Beckford PL, Robotham H (1992). Effects of *Meloidogyne incognita* e *Rotylenchulus reniformis* on growth and nutrient content of *Vigna unguiculata* e *Zea mays*. Nematropica 22:139-148.

Huber DM (1991). The use of fertilizers and organic amendments in the control of plant disease. In: Pimentel D, Hanson AA (ed) Handbook of pest management in agriculture. Flórida: CRC, pp. 357-394.

Huber DM, Arny DC (1985). Interactions of potassium with plant disease. In: Munson RD (ed) Potassium in Agriculture. Madison: ASA, CSSA, SSA. pp. 467-488.

Huber DM, Wilhelm NS (1988). The role of manganese in resistance to plant diseases. In: Graham RD, Hannan RJ, Uren NC (eds) Manganese in soils and plants. Dordrecht: Kluwer Academic. pp. 155-173.

Huebner RA, Rodríguez-Kábana R, Patterson RM (1983). Hemicellulosic waste and urea for control of plant parasitic nematodes: Effect on soil enzyme activities. Nematropica 13:37-54.

Hurchanik D, Schmitt DP, Hue NV, Sipes BS (2003). Relationship of *Meloidogyne konaensis* population densities to nutritional status of coffee roots and leaves. Nematropica 33:55-64.

Hurchanik D, Schmitt DP, Hue NV, Sipes BS (2004). Plant nutrient partitioning in coffee infected with *Meloidogyne konaensis*. J. Nematol. 36:76-84.

Jackson TJ (2000). Action of the fungicide phosphite on *Eucaliptus marginata* inoculated with Phytophthora cinnamomi. Plant. Pathol. 49:147-154.

Kandeler E, Gerber H (1988). Short-term assay of soil urease activity using colorimetric determination of ammonium. Biol. Fert. Soils 6:68-72.

Marschner H (1997). Mineral nutrition of higher plants. London: Academic Press. P. 889.

McGuire RG, Kelman A (1986). Calcium in potato tuber cell walls in relation to tissue maceration by *Erwinia carotovora* pv. *atroseptica*. Phytopathology 76:401-406.

Merwin IA, Stiles WC (1989). Root-lesion nematodes, potassium deficiency, and prior cover crops as factors in apple replant disease. J. Am. Soc. Hon. Sci. 114:724-728.

Mohamed MM, Youssef MMA (2009). Efficacy of calcium carbide for managing *Meloidogyne incognita* infesting squash in Egypt. Int. J. Nematol. 19:229-231.

Obici LV, Dias-Arieira CR, Klosowski ES, Fontana LF, Cunha TPL, Santana SM, Biela F (2011). Efeito de plantas leguminosas sobre *Pratylenchus zeae* e *Helicotylenchus dihystera* em solos naturalmente infestados. Nematropica 41:215-222.

Oka Y, Tkachi N, Mor M (2007). Phosphite inhibits development of the nematode *Heterodera avenae* and *Meloidogyne marylandi* in cereals. Nematology 97:396-404.

Oteifa BA, Elgindi AY (1976). Potassium nutrition of cotton, *Gossypium barbadense*, in relation to nematode infection by *Meloidogyne incognita* and *Rotylenchulus reniformis*. Proc. 12th Colloq. Int. Potash Inst. Bern. pp. 301-306.

Perrenoud S (1990). Potassium and plant health. Bern: International Potash Institute, 2. ed. P. 363.

Pettigrew WT, Meredith Jr WR, Young LD (2005). Potassium fertilization effects on cotton lint yield, yield components, and reniform nematode populations. Agron. J. 97:1245-1251.

Pinheiro JB, Pozza EA, Pozza AAA, Moreira AS, Campos VP (2009). Estudo da influência do potássio e do cálcio na reprodução do nematoide do cisto da soja. Nematol. Bras. 33:17-27.

Quimette DG, Coffey MD (1990). Symplastic entry and phloem translocation of phosphonate. Pestic. Biochem. Phys. 38:18-25.

Rocha MR, Carvalho I, Corrêa GC, Cattini GP, Paolini G (2006). Efeito de doses crescentes de calcário sobre a população de Heterodera glycines em soja. Pesq. Agropec. Trop. 36:89-94.

Rodrigues FA, Datnoff LE (2005). Silicon and rice disease management. Fitopatol. Bras. 30:487-469.

Rodríguez-Kábana R (1986). Organic and inorganic nitrogen amendments to soil as nematode suppressants. J. Nematol. 18:129-135.

Rodríguez-Kábana R, King PS (1980). Use of mixtures of urea and blackstrap molasses for control of root-knot nematodes in soil.

Nematropica 10:38-44.

Rodríguez-Kábana R, King PS, Pope MH (1981). Combinations of anhydrous ammonia and ethylene dibromide for control of nematodes parasitic on soybeans. Nematropica 11:27-41.

Salgado SML, Resende MLV, Campos VP (2007). Efeito de indutores de resistência sobre *Meloidogyne exigua* do cafeeiro. Ciênc. Agrotecnol. 31:1007-1013.

Santos JM (1978). Efeitos de fertilizantes sobre *Meloidogyne exigua* e influência de seu parasitismo sobre a absorção e translocação de nutrientes em mudas de *Coffea arabica* L. 73. Dissertação (Mestrado em Fitopatologia), Universidade Federal de Viçosa, Viçosa.

Shaukat SS, Siddiqui IA (2003). Zinc improves biocontrol of Meloidogyne javanica by the antagonistic rhizobia. Pak. J. Biol. Sci. 6:575-579.

Siddiqui IA, Shaukat SS (2002). Zinc and glycerol enhance the production of nematicidal compounds in vitro and improve the biocontrol of *Meloidogyne javanica* in tomato by fluorescent pseudomonads. Lett. Appl. Microbiol. 35:212-217.

Siddiqui IA, Shaukat SS, Hamid M (2002). Role of zinc in rhizobacteria-mediated suppression of root-infecting fungi and root-knot nematode. J. Phytopath. 150:569-575.

Silva JFV, Garcia A, Pereira JE, Hiromoto D (1997). Nematoide de cisto da soja (*Heterodera glycines* Ichinohe). Londrina: Embrapa Soja, Documentos, P. 104.

Silva RV, Oliveira RDL, Nascimento KJT, Rodrigues FA (2010). Biochemical responses of coffee resistance against *Meloidogyne exigua* mediated by silicon. Plant Pathol. 59:586-593.

Streeter TC, Rengel Z, Neate SM, Graham RD (2001). Zinc fertilization increases tolerance to *Rhizoctonia solani* (AG 8) in Medicago truncatula. Plant Soil 228:233-242.

Terry LA, Joyce DC (2004). Elicitors of induced disease resistance n postharvest horticultural crops: a brief review. Postharv. Biol. Tecn. 32:1-13.

Wang ELH, Bergeson GB (1974). Biochemical changes in root exsudate and xylem sap of tomato plants infected with *Meloidogyne incognita*. J. Nematol. 6:194-202.

Wilhelm MS, Fisher JM, Graham RD (1985). The effect of manganese deficiency and cereal cyst nematode infection on the growth of barley. Plant Soil 85:23-32.

Zambolim L, Costa H, Vale FXR (2001). Efeito da nutrição mineral sobre doenças de plantas causadas por patógenos de solo. In: Zambolim L (ed) Manejo integrado fitossanidade: cultivo protegido, pivô central e plantio direto. Viçosa: Editora UFV. pp. 347-408.

Zambolim L, Rodrigues FA, Capucho AS (2005). Resistência a doenças de plantas induzida pela nutrição mineral. In: Venzon M, Júnior TJP, Pallini A (eds) Controle alternativo de pragas e doenças. Viçosa: EPAMIG. pp. 185-219.

3

Reproductive performance of different crossbred cows of Bangladesh

F. Kabir and J. J. Kisku

Faculty of Animal Science and Veterinary Medicine, Patuakhali Science and Technology University, Babugonj, Barisal, Bangladesh.

The study was conducted at Central Cattle Breeding and Dairy Farm in Savar, Dhaka to evaluate the reproductive performance of different crossbred cows in terms of gestation length, service per conception, postpartum heat period and calving interval. The genotypes Australian Friesian Sahiwal (AFS), Sahiwal × Friesian (SL×F), Local × Friesian (L×F), Local × Friesian × Friesian (LF$_1$×F) and Local × Friesian × Friesian × Friesian (LF$_2$×F) were considered. The highest performance for the trait service per conception was found in AFS (1.40±0.69) and the lowest performance was found in SL×F (1.80±0.63). The longest gestation length and calving interval were observed in SLxF, 281.0±3.26 and 542.0±9.87 days, respectively. The longest postpartum heat period (201.7±17.40 days) was found in LF$_1$×F cows. On the other hand, the shortest gestation length (277.0±5.21 days), postpartum heat period (135.5±10.58 days) and calving interval (436.07±9.87 days) were observed in L×F crossbred cows. From the above perspective it is concluded that L×F crossbred cows are more suitable for Bangladesh.

Key words: Reproductive, performance, crossbred, cows.

INTRODUCTION

Bangladesh is agriculture based subtropical country. Livestock is an important sub-sector of agriculture which plays an important role to promote human health and poverty alleviation. About 20% of the people directly depends on the livestock sector and thus contributes around 16.5% to the country GDP. The cattle production is an important part of livestock. Farmers rear cattle mainly for draft purpose but also as a means of economic upliftment from sale of milk. Majid et al. (1995) observed that the highest gestation was found in ¼ local - ¼ Friesian - ½ Sahiwal and shortest was in Friesian cows. Sultana (1995) found that the gestation period of local (L), Sahiwal (SL), ½ F- ½ SL, Jersey (J), ½ L- ½ J and ½ F was 274.98±1.76, 276.29±4.71, 275.5±2.42 and

274.72±1.48 days respectively. Islam and Bhuiyan (1997) observed that the breed type had no significant effect on number of services per conception. Khan (1990) found that the genotypic mean independent of parity had non-significant difference for service per conception.

The native cattle of Bangladesh have low productivity but disease resistance capacity was higher than that of exotic breeds. To develop the performance of native cattle, up gradation is necessary. Livestock development depends mainly on genetic potential of the animal. Native ruminant animals are non-descriptive and their genetic potential have not yet been recognized. Conservation and improvement of native animal germplasm are essential for profitable livestock farming to meet the increasing demand of milk and meat. Optimum nutrition, disease control and management practices permit better expression of genetic potential. Climatic stresses in the form of high ambient temperature, high humidity and erratic or inadequate rainfall affect the productivity of dairy cattle in the tropics. Reproductive efficiency is a major factor in the profitability of a dairy enterprise through its effect on the annual milk production of the herd and the cost of herd depreciation. In this study an attempt was made to evaluate the reproductive performance of different crossbred cows and to find out the suitable crossbred animals.

MATERIALS AND METHODS

Study area

The study was conducted at Central Cattle Breeding and Dairy Farm, Savar, Dhaka. The data collected during the period from 2005 to 2009.

Animals and data collection

The information on the reproductive performance of 54 cows of different genotypes was collected at Central Cattle Breeding and Dairy Farm, Savar, Dhaka. The experimental animals were divided into five genetic groups according to their genetic composition, that is, Australian Friesian Sahiwal (AFS), Sahiwal × Friesian (SL×F), Local × Friesian (L×F), Local × Friesian × Friesian (LF$_1$×F) and Local × Friesian × Friesian × Friesian (LF$_2$×F). A total of 270 observations on the reproductive parameters were recorded and statistically analyzed by following slight modification method of Majid et al. (1995).

Feeding and management

Feeding and management system in the farm was uniform throughout the year. Stall-feeding was practiced regularly. Concentrate feeds were included wheat bran, sesame oil cake, rice bran, grass pea and salt. Green grasses were supplied daily. Different types of green grasses, that is, Napier, Para, Maize, German and Oats were cultivated in the field near the farm. The grass after collection were ensiled in the pits and fed to cows both as fresh and ensiled. Records on date of birth, birth weight, date of first service, date of calving, daily milk yield, date of abortion, date of death and sold and lactation length were kept.

Traits studied

The following characteristics were used to measure reproductive performance of different crossbred animals that is gestation length, service per conception, postpartum heat period and calving interval.

Data collection

Gestation length

Rectal palpation technique was used for diagnosis of pregnancy. The period of intra-uterine development of embryo and fetus was considered as gestation length. It was calculated as the interval

from fertile service to parturition. The duration of gestation was determined in days. The difference in gestation period was associated with twinning, sex of calf and parity of cow.

Service per conception

This is defined as the average number of services or insemination required per conception and is a simple method of assessing fertility (Payne, 1970). Experimental animals were serviced by using artificial insemination (A.I.) technique.

Postpartum heat period

Postpartum heat period was calculated as the interval between parturition to next heat that was observed after a certain period of parturition. The period was considered in days.

Calving interval

The calving intervals were recorded on the basis of interval between the dates of one calving to the date of next calving. The calving intervals were recorded in days.

RESULTS AND DISCUSSION

The reproductive parameters of different crossbred animals and the average values were presented in the Table 1. The results are discussed under the following sub-headings.

Gestation period

Gestation period of different genetic groups were found to be 274.5±6.83, 281.0±3.26, 277.0±5.21 279.3±4.54 and 277.2±3.93 days under the genotypes AFS, SL×F, L×F, LF$_1$×F, LF$_2$×F, respectively. The mean gestation length was highest in SL×F (281.0±3.26 days) and lowest in AFS (274.5±6.83 days). The gestation length of present findings are more or less similar with the findings of Maarrof et al. (1987) who analyzed the data of 85 Jenubi cattle in dairy farms of central Iraq, where average gestation length was 283±1.5 days. Variation in gestation length within the species may be contributed mainly by maternal and fetal factors. The maternal factors include age of the dam, nutritional status and body condition of the dam. Fetal factors include the sex of the fetus, twinning and hormonal functions of the fetus. Environment such as season, feeding, and management also contribute to some extent (Hafez, 1993).

Service per conception

Service per conception of different genetic groups were found to be 1.40±0.69, 1.80±0.63, 1.64±0.74, 1.70±0.67 and 1.50±0.70 under the genotypes AFS, SL×F, L×F,

Table 1. Reproductive performance of different cross bred cows.

Parameter	Genotypes					Overall (Mean±SD)
	AFS (Mean±SD)	SL×F (Mean±SD)	L×F (Mean±SD)	LF$_1$×F (Mean±SD)	LF$_2$×F (Mean±SD)	
Gestation length (days)	274.5±6.83	281.0±3.26	277.0±5.21	279.3±4.54	277.2±3.93	277.8±2.46
Service per conception	1.40±0.69	1.80±0.63	1.64±0.74	1.70±0.67	1.50±0.70	1.60±0.15
Postpartum heat period (days)	177.6±10.24	182.0±12.40	135.5±10.58	201.7±17.40	181.1±11.74	175.58±24.3
Calving interval (days)	478.6±8.52	542.0±9.87	436.07±9.87	530.4±16.19	508.6±12.01	499.13±42.7

LF$_1$×F, LF$_2$×F, respectively. The highest performance in AFS (1.40±0.69) and lowest performance in SL×F (1.80±0.63) cows were recorded in terms of services required per conception. The results of this study are similar with the findings of Majid et al. (1995) who observed almost similar service per conception for different genotypes where the value of service per conception of Sahiwal (SL), Friesian (F), Local (L), ½ L-½ F (F$_1$), ½ SL- ½ F (F$_1$), ¼ L- ¾ F (F$_2$), ½ L- ½ F (F$_2$), 1/4 L- ¼ SL- ½ F (F$_2$), ¼ L- ¼ F- ½ SL and L- ½ F- SL (F$_3$) were 1.90±0.12, 1.27±0.19, 1.76±0.08, 2.20±0.49, 2.21±0.23, 2.00±0.37, 1.73±0.18, 2.00±0.39, 1.53±0.19 and 1.25±0.25, respectively. Islam and Bhuiyan (1997) found that service per conception was 1.23±0.17 in JR, 1.46±0.19 in JR×SN, 1.45±0.12 in SL×PMC and 1.23±0.10 in ¼ PMC × ¾ SL cows at Baghabarighat milk shed area. Hossen et al. (2012) observed the lowest service per conception (1.22) in PMC cows. A number of other factors, which influences service per conception are the quality and quantity of semen used in artificial insemination, improper detection of heat, failure to inseminate at appropriate time and skill of the inseminator. The other related factors are the level of fertility, which may be influence by diseases, semen handling techniques and other environmental factors.

Postpartum heat period

Postpartum heat period of different genetic groups were found 177.6±10.24, 182.0±12.40, 135.5±10.58, 201.7±17.40, 181.1±11.74 days under the genotypes AFS, SL×F, L×F, LF$_1$×F, LF$_2$×F, respectively. The highest average postpartumm heat period was obtained in LF$_1$×F (201.7±17.40) and lowest in L×F (135.5±10.58) crossbred cows. The results of the study corroborates with the findings of Majid et al. (1995) who found longest average postpartum heat period (223.5±40.14 days) in ¼ Local-Friesian crossbreed and the lowest (117.24±7.2 days) in ½ Local – ½ Friesian cows at the Central Cattle Breeding and Dairy Farm, Savar, Dhaka. Postpartum heat period is an important economic reproductive trait in a dairy herd. Hafez (1993) suggested that the postpartum breeding delayed up to 60 to 70 days after parturition, when the uterus under goes recovery and preparation for the next

conception. Chowdhury et al. (1994) was found the postpartum heat period 154.8 days in FN×SL crossbred cows. Hossen et al. (2012) observed that the shortest postpartum heat period (133.23 days) was in PMC cows. A period after calving to next heat and ovulating is considered anestrous or postpartum heat interval of cows. The length of the postpartum interval is influenced by nutrition, body condition, age, genetics and presence of the calf.

Calving interval

Calving interval of AFS, SL×F, L×F, LF$_1$×F and LF$_2$×F cows were 478.6±8.52, 542±9.87, 436.07±9.87, 530.4±16.19 and 508.6±12.01 days, respectively. The highest value of calving interval was observed in SL×F (542.0±9.87 days) cows and the lowest value was in L×F (436.07±9.87 days) cows. These results are more or less similar with the findings of Ghose et al. (1997) who recorded calving interval of 489.52 days for Pabna, 524.00 days for Dhaka, 430.86 days for Red Chittagong, 491.16 days for Sahiwal, 490.00 for Sindhi, 571.00 days for Sindhi × Pabna, 457.00 days for Sindhi × Local and 485.25 days for Sahiwal × Local cows. Hossen et al. (2012) found the shortest calving interval (414.90 days) in PMC cows. The differences in calving interval observed in the present study may be due to different environment, feeding, management and also due to irregularity in estrous.

Conclusion

In this study the considered genotypes were Australian Friesian Sahiwal (AFS), Sahiwal × Friesian (SL×F), Local Friesian (L×F), Local Friesian (LF$_1$×F) and Local × Friesian × Friesian (LF$_2$×F). The length of gestation period was more or less close to each other. The lowest gestation period was AFS cows followed by and highest gestation length SL×F. Minimum number of services was required for the conception of AFS cows. Maximum services were required for SL×F. The highest value of calving interval was obtained from SL×F cows and the lowest value was obtained from L×F cows.

The highest postpartum heat period was observed in LF$_1$×F and the lowest was obtained from L×F crossbred cows. The pure exotic breed (e.g. Holstein Friesian) is not suitable in context of Bangladesh in terms of environmental condition. It may require low temperature, better feeding and management. The disease prevention capacity is also lower than that of native cattle. On the other hand, the local cattle are well adapted as well as high disease resistance than exotic pure breed. The crossbred cattle performed better than that of exotic and native cattle in terms of adaptability and production. So, it is necessary to improve native cattle by selective breeding to increase the productive and reproductive performance. Considering the above perspective it is concluded that L×F crossbred cows are most suitable for Bangladesh.

REFERENCES

Chowdhury MZ, Tahir MJ, Rafique M (1994). Production performance and milk producing efficiency in different filial groups of Holstein Friesian × Sahiwal half-breds. Asian-Aust. J. Anim. Sci., 7: 383-387.

Ghose SC, Haque M, Rahman M, Saadullah M (1997). A comparative study of age at first calving, gestation period and calving of different breeds of cattle. Bangl. J. Vet. Med., 11: 9-14.

Hafez ESE (1993). Reproduction in Farm Animals. 6th eds. Lea and Febiger. USA.

Hossen MS, Hossain SS, Bhuiyan AKFH, Hoque MA, Talukder MAS (2012). Comparison of some important dairy traits of crossbred cows at Baghabarighat milk shed area of Bangladesh. Bangl. J. Anim. Sci., 41: 13-18.

Islam SS, Bhuiyan AKFH (1997). Performance of crossbred Sahiwal cattle of at the Pabna milk shed area in Bangladesh. Asian-Aust. J. Anim. Sci., 10: 581-586.

Khan AA (1990). A comparative study on the reproductive efficiency of native and crossbred cows. MSc Thesis, Bangladesh Agricultural University, Mymensingh.

Maarrof MN, Al-ani LM, Raseed ST (1987). Performance of Jersey cattle. Indian J. Anim. Sci., 57: 719-727.

Majid MA, Nahar TN, Talukder AI, Rahman MA (1995). Factors affecting the reproductive efficiency of crossbred cows. Bang. J. Anim. Sci., 2: 18-22

Payne WKA (1970). Cattle production in the tropics. Vol 1, Longman, London.

Sultana R (1995). Quantitative analysis of reproductive performance of purebred and their crossbred in the Savar Dairy Farm. MSc Thesis, Bangladesh Agricultural university, Mymensingh.

4

Effect of dietary lysine on performance and immunity parameters of male and female Japanese quails

Hajkhodadadi*, M. Shivazad, H. Moravvej and A. Zare-shahneh

Department of Animal Science, Faculty of Agriculture and Natural Resources, Tehran University, Karaj, Iran.

The present study was conducted to investigate the effects of dietary lysine (Lys) and sex on growing quail performance and immunity parameters. The six dietary Lys: 10, 11.5, 13.0, 14.5, 16.0 and 17.5 g/kg were provided. This experiment was carried out in a completely randomized design arrangement from 3 to 24 days. Each treatment was consisted of 5 floor pens (50 quail chicks). Body weight (BW) and body weight gain (BWG) increased significantly (P<0.05) with dietary Lys supplementation up to 14.5 g/kg. The quails that consumed 14.5 g/kg Lys had the best feed conversion ratio (FCR). The carcass weight, breast weight and yield, thigh weight and yield, mortality and response to sheep red blood cell (SRBC) were significantly (P<0.05) affected by dietary Lys. The breast yield, thigh weight and yield, white blood cells count, monocyte percentage and response to SRBC significantly (P<0.05) were influenced by sex. Neither age nor sex had any significant effect on lymphocyte percentage, heterophile percentage and heterophile to lymphocyte ratio at 24 days of age. In this study, an interaction of sex with Lys were not significant (P>0.05) for any traits. It concluded that during the 3 to 24 days of age optimum BW, BWG, FCR and immunity status could be obtained with 14.5 g/kg Lys, so these levels is more than NRC 1994 recommendation.

Key words: Lysine, Japanese quail, performance, immunity.

INTRODUCTION

Japanese quail (*Coturnix coturnix Japonica*) belongs to the order Galliformes and the family Phasianidae like the chicken (Karaalp, 2009). The Japanese quail possesses several advantages, such as rapid growth, early sexual maturity, high rate of egg production, and a short generation interval (Mandel et al., 2006).

Essential amino acid recommendations by NRC 1994 for Japanese quail are largely based on researches that conducted at least 5 to 6 years before 1994's but, the meat production performance of Japanese quails has also been improved during recent years due to genetic selection. Therefore, there is need of updating optimal nutritional requirements of Japanese quails with the improvement in genetic makeup to exploit production

potentiality (Kaur et al., 2008).

Lysine (Lys) is the second limiting amino acid after methionine in a maize–soybean based diet and most scientists use Lys as the basis to which all other amino acids are related when generating an "ideal balance", furthermore lysine and sulfur amino acids (SAA) are known to exhibit specific effects on carcass composition (Corzo et al., 2002).

In chickens as well as in mammals, it has been shown that deficiency or excess of dietary amino acids (Augspurger and Baker, 2007; Dozier et al., 2008) alters performance and immune responses but information on the effects of lysine on performance and immunity of Japanese quails is scanty (Kaur et al., 2008). The literature is also silent on the amino acid values in different feedstuffs and their requirements for Japanese quails. Therefore, the present study was conducted to elucidate the response of growing Japanese quail to different levels of Lys for optimum growth performance, feed utilization efficiency and immunity.

*Corresponding author. E-mail: ihkhodadadi@ut.ac.ir, imanhkhodadadi@yahoo.com.

Table 1. Ingredients and calculated composition of experimental diets (percentage as-fed basis).

Item	Graded level of lysine (g/kg)					
	10.0	11.5	13.0	14.5	16.0	17.5
Ingredient (%)						
Ground corn	52.34	52.34	52.34	52.34	52.34	52.34
Soybean meal(44% cp)	26.24	26.24	26.24	26.24	26.24	26.24
Corn gluten meal	15.16	15.16	15.16	15.16	15.16	15.16
Calcium carbonate	1.36	1.36	1.36	1.36	1.36	1.36
Dicalcium phosphate	0.88	0.88	0.88	0.88	0.88	0.88
Sodium chloride	0.16	0.16	0.16	0.16	0.16	0.16
Sodium bicarbonate	0.19	0.19	0.19	0.19	0.19	0.19
Vitamin premix[1]	0.25	0.25	0.25	0.25	0.25	0.25
Mineral mineral[2]	0.25	0.25	0.25	0.25	0.25	0.25
L-Lys·HCl	0	0.19	0.39	0.59	0.80	1.00
Gln	0.75	0.60	0.45	0.30	0.15	0
L-Thr	0.17	0.17	0.17	0.17	0.17	0.17
DL-Met	0.07	0.07	0.07	0.07	0.07	0.07
Inert filler[3]	2.18	2.14	2.09	2.04	1.98	1.93
Total	100	100	100	100	100	100
Calculated composition						
AME (kcal/kg)	2900	2900	2900	2900	2900	2900
CP (%)	24	24	24	24	24	24
Lys (%)	1.00	1.15	1.3	1.45	1.6	1.75
Met (%)	0.55	0.55	0.55	0.55	0.55	0.55
Met + Cys (%)	0.82	0.82	0.82	0.82	0.82	0.82
Thr (%)	1.12	1.12	1.12	1.12	1.12	1.12
Nonphytate P (%)	0.3	0.3	0.3	0.3	0.3	0.3
Calcium (%)	0.8	0.8	0.8	0.8	0.8	0.8
Sodium (%)	0.15	0.15	0.15	0.15	0.15	0.15

[1]Vitamin premix include per kilogram of diet: vitamin A (vitamin A acetate), 3600 IU; cholecalciferol (D3), 800 ICU; vitamin E (DL-α-tocopheryl acetate) , 7.7 IU; menadione, 0.8 mg; vitamin B12 0.01 mg; folic acid, 0.4 mg; choline chloride, 170 mg; D-pantothenic acid, 12 mg; riboflavin, 2.6 mg; niacin, 4 mg; biotin, 0.2 mg; thiamin, 0.7 mg; pyridoxine. [2]Mg; butylated hydroxytoluene, 125 mg. Mineral premix supplied the following per kilogram of diet: manganese, 16 mg; zinc, 15 mg; iron, 8 mg; copper, 4 mg; iodine, 1.6 mg; selenium, 0.08 mg; butylated hydroxytoluene, 125 mg. [3]Filler represents inert space (builder s sand) in the diet to which L-Lys-HCl was added to derive the projected Lys level.

MATERIALS AND METHODS

Birds

Day-old Japanese quail chicks (*C. cotornix Japonica*) were provided from commercial hatchery and reared on a common starter diet during the first 3 days of age. At 3 days of age, chicks were weighed individually and based on similar body weight, 1500 chicks selected and randomly allocated to 30 floor pens (50 chicks per pen) so average initial body weight and variance were similar between all pens. Each of the experimental diet was then randomly assigned to 5 pens.

All birds reared on floor pens (80 × 150 × 100 cm) on wood shavings in temperature-controlled house during 3 to 24 days of age. The temperature was 35ºC in the first week and reduced by 2.5ºC/week up to 3 weeks of age. The birds received 24 h of light/day during the 24 days of age. Food and water were available *ad libitum*. All chicks received feeds from placement until 24 days of age in mash form, according to its treatment.

Diets

A basal diet was deficient in lysine content (1%) and formulated mainly based on corn, corn gluten meal, and soybean meal (Table 1). Crystalline amino acids were included to assure the minimum levels of all other essential amino acids in a manner that would meet or exceed NRC (1994) recommendations (Corzo et al., 2008). The calculated content of AME was 2,900 kcal/kg (12.1 MJ/kg), and the analyzed concentrations of CP and lysine were 261 and 11 g/kg, respectively. The lysine level was gradually increased in 6 further isonitrogenous diets by stepwise inclusion of L-lysine •HCl at the expense of L-glutamine to achieve the following lysine concentrations (g/kg): 10, 11.5, 13.0, 14.5, 16.0 and 17.5 (Table 1). L-lysine•HCl, DL-methionine and L-threonine were feed grade quality. All ingredients with the exception of the variable ones (L-glutamine, L-lysine•HCl and inert filler) were mixed as a single lot and the mix were divided into 6 parts. L- Glutamine, L-lysine •HCl and inert filler were added separately in the respective proportions to each diet then mixed again.

Table 2. Dietary Lys effect on mixed-Japanese quails' performance at 3 to 24 d of age[1].

Variable	Dietary lysine (g/kg)						SEM[2]	CV[3]	P-value
	10.0	11.5	13.0	14.5	16.0	17.5			
Initial body weight	17.76	17.62	17.68	17.56	17.53	17.64	0.42	2.38	0.970
Body weight									
BW, 10 days (g)	27.91[b]	28.07[b]	28. 76[b]	31.56[a]	31.15[a]	30.16[ab]	1.46	4.95	0.0006
BW, 17 days (g)	50.96[c]	56.96[b]	57.25[b]	62.22[a]	60.15[ab]	61.88[a]	2.78	4.77	0.0002
BW, 24 days (g)	93.83[c]	97.64[c]	100.01[bc]	110.62[a]	107.48[ab]	110.51[a]	5.12	4.96	0.0005
Body weight gain									
BWG, 3-10 days (g)	10.23[c]	10.56[bc]	11.13[bc]	13.99[a]	13.61[a]	12.52[ab]	1.29	10.75	0.001
BWG,10-17 days (g)	23.05[c]	26.89[b]	27.49[b]	30.66[ab]	29.00[ab]	31.71[a]	2.35	8.38	0.001
BWG,17-24 days (g)	39.86[b]	42.76[ab]	46.67[ab]	48.40[a]	47.32[a]	48.63[a]	4.54	9.95	0.076
BWG, 3-24 days (g)	72.27[c]	81.83[b]	84.13[ab]	93.06[a]	89.94[ab]	92.87[a]	5.84	6.82	0.0005

[1]Means with different superscripts within the same row differ significantly (P<0.05). [2]Standard error of means. [3]Coefficient of variation.

Measurements

Performance parameters

The body weight, feed consumption of all birds were recorded at 3, 10, 17, 24 day of age, the average body weight gain was calculated weekly. The mortality was recorded daily. Feed conversion was corrected for mortality and represents grams of feed consumed by all birds in a pen divided by grams of body weight gain per pen (Fisher, 1998).

At 24 days of age, 10 female and 10 male quails from each treatment were selected and weighed individually before slaughter. The quails were starved for 4 h, but drinking water was supplied. The quails were slaughtered by severing the jugular vein, blood samples were collected into 1.5 ml tubes immediately thereafter.

The quails were plucked mechanically after hot water scalding then carcasses was eviscerated by hand. Weights were individually recorded for the carcass, thigh, breast, liver and total alimentary tract. The yields of different carcass traits were expressed as percentage to final body weight (Fatufe et al., 2004).

Blood parameters

At 21 days of age blood samples were collected from two birds in each pen into the heparinzed tubes from the jugular vein. The samples were transferred to laboratory immediately then blood erythrocyte count, blood leukocyte count, differential leukocyte count; lymphocyte (Lym), monocyte (Mon) and heterophile (Het) was measured.

At 24 days of age, four quails from each pen (2 from each sex) were bled to collect serum. After overnight clotting at 4°C, the samples were centrifuged for 20 min at 4,000 × g then blood serum was frozen until analysis.

Immunity response parameters

At 14 days of age, 2 female and 2 male quails were selected from each pen (that is, 20 birds/dietary treatment providing 100 birds in all) and were inoculated intramuscularly with 0.1 ml of a 0.5% suspension of sheep red blood cell (SRBC). For measuring the primary immunity response blood samples were obtained from all SRBC injected birds at 0 and 5 days post-inoculation.

All the samples were incubated at 37ºC for one hour to aid clotting and retraction then centrifuged at 15000 ×g for 5 min for collection of sera. Blood serum was frozen until analysis for antibody titers could be performed. The all serum samples after thawing were used for analysis. All the microtitre plates (U-bottomed) were rinsed with phosphate-buffered saline (PBS; pH 7 to 6) then dried before the haemagglutination antibody (HA) titre was estimated by a micro-haemagglutination method using twofold serial dilutions of sera (Biswas et al., 2006).

Statistical analysis

Data in this experiment were evaluated by analysis of variances (ANOVA) in a completely randomized design. Pen was used as the experimental unit for analysis. All data were examined for normality distribution for ANOVA analysis was then analyzed by the general linear models (GLM) procedure of SAS (1996). Lysine and sex effects (P<0.05) were separated using the Duncan multiple range test option of SAS (1996) with an alpha (α) of 0.05.

RESULTS AND DISCUSSION

Performance

The performance results are presented in Tables 2 and 3. There was no significant difference (P>0.05) in initial body weight, between experimental treatments. Dietary Lys significantly (P<0.05) affected BW at 10, 17 and 24 days of age. BW was increased by Lys increasing, so the quails that consumed at least 14.5 g/kg Lys was heavier than other Lys treatments and beyond this level there was no significant difference in BW between treatments.

Body weight gain (BWG) within 3 to 10, 10 to 17, 17 to 24 and 3 to 24 days was significantly (P<0.05) affected by dietary Lys. BWG was in line with BW so the quails that consumed at least 14.5 g/kg Lys had highest BWG than others. The BWG of quails that consumed dietary Lys more than NRC recommendation had significant

Table 3. Dietary Lys effect on mixed-Japanese quail feed intake and feed conversion at 3-24 d of age[1].

Variable	Dietary lysine (g/kg)						SEM[2]	CV[3]	P-value
	10.0	11.5	13.0	14.5	16.0	17.5			
Daily feed intake									
FI, 3-10 days (g)	5.36	5.97	5.58	5.55	5.58	6.20	0.54	9.53	0.293
FI, 10-17 days (g)	11.44	12.02	11.12	11.59	10.51	11.96	1.11	9.7	0.423
FI, 17-24 days (g)	25.47[a]	25.66[a]	24.56[ab]	23.25[bc]	23.34[bc]	22.20[c]	1.27	5.31	0.007
FI, 3-24 days (g)	42.27[ab]	43.66[a]	41.28[ab]	40.39[b]	40.37[b]	39.43[b]	1.88	4.53	0.059
Feed conversion ratio									
FCR, 3-10 days (g:g)	3.15[a]	3.22[a]	3.30[a]	2.38[b]	2.46[b]	2.99[a]	0.296	10.14	0.0007
FCR, 10-17 d (g:g)	2.66[ab]	2.55[abc]	2.89[a]	2.27[bc]	2.18[c]	2.27[bc]	0.252	10.21	0.0056
FCR, 17-24 days (g:g)	3.89[a]	3.66[ab]	3.17[bc]	2.97[c]	2.89[c]	2.75[c]	0.399	12.38	0.0039
FCR, 3-24 days (g:g)	3.23[a]	3.14[a]	3.12[a]	2.51[b]	2.54[b]	2.67[b]	0.15	5.24	<0.0001

[1]Means with different superscripts within the same row differ significantly (P<0.05). [2]Standard error of means. [3]Coefficient of variation.

Table 4. Dietary Lys and sex effects on slaughter related traits and mortality of quails at 24 days of age[1].

Variable	Dietary lysine (g/kg)						Sex		Effects			CV[2]	SEM[3]
	10.0	11.5	13.0	14.5	16.0	17.5	Female	Male	T	S	T×S		
Carcass weight (g)	89.89[c]	85.01[bc]	96.47[abc]	103.13[a]	102.92[a]	99.25[ab]	98.75	96.73	*	NS	NS	6.96	6.80
Breast weight (g)	26.26[c]	28.24[bc]	29.66[ab]	30.42[ab]	31.51[a]	30.65[ab]	29.27	29.64	*	NS	NS	8.51	2.50
Breast yield (%)	23.14[c]	23.45[bc]	24.84[abc]	24.49[abc]	25.99[a]	25.20[ab]	23.93[b]	25.11[a]	*	*	NS	7.37	1.80
Tight weight (g)	16.45[b]	18.28[ab]	18.23[ab]	18.72[a]	19.74[a]	19.52[a]	17.57[b]	19.41[a]	*	*	NS	9.62	1.77
Tight yield (%)	14.44[c]	15.12[bc]	15.22[bc]	14.99[bc]	16.32[a]	15.99[ab]	14.29[b]	16.40[a]	*	*	NS	6.55	1.00
Liver weight (g)	3.47	3.75	3.42	3.47	3.56	3.38	3.52	3.50	NS	NS	NS	12.10	0.42
Liver yield (%)	3.06	3.92	2.86	2.78	2.95	2.79	2.89	2.97	NS	NS	NS	10.24	0.30
GI weight (g)	14.48	14.79	14.23	14.89	14.43	14.11	14.61	14.36	NS	NS	NS	11.29	1.63
GI yield (%)	12.76	12.43	11.93	11.92	11.99	11.64	12.01	12.21	NS	NS	NS	10.57	1.28
Mortality (%)	3.2[a]	2.5[a]	1.9[ab]	1.4[b]	0.98[b]	1.3[b]	2.43	2.86	*	NS	NS	14.56	0.45

[1]Means with different superscripts within the same row differ significantly (P<0.05). [2]Coefficient of variation. [3]Standard error of means.

(P<0.05) difference with other lower levels. This findings was agree with Dozier et al. (2008) in broilers, who concluded that the NRC (1994) Lys requirement of 0.85% (total basis) is not adequate from 6 to 8 weeks of age for Ross × Ross 708 broilers.

There were no significant differences (P>0.05) in feed intake at 3 to 10 d, 10 to 17 days but Lys significantly (P<0.05) affected feed intake at 17 to 24 and 3 to 24 days. The significant difference in feed intake at 17 to 24 day may occurred for Lys deficiency diets, in the other hand the quails with inadequate Lys intake consumed more feed than other treatments for compensating the deficient Lys uptake.

Dietary lysine significantly (P<0.05) affected feed conversion ratio (FCR) at 3 to 10, 10 to 17, 17 to 24 and 3 to 24 days of age (Table 3). With increasing dietary Lys supplementation FCR showed significantly (P<0.05) improvement. However, the quails that consumed 14.5 g/kg dietary Lys showed the best FCR in compared with the other dietary Lys treatments.

Some studies reported that the Lys has positive effects on broiler body weight gain of increasing dietary Lys is obtained through the improvement of both feed intake and feed conversion (Fatufe et al., 2004; Sterling et al., 2005; Tesseraud et al., 2009) whereas other reported that increasing Lys only improved feed conversion without affecting feed intake (Corzo et al., 2002; Corzo et al., 2003). In the present study with Japanese quails we found that effect of Lys on body weight by means of affecting on the both feed intake and feed conversion.

Carcass constitutes

The effects of dietary Lys and sex on carcass constitutes at 24 days is presented in Table 4. Dietary Lys significantly (P<0.05) affects carcass weight, breast weight and yield, thigh weight and yield. Carcass weight increased up to 103.13 g with 14.5 g/kg dietary Lys although this Lys level has no significant difference with 13 g/kg Lys.

Table 5. Dietary Lys and sex effects on hematology and immunity parameters of Japanese quails at 24 days of age[1]

| Variable | Dietary lysine (g/kg) | | | | | | Sex | | Effects | | | CV[2] | SEM[3] |
	10	11.5	13	14.5	16	17.5	Female	Male	T	S	T×S		
Packed cell volume (%)	33.87	33.48	34.51	36.41	33.61	35.26	31.54[b]	37.51[a]	NS	*	NS	10.85	3.76
Red blood cell (n×10^6/μl)	2.97[b]	3.08[ab]	3.12[ab]	3.30[ab]	3.11[ab]	3.41[a]	3.14	3.19	*	NS	NS	9.28	0.29
White blood cell (n/μl)	23863	20013	22100	24538	23050	21075	20179[b]	24700[a]	NS	*	NS	20.33	4563
Differential leucocytes count													
Lymphocyte (%)	72.00	70.62	73.00	74.87	72.62	75.62	72.12	74.12	NS	NS	NS	6.36	4.65
Monocyte (%)	3.37	2.75	2.75	2.87	2.62	2.37	2.37[b]	3.20[a]	NS	*	NS	45.03	1.32
Heterophil (%)	20.25	20.12	21.62	22.12	21.12	19.00	19.91	21.50	NS	NS	NS	20.26	4.19
Hetro/Lympho ratio	0.280	0.273	0.296	0.311	0.296	0.251	0.290	0.279	NS	NS	NS	27.2	0.07
Humeral immunity													
Response to SRBC	5.62[b]	7.25[ab]	7.37[ab]	8.12[a]	7.62[a]	6.87[ab]	7.16	7.12	*	NS	NS	23.17	1.65

[1]Means with different superscripts within the same row differ significantly (P<0.05). [2]Coefficient of variation. [3]Standard error of means.

The breast weight and yield showed the best values between 13 to 16 g/kg dietary Lys level. The quails that consumed 16 g/kg dietary Lys had the highest amount of thigh weight and yield. This result was agree with other finding in broiler or turkey, that concluded Lys need is higher when breast and thigh weight was included as response of requirement determination (Coleman et al., 2003; Corzo et al., 2002; Fatufe et al., 2004; Kaur et al., 2008; Wang et al., 2006).

Sex had no significant (P>0.05) effect on the carcass and, breast weights but breast yield, thigh weight and yield were significantly (P<0.05) affected by sex, so in all mentioned traits male quails presented the higher amount than female quails. Neither Dietary Lys nor sex had any significant (P>0.05) effect on liver, gastrointestinal tract weights and yields.

Supplementation of diets with Lys significantly (P<0.05) reduced mortality during 3 to 24 d of age, so quails, which consumed diet with 16 g/kg Lys content, represented minimum mortality in compared with other treatments. Mortality of quails during 3 to 24 days of age did not differ significantly (P>0.05) between male and female quails.

Blood parameters

The effects of dietary Lys and sex on hematology parameters at 24 days were presented in Table 5. Dietary Lys had no significant (P>0.05) effect on packed cell volume (PCV) at 24 days of age, but packed cell volume was differed significantly (P<0.05) between male and female quails, so male quails had more PCV than female ones. The sex has no significant (P>0.05) effect on red blood cell (RBC), but Lys had significant (P<0.05) effect on red blood cell, it was increased numerically with increasing dietary Lys level but to our knowledge, there

are no similar reports in the literature on Japanese quails for comparing our results.

Immunity response

The effects of dietary Lys and sex on white blood cell (WBC) count, differential leukocyte count; lymphocyte (Lym), monocyte (Mon), heterophile (Het) and humoral immunity (response to SRBC injection) were shown in Table 5. Lys had no significant (P>0.05) effect on WBC, but sex had significant (P<0.05) effect on WBC so the male quails (24700 n/μl) had higher WBC than the female quails (20179 n/μl).

Neither Lys nor sex had any significant (P>0.05) effect on blood lymphocyte, heterophile and heterophile to lymphocyte ratio at 24 days of age. Sex has significant (P<0.05) effect on blood monocyte so the male quails (3.2%) had more monocyte than the female quails (2.37%). Lys had significant (P<0.05) effect on response to SRBC injection as humoral immunity of quails but sex had no significant (P>0.05) influence on response to SRBC.

Our finding indicated that dietary supplementation of Lys improved humoral immunity of growing quails at 24 days of age so the quails who consumed diet with 14.5 g/kg dietary level showed the highest amount of immunity response, these finding was agree with (Biswas et al., 2006; Scholtz et al., 2009) who concluded that SRBC injection could use in growing quail as immunity response and nutrient supplementation has a beneficial effect on immune responses in growing Japanese quail.

Conclusion

This study indicated that a better response can be

obtained with 14.5 g/kg dietary Lys, when body weight, feed conversion ratio were considered and may be at least 16.0 g/kg when breast and thigh meat properties are taken into account in Japanese quail from 3 to 24 d of age, so these levels are more than NRC (1994) recommendation for Japanese quail.

Furthermore, humeral immune system functions of healthy quails were improved numerically by fortifying diets with Lys in excess of NRC (1994) recommendations, although it did not differ statistically with NRC 1994 recommendation level. Future research, however, should compare dietary Lys effects on immune system functions of quails reared in conventional environments or during an infectious challenge.

ACKNOWLEDGEMENTS

This study was funded by Department of Animal Science, Faculty of Agriculture and Natural Resources, Tehran University, Iran. The author wishes to acknowledge Degussa AG (Germany) for doing amino acid analyses in maize, corn gluten meal and soybean meals. The authors greatly appreciated Mrs. Mohammad Reyahi and Babake Sajedi for their practical assistance.

REFERENCES

Augspurger NR, Baker DH (2007). Excess dietary lysine increases growth of chicks fed niacin-deficient diets, but dietary quinolinic acid has no niacin-sparing activity. Poult. Sci. 86:349-355.

Biswas A, Mohan J, Sastry KVH (2006). Effect of higher levels of dietary selenium on production performance and immune responses in growing Japanese quail. Br. Poult. Sci. 47:511-515.

Coleman RA, Bertolo RF, Moehn S, Leslie MA, Ball RO, Korver DR (2003). Lysine requirements of pre-lay broiler breeder pullets: Determination by indicator amino acid oxidation. J. Nutr. 133:2826-2829.

Corzo A, Moran Jr ET, Hoehler D (2002) Lysine need of heavy broiler males applying the ideal protein concept. Poult. Sci. 81:1863-1868.

Corzo A, Moran Jr ET, Hoehler D (2003) Arginine need of heavy broiler males: Applying the ideal protein concept. Poult. Sci. 82:402-407.

Corzo A, Dozier Iii WA, Kidd MT (2008) Valine nutrient recommendations for Ross x Ross 308 broilers. Poult. Sci. 87:335-338.

Dozier WA, Corzo A, Kidd MT, Schilling MW (2008) Dietary digestible lysine requirements of male and female broilers from forty-nine to sixty-three days of age. Poult. Sci. 87:1385-1391.

Fatufe AA, Timmler R, Rodehutscord M (2004) Response to lysine intake in composition of body weight gain and efficiency of lysine utilization of growing male chickens from two genotypes. Poult. Sci. 83:1314-1324.

Fisher C (1998) Amino Acid Requirements of Broiler Breeders. Poult. Sci. 77:124-133.

Karaalp M (2009) Effects of decreases in the three most limiting amino acids of low protein diets at two different feeding periods in Japanese quails. Br. Poult. Sci. 50:606-612.

Kaur S, Mandal AB, Singh KB, Kadam MM (2008) The response of Japanese quails (heavy body weight line) to dietary energy levels and graded essential amino acid levels on growth performance and immuno-competence. Livestock Sci. 117:255-262.

Scholtz N, Halle I, Flachowsky G, Sauerwein H (2009) Serum chemistry reference values in adult Japanese quail (Coturnix coturnix japonica) including sex-related differences. Poult. Sci. 88: 1186-1190.

Sterling KG, Vedenov DV, Pesti GM, Bakalli RI (2005) Economically optimal dietary crude protein and lysine levels for starting broiler chicks. Poult. Sci. 84:29-36.

Tesseraud S, Bouvarel I, Collin A, Audouin E, Crochet S, Seiliez I, Leterrier C (2009) Daily variations in dietary lysine content alter the expression of genes related to proteolysis in chicken Pectoralis major Muscle. J. Nutr. 139:38-43.

Wang Y, Hou S, Huang W, Zhao L, Fan HP, Xie M, Wang Lh (2006) Lysine, Methionine and Tryptophan Requirements of Beijing Ducklings of 0-2 Weeks of Age. Agric. Sci. China 5:228-233.

5

Superovulation and embryo quality with porcine follicle stimulation hormone (pFSH) in Katahdin hair sheep during breeding season

Fernando Sánchez[1], Hugo Bernal [1], Alejandro S. del Bosque [1], Adán González [1], Emilio Olivares[1], Gerardo Padilla[3], Rogelio A.Ledezma[2]

[1]Universidad Autónoma de Nuevo León, Department of Agronomy, Laboratory of Animal Reproduction, Agricultural Sciences Campus UANL, Francisco Villa S/N, Ex Hacienda Canadá, P.O. Box 66050 General Escobedo, Nuevo León, México.
[2]Department of Veterinary Medicine, Laboratory of Reproduction, Agricultural Sciences Campus UANL, Francisco Villa S/N, Ex Hacienda Canadá, P.O. Box 66050 General Escobedo, Nuevo León, México.
[3]School of Medicine,Agricultural Sciences Campus UANL, Francisco Villa S/N, Ex Hacienda Canadá, P.O. Box 66050 General Escobedo, Nuevo León, México.

Estrus synchronization of 21 ewes was carried out using intravaginal sponges (60 mg of medroxyprogesterone acetate, Sincrogest®-Sanfer, Mexico) for 14 days. 7 ewes were randomly assigned to one of the three treatments: T_1, 80 mg; T_2, 120 mg; T_3, 140 mg of porcine follicle stimulating (pFSH) (Folltropin-p®-Bioniche, Canada) administered on day 12 after sponge insertion. The dose was divided in decreasing doses; second and third day representing 50 to 75% and 25 to 33% of the first day dose, respectively. At the beginning of superovulation treatment, each ewe received 5 mg of dinoprost tromethamine (1 ml of Lutalyse® Pfizer, Mexico). An effect of pFSH level was observed on the interval from sponge removal and the onset of estrus (REI) (P < 0.01), with a difference of 13.5 h between ewes from treatment 2 and 3. The number of follicles (F), number of collected embryos (CE) and transferable embryos (TE) (P < 0.05) was also different among treatments, with highest values observed in T3 (3.0 ± 0.6 for CE and 2.3 ± 0.5 for TE). Embryo recovery rate was in average 66.5%, (P > 0.05) of CL observed. Further investigation is guaranteed to evaluate the effect of pFSH doses on estrus synchronization for selected breeding seasons.

Key words: Embryo transfer, hair sheep, porcine follicle-stimulating hormone (pFSH), breeding season.

INTRODUCTION

Sheep production in Mexico has increased during the last years due to high demand of ovine meat (AMCO, 2008). Currently, there is a sheep population of 8.5 million of heads, with a trend to reach 10 million animals in the next years, of which 33% is hair sheep (FAO, 2008). At present, sheep breeders are enforced to improve the genetic potential of their animals, and in this regard they also look for opportunities to spread the genetic material, using several reproductive techniques, like artificial insemination, as well as assisted reproductive techniques such as MOET (multiple ovulation and embryo transfer) integral program, where the aim is to speed up via mothers' genetic progress (Baldasarre, 2007). Since its beginning, this technique has had varied responses in

Table 1. Procedure of experimental superovulation treatments applied in the present study to Katahdin hair sheep.

Treatment pFSH	Administration shift	Administration protocol (mg of pFSH per dose)		
		First day	Second day	Third day
1 (80 mg)	a.m.	20	15	5
	p.m.	20	15	5
2 (120 mg)	a.m.	30	20	10
	p.m.	30	20	10
3 (140 mg)	a.m.	40	20	10
	p.m.	40	20	10

goats and sheep in accordance with embryo recovery rate and depending on each of the protocols of superovulation that have been used on different breeds worldwide (Cognie et al., 2003; Gonzalez-Bulnes et al., 2004; Baldasarre, 2007; Paramio, 2010). Among the limitations recently found, one concerns optimal response of the different superovulation protocols that have been used, where dosage of the follicle-stimulating hormone, (FSH) may vary in accordance with body size and breed affecting follicular dynamic (Baldasarre, 2007; Veiga-Lopez et al., 2008; Menchaca et al., 2009, 2010; Paramio, 2010). Most of the studies have been done on wool sheep (Cognie et al., 2003). However, in hair sheep, great part of own features are due to their adaptation to diverse climatic conditions (Ortega-Abasolo, 2006), high prolificacy rates (Lopez-Junior et al., 2006; Sánchez et al., 2011) and too low seasonal anoestrus presence, as well as its resistance to parasites (Santos et al., 2007; Tabarez-Rojas et al., 2009; Ungerfeld and Sánchez, 2012); there is great interest in propagating hair breeds in Mexico, either from paternal or maternal side (embryo transfer). Since the year 2000, Katadhin breed started being introduced to the United States of America, because of its capacity for adaptation to diverse climates and for being a dual-purpose breed (AMCO, 2008). Currently, in Mexico, breeding programmes related with artificial insemination and embryo transfer have been developing; for which, in regard to Katadhin breed, there have been no reports of an embryo transfer programme. The aim of this study was to evaluate the effect of 3 doses of porcinefollicle-stimulating hormone (pFSH) on the response of embryo recovery and ovary structures in hair sheep, during the breeding season.

MATERIALS AND METHODS

Location

The present study was carried out during the natural breeding season (autumn) at the "Mary" sheep flock, located in Higueras, Nuevo Leon, at 25° 54' North latitude and 99° 58' West longitude, at an altitude of 451 masl. The weather is extreme, with an annual

mean temperature of 25°C. Twenty one Katahdin ewes, (initial average age = 22 months, body weight = 47.5 ± 3.7 kg and average body condition (BC) of 2.75 were used. Embryo recovery was carried out at the Laboratorio de Reproducción Animal of the Departament of Agronomy of the Universidad Autónoma de Nuevo León, located at the km 17.5 of the Zuazua-Marin road in Marír, Nuevo León, México.

Selection and management of ewes

Ewes that had given birth during the last breeding season were selected, with a postpartum period of 3 months before initiating the superovulation treatment. Sanitary management was provided to the females, which consisted of administration of vitamins A, D and E at a rate of 1 ml/animal with a concentration of 500 000, 75 000 and 500 IU, respectively. Likewise, internal deworming was carried out using a combination of ivermectin (5 mg, ®Pfizer, Mexico) and closantel (125 mg, ®Bayer, Mexico) by subcutaneous via at a rate of 1 ml/25 kg of body weight. A month before hormonal treatment nutritional supplementation was initiated; daily concentrate allowance was 0.81 kg, containing 18% of crude protein (CP) and 2.5 Mcal/kg of body weight. All ewes had access to green forage and were kept in confinement pens during the experimental period.

Synchronization and superovulation protocol

Estrus synchronization of the 21 ewes was carried out by the administration of intravaginal sponges, containing 60 mg of medroxyprogesterone acetate (MPA), for 14 days (Sincrogest®-Sanfer, Mexico). Superovulatory treatment consisted in intramuscular administration of porcine follicle-stimulating hormone (pFSH, Folltropin-p®-Bioniche, Canada). The administration of pFSH started on day 12 after sponge insertion. The pFSH dose was divided into 6 applications at intervals of 12 h (8 a.m. and 8 p.m., for 3 days) in decreasing doses (Table 1). Each ewe received 5 mg of dinoprost tromethamine (1 ml of Lutalyse® Pfizer, Mexico) at the moment superovulation treatment was initiated. 7 ewes were randomly assigned to each of the 3 treatments described in Table 1. Upon removing the sponges, estrus was monitored, twice a day during 3 days, with a fertile male ram. After estrus detection, ewes were fertilized by natural mating.

Embryo recovery

Embryo recovery was carried out via laparotomy technique at day 7 after vaginal sponge removal, using uterine washing enriched with PBS (phosphate buffered saline, Biolife, Agtech, KS, USA) (40 ml

Table 2. Responses to superovulatory treatments (mean ± SEM) in characteristics of observed follicles and collected embryos in Katahdin hair sheep.

pFSH dose (mg)	No. females	REI (h)REI	No. follicles/ ewe (F)	Corpus luteum/ ewe (CL)	Collected embryos/ewe (CE)	Transferable embryos/ewe(TE)
T1 = 80	7	20.1 ± 2.1^b	0.1 ± 0.6^b	1.4 ± 1.1^{ns}	1.0 ± 0.6^b	1.0 ± 0.5^b
T2 = 120	7	14.9 ± 2.3^c	1.8 ± 0.6^a	4.5 ± 1.2^{ns}	3.0 ± 0.7^a	2.0 ± 0.6^a
T3 = 140	7	28.4 ± 2.5^a	2.1 ± 0.6^a	4.6 ± 1.1^{ns}	3.0 ± 0.6^a	2.3 ± 0.5^a

Values with different letters within each column are significantly different in each variation source ($P < 0.05$); [ns], non-significant.

Table 3. Responses to superovulatory treatments (mean ± SEM) in characteristics of collected and transferable embryos in Katahdin hair sheep.

pFSH dose (mg pFSH)	% Embryo recovery	% Transferable embryos
$T_1 = 80$	$1/1.4 (67.4)^{ns}$	$1.0/1.0 (100)^a$
$T_2 = 120$	$3/4.5 (66.6)^{ns}$	$2.0/3.0 (66.6)^c$
$T_3 = 140$	$3/4.6 (65.6)^{ns}$	$2.3/3.0 (76.3)^b$

Values with different letters within each column are different ($P < 0.05$); [ns], non-significant.

for each uterine horn). The technique described by Baril et al. (1995), where ewes were subjected to a nutritional and liquid diet of 24 h before surgery, was used. Once the washed uterine liquid was collected, identification of ovary structures was carried out, as well as search and classification of embryos based on the description of Robertson and Nelson (1999). Embryos showing a symmetric and spherical mass, with individual and uniform blastomeres in regard to size, color and density, were considered able to be transfered. Parallel with evaluation of collected embryos, at the moment uterine flushing, the number of follicles and corpus lutea were visually counted, without considering type and size of these 2 ovarian structures.

Statistical analysis

The evaluated variables were: estrus percentage, interval from sponge removal to the onset of estrus (REI), follicle number (F), corpus lutea number (CL), collected embryo number (CE), and transferable embryo number (TE). The influence of treatment was analyzed using a lineal model after normalizing the data through the square root transformation, according to Snedecor and Cochran (1980). When significant differences occurred, the minimum significant difference (MSD) was used for post-hoc comparisons. Data were analyzed using SPSS 17 (2008) and are presented as mean ± SEM.

RESULTS

In the 3 treatments, 100% of estrus was observed in the treated ewes. A significant effect ($P < 0.05$) of pFSH level on REI, as well as on the number of follicles (F), number of collected embryos (CE) and transferable embryos (TE) was detected. There was no effect of treatments on the number of corpus luteum (CL).

Table 2 shows the results for every pFSH dose used. Overall REI average was 21.1 ± 2.1 h in average for all ewes. REI was longest for T3, shortest for T2 and

intermediate for T1 ($P < 0.05$). Ewes from T3 and T2 showed the highest values for F, CE and TE ($P < 0.05$); the lowest response with regard to the evaluated variables was shown in ewes receiving 80 mg of pFSH (T1) ($P < 0.05$).

There were no differences among treatments ($P > 0.05$) in the percentages of embryo recovery (Table 3). The lowest ($P < 0.05$) values of transferable embryos (66.0%) were found in ewes receiving 120 mg of pFSH (T2), followed by T3 (76.3%) and T1 (100%), respectively.

DISCUSSION

In the present study, estrus percentage was higher than the one reported by Lopes-Junior et al. (2006) in Morada Nova native ewes, from the north of Brazil, where they obtained 88.9 and 90.9% presence of estrus in young and adult females, respectively. This may be due to good body condition of sheep, as well as having been carried out during the breeding season, which guarantees a better response than if conducted outside their reproductive time. Likewise, the sponge used contained 60 mg of medroxyprogesterone, which ensures high progesterone concentration at the moment of device removal and therefore, estrus rate is higher, since there are sponges containing 30 and 40 mg medroxyprogesterone available in the market.

However, up until 90% of estrous can be achieved in hair sheep, by using new CIDR, comparing with 80.5% achieved by using intravaginal sponges, or 61.8% by previously using CIDR (Godfrey et al., 1999; Kohno et al., 2005; Ortega-Abasolo, 2006). According to González-Bulnes et al. (2004), an additional alternative for increasing estrus percentage can be the use of a second sponge inserted 7 days apart from the first one. With this protocol it is possible to record up to 95% estrus percentage, which represents 18% improvement in results, compared with the figures obtained by using a single sponge over 14 days. This improvement supposes application of a prostaglandin analogue at the beginning of the superovulation (González-Bulnes et al., 2004).

Lopes-Junior et al. (2006) reported a greater variation in the onset of estrus when progesterone concentration is low at the time of removing the intravaginal device. New short synchronization protocols (5 to 7 days) are being used in order to obtain higher progesterone concentrations at the moment of removing the intravaginal device (Menchaca et al., 2009). With both alternatives, it is possible to achieve a maximum concentration of serum progesterone, 48 h after insertion of the intravaginal device (Lopez, 2004) and to avoid decreasing progesterone concentrations afterwards, which can turn insufficient to stimulate the corpus luteum activity at the end of the long synchronization protocol (Menchaca et al., 2010). Acording to Lopez (2004), this issue may be responsible for the great variability in estrus appearance as a result of conventional long term synchronization protocols.

In this study, body condition score of the ewes was 2.75, and hence better than the average of 2.5 in a scale of 0 to 5 (Mendizabal et al., 2011). Thus it can be assumed that females used in the present experiment were in positive energy balance, this influencing positively blood estrogen levels and consequently the estrous rate (Scaramuzzi et al., 2006; Sosa et al., 2009). Moreover, the rate of device losses was zero (Wildeus, 2000; Avendaño-Reyes et al., 2007). Also, it must be considered that the present study was carried out during the breeding season, when better rate of estrus in contrast to anoestrus season has been reported (Wildeus, 2000). Regarding the interval from sponge removal to the onset of estrus (REI), the results were lower than those reported by Godfrey et al. (1999), Lida et al. (2004) and Kohno et al. (2005) in ewes synchronized with CIDR (26.5 ± 2.3, 23 ±1.8 and 36.3 ± 15.7 h, respectively). Differences in REI can be due to genetics, as well as to great variability of the onset of estrus, and its relation to ovulation onset (Lopes et al., 2006; Bartlewski et al., 2011). Lopez (2004) stated that once the intravaginal device is applied, the maximum progesterone concentrations are reached within the following 48 h, decreasing thereafter to very low concentrations, which are insufficient for stimulating corpus luteum activity at the end of the treatment. In the case of this study, REI was shorter than the one reported by Lopez (2004); however, it was similar to that reported by Chagas e Silva et al. (2003) in native Saloia de Portugal ewes (25.3 ± 0.5 h). Hair sheep have a shorter REI than wool breed sheep and this effect is even stronger if superovulation treatment is performed during the normal breeding season (Bartlewski et al., 2011). The majority of the studies in sponge and CIDR report CIDR values between 28 and 35 h, considering that the device administration, in the present study, works in Katadhin sheep, according to the conditions in which it is exploited in northeast of Mexico.

The number of corpus lutea, was smaller than that reported by Lopes-Junior et al. (2006) in young (10.2 ± 1.2) and adult females (5 ± 0.8) Morada Nova ewes using a dose of 200 IU of pFSH, distributed in 6 applications. Chagas e Silva et al. (2003), found CL values of 8.3 ± 0.8 in summer and 9.3 ± 1.1 in autumn with no differences between the 2 seasons (P > 0.05) Variations in corpus luteum are due, among other factors to the age of the ewe (Lopes-Junior et al., 2006), breed prolificity (Gonzalez-Bulnes et al., 2004) and nutritional status of the ewe donors (Sosa et al., 2009). In the present study pFSH-doses were lower than those used by others. Considering that there are no pFSH doses reported in Katadhin sheep, lower doses to evaluate the response under environmental and management exploitation in Mexico were applied, for instance, Herrera-Camacho et al. (2008) applied a total dose of 200 mg pFSH and obtained a greater number of CL per superovulated Pelibuey ewe, when they were supplemented with high energetic ingredients (corn oil) (14.7 ± 1.9) in contrast to unsupplemented control group (10.7 ± 1.4). A main aspect of the region is the high environmental temperature that could have affected the ovulatory response, since according to Tabarez-Rojas et al. (2009), exposure of superovulated ewes to heat stress may have an influence on incidence of premature CL regression, which should be considered in MOET programs, and with greater caution in hair sheep that are constantly exploited in warm and semi-desert regions, with high environmental temperatures. In this study, the number of follicles (F) was smaller than that reported by Chagas e Silva et al. (2003) (average of 2.7 follicles larger than 5 mm in Saloia de Portugal ewes), who evaluated 2 hormone sources in superovulated ewes, and found that animals treated with equine chorionic gonadotropin (eCG) showed a greater number of follicles (4.7 ± 0.9) compared with those receiving FSH in decreasing doses (1.3 ± 0.2 follicles); thus pFSH-dose used in the present study may be increased in future evaluations. Menchaca et al. (2010) reported that the administration of eCG causes high peak estradiol levels compared to FSH, which may interfere with fimbria ovum capture or with oviductal ovum transport. In this study, the presence of more follicles at the end of Treatment 1 was due to lower concentration of pFSH. Compared to eCG, the administration of decreasing doses of FSH may have a better effect on the number of developing follicles (Gonzalez-Bulnes et al., 2004; Bartlewski et al., 2008), based on a better emulation of naturally observed FSH decreasing concentrations after luteolysis, primarily due to the inhibitory effect of estradiol and inhibin released by preovulatory follicles.

In this study, the number of collected transferable embryos was smaller than those reported by several groups of researchers. Gonzalez-Bulnes et al., (2004), found values up to 11.9 of CE and 6.1 of TE. Administration of 160 mg of FSH in decreasing doses promoted better CE (5.8 ± 0.8 collected embryos/ewe), and TE response (4.0 ± 0.7 transferable embryos/ewe), in contrast to ewes which received 1500 IU of eCG, (CE = 3.6 ± 0.6 collected embryos/ewe, and TE = 2.6 ± 0.5 transferable embryos/ewe) (Chagas e Silva et al., 2003). It is necessary to consider the lower body weight of the animals used in this study, than in former ones. Rate of transferable embryos can improve, when young hair sheep are used, with values up to 5.6 ± 1.1/ewe (Lopes-Junior et al., 2006). With pFSH doses used in this work, quality and quantity of embryos in hair sheep are not compromised, considering that higher doses can be used to achieve better embryo response. The percentage of collected and transferable embryos is within the range obtained by several researchers which have been working in MOET programs (Bari et al., 2000; Cordeiro et al., 2003; Gonzalez-Bulnes et al., 2004; Lopes- Junior et al., 2006). However, doses can be increased considering other factors for the administration of pFSH, as the use of different progestogen sources, presence of a dominant follicle, as well as CL, which leads to the adequate starting point of the ovary to initiate the superovulation treatment (Gonzalez-Bulnes et al., 2004; Baldasarre, 2007). Hair sheep of Katahdin breed are being used as males in intensive meat producing systems. In Mexico, the use of this technique depends on the evaluation of different factors, studied by other researchers, and on the adaptation of this information to the prevailing local conditions related to environment, health, and nutrition. Considering that recent interest in embryo transfer for animal production systems is growing, this technique assists the genetic improvement of animals as pursued by genomic selection. For this reason, efforts to improve this biotechnology will make it easier for farmers to have access to animals with higher genetic merit.

Conclusion

The results show that the doses of pFSH used in this work may be modified to obtain greater embryo recovery rate without compromising their quality, for which it is suggested that higher doses should be evaluated, as well as out of breeding season, to be able to implement a MOET program, under breeding conditions of hair sheep in Mexico.

ACKNOWLEDGEMENTS

This study was carried out with the internal support of Programa Integral de Fortalecimiento Institucional (PIFI) through the Subdirección de Estudios de Posgrado e Investigación of the Facultad de Agronomía of the Universidad Autónoma a de Nuevo León (FAUANL). Under Mexican law, published in the Official Standard NOM-062-ZOO-1999, it has been complied with animal care throughout the development of this research in goats owned by Department of Agronomy of the UANL.

REFERENCES

AMCO, Asociación Mexicana de criadores de ovinos (2008). Situación de la ovinocultura en México. Pachuca, Hidalgo, P. 83.

Avendaño-Reyes L, Álvarez-Valenzuela FD, Molina-Ramírez L, Rangel-Santos R, Correa-Calderón A, Rodríguez-García J, Cruz-Villegas M, Robinson PH, Fámula TR (2007). Reproduction performance of Pelibuey ewes in response to estrus synchronization and artificial insemination in Northwestern Mexico. J. Anim. Vet. Adv. 6:807-812.

Baldasarre H (2007). Reproducción asistida en la especie caprina: Inseminación artificial a clonación. Revista Brasileña Reproducción Animal 31:274-282.

Baril G, Brebion P, Chesné P (1995). 115 Manual de formación práctica para el trasplante de embriones en ovejas y cabras. Italia. FAO, P. 175.

Bari F, Khalid M, Haresign W, Murray A, Merrell B (2000). Effect of mating system, flushing procedure, progesterone dose and donor ewe age on the yield and quality of embryos within a MOET program in sheep. Theriogenology 53:727-742.

Bartlewski PM, Alexander BD, Rawlings NC, Barrett DMW, King WA (2008). Ovarian Responses, Hormonal Profiles and Embryo Yields in Anoestrous Ewes Superovulated with Folltropin®-V after Pretreatment with Medroxyprogesterone Acetate-releasing Vaginal Sponges and a Single Dose of Oestradiol-17β. Repr. Dom. Anim. 43:299-307.

Bartlewski PM, Tanya EB, Jennifer GG (2011). Reproductive cycles in sheep. Anim. Repr. Sci. 124:259-268.

Chagas e Silva LD, Cidada JC, Robalo-Silva J (2003). Plasma progesterone profiles, ovulation rate, donor embryo yield and recipient embryo survival in native Saloia sheep in the fall and spring breeding seasons. Theriogenology 60:521-532.

Cognié Y, Baril G, Poulin N, Mermillod P (2003). Current status of embryo technologies in sheep and goat. Theriogenology 59:171-188.

Cordeiro MF, Lima-Verde JB, Lopes-Júnior ES, Teixeira DIA, Farias LN, Salles HO, Simplício AA, D. Rondina D, Freitas VJF (2003). Embryo recovery rate in Santa Inês ewes subjected to successive superovulatory treatments with pFSH. Small Rum. Res. 49:19-23.

FAO (2008). Anuario estadístico 2005-2006 de ganadería. FAO-STAT, Italia. P. 339.

Godfrey RW, Collins JR, Hensley EL, Wheaton JE (1999). Estrus synchronization and artificial insemination of hair sheep in the tropics. Theriogenology 51:985-987.

González-Bulnes A, Baird DT, Campbell BK, Cocero MJ, Garcia GRM, Keith I, Lopez Sebastian A, McNeilly AS, Santiago-Moreno J, Souza, CJH, Veiga-López A (2004). Multiple factors affecting the efficiency of multiple ovulation and embryo transfer in sheep and goats. Repr. Fert. Dev. 16:421–435.

Herrera-Camacho J, Aké-López JR, Ku-Vera JC, Williams GL, Quintal-Franco JÁ (2008). Respuesta ovulatoria, estado de desarrollo y calidad de embriones de ovejas Pelibuey superovuladas suplementadas con ácidos grasos poliinsaturados. Téc. Pec. Méx. 46:107-117.

Lida K, Kobayashi N, Kohno H, Miyamoto A, Fukui Y (2004). A comparative study of induction of estrus and ovulation by 3 different intravaginal devices in ewes during the non-breeding season. J. Repr. Dev. 50:63-69.

Kohno H, Okamoto C, Lida K, Takeda T, Kaneko E, Kawashima C, Miyamoto A, Fukui Y (2005). Comparison of estrus induction and subsequent fertility with 2 different intravaginal devices in ewes during the non-breeding season. J. Repr. Dev. 51:805-812.

Lopez AC (2004). Supresión del efecto de dominancia folicular en protocolos de estimulación ovárica en ganado ovino mediante la administración de una dosis única de antagonista de GnRh. Tesis Doctoral, Universidad Complutense Madrid, P. 139.

Lopes-Junior ES, Maia EL, Paula NR, Teixeira DIA, Villarreal ABS, Rondina D, Freitas VJF (2006). Effect of age of donor on embryo production in Morada Nova (white variety) ewes participating in a conservation programme in Brazil. Trop. Anim. Health Prod. 38:555-561.

Menchaca A, Vilariño M, Pinczak A, Kmaid S, Saldaña JM (2009). Progesterone treatment, FSH plus eCG, GnRH administration and Day 0 protocol for MOET programs in sheep. Theriogenology 72:477-483.

Menchaca A, Vilariño M, Crispo M, de Castro T, Rubianes E (2010). New approaches to superovulation and embryo transfer in small ruminants. Repr. Fert. Dev. 22:113-118.

Mendizabal JA, Delfa R, Arana A, Purroy A (2011). Body condition score and fat mobilitazation as management tools for goats on native pastures. Small Rum. Res. 98:121-127.

Ortega-Abasolo JC (2006). Comparación de dos métodos de sincronización del estro en ovinos de pelo. Tesis Maestría. Facultad Zootecnia Universidad Autónoma de Chihuahua, México. P. 62.

Paramio MT (2010). In vivo and in vitro embryo production in goats. Small Rum. Res. 89:144-148.

Robertson I, Nelson R (1999). Certification and identification of embryos. In: D.A. Stringfellow and S.M. Seidel (eds), Guide of International Embryo Transfer Society. pp. 109–122.

Sánchez DF, Bernal H, Negrete JC, Olivares E, del Bosque AS, Ledezma R, Ungerfeld R (2011). Environmental factors and interval from the introduction of rams to estrus in postpartum Saint Croix sheep. Trop. Anim. Health Prod. 43:887-891.

Santos MHB, Gonzalez CIMBezerra FQG, Neves JP, Reichenbach HD, Lima PFOliveira MAL (2007). Sexing of Dorper sheep fetus derived from natural mating and embryo transfer by ultrasonography. Repr. Fert. Dev. 19:366-369.

Scaramuzzi RJ, Campbell BK, Downing JA, Kendall NR, Khalid M, Muñoz-Gutierrez M, Somchit A (2006). A review of the effects of supplementary nutrition in the ewe on the concentrations of reproductive and metabolic hormones and the mechanisms that regulate folliculogenesis and ovulation rate. Nutr. Repr. Dev. 46:339-354.

Snedecor H, Cochran G (1980). Diseños estadísticos para especies animales. Ed. Mc. GrHill. México. pp. 34-56.

Sosa CA, González-Bulnes JA, Abecia F, Forcada F, Meikle A (2009). Short-term undernutrition affects final develoment of ovulatory follicles in sheep synchronized for ovulation. Repr. Dom. Anim. 45:1033-1038.

SPSS 17 (2008). Statistical Package for Social Sciences Windows version 12.0. Chicago, Illinois, P. 323.

Tabarez-Rojas A, Porras-Almeraya A, Vaquera-Huerta H, Hernández-Ignacio J, Valencia J, Rojas-Maya S, Hernández-Cerón, J (2009). Desarrollo embrionario en ovejas Pelibuey y Suffolk en condiciones de estrés calórico. Agrociencia 43:671-680.

Ungerfeld R, Sánchez F (2012). Oestrus synchronization in postpartum autumn-lambing ewes: effect of postpartum time, parity, and early weaning. Sp. J. Agr. Res. 10:62-68.

Veiga-Lopez A, Encinas T, McNelly AS, Gonzalez-Bulnes A (2008). Timing of preovulatory LH surge and ovulation in superovulated sheep are affected by follicular status at start of the FSH treatment. Repr. Dom. Anim. 43:92-98.

Wildeus S (2000). Current concepts in synchronization of estrus: Sheep and goats. J. Anim. Sci. 77:1-14.

6

Life table of *Phenacoccus solenopsis* Tinsley (Pseudococcidae: Hemiptera) on various phenological stages of cotton

Kumar S.[1], Kular J. S.[2], Mahal M. S.[2] and Dhawan A. K.[2]

[1]Regional Research Station, Bathinda Punjab Agricultural University, Ludhiana, Punjab, India-141004, India.
[2]Department of Entomology, Punjab Agricultural University, Ludhiana, Punjab, India-141004, India.

A study of the life table of *Phenacoccus solenopsis* Tinsely on cotton (*Gossypium hirsutum* L.) for two constitutive seasons at Entomological Farm, Punjab Agricultural University (PAU) Ludhiana revealed that among three phenophases (vegetative, flowering and maturation stage), gross reproductive rate (GRR), net reproductive rate (Ro), intrinsic rate of increase (rm), precise value of rm and finite rate of increase (λ) were maximum (329.83 and 349.10, 65.29 and 278.46, 0.169 and 0.242, 0.183 and 0.246, 1.184 and 1.274) at vegetative stage during 2008 and 2009, respectively. The total life cycle was completed in 32 days at vegetative stage while took maximum (39.00) days at maturation stage. During 2008, apparent mortality (100qx) was maximum (74.33 and 77.31%) at vegetative and flowering, respectively. Survival rate (Sx) was (0.26 and 0.23) in second instar at vegetative and flowering stage, whereas, at maturation stage the corresponding values for apparent mortality and survival rate were 40.69 and 0.59 in third instar. During 2009, maximum apparent mortality (76.35 and 56.91) and minimum survival rate (0.24 and 0.43) were recorded in adults at vegetative and maturation stage respectively. Heavy rainfall, *Aenasius* sp. (Hymenoptera: Encyrtidae) and *Coccinella* sp (Coleoptera: coccinellidae) contributed maximum mortality at vegetative and flowering stage, while rain fall and low temperature were major factors for mortality during maturation stage. The study suggested that *P. solenopsis* population is more affected by the biotic and abiotic factors in the vegetative stage. Therefore, the application of control measures at this stage could be drastically further reduces the population and thus prevents the buildup of population at later in the season.

Key words: Life table, *Phenacoccus solenopsis*, phenological stages, net reproductive rate (Ro), finite rate of increase (λ), intrinsic of increase (rm).

INTRODUCTION

Phenacoccus solenopsis Tinsley (Hemiptera: Pseudococcidae) had emerged as a major threat to cotton, vegetables and fruits in the world, causing huge losses (779.43 US$/ha) and reduced average seed cotton yield by 44% (Dhawan et al., 2007). *P. solenopsis* had caused 14% loss to cotton crop during 2005 in Pakistan (PWQCP, 2005). *P. solenopsis* originally reported on ornamentals and fruit crops in New Mexico (Tinsley, 1898).

Then it spread to other parts of world like Caribbean and Ecuador (Ben-Dov, 1994), Chile (Larraín, 2002), Argentina (Granara de willink, 2003), Brazil (Culik and Gullan, 2005), Pakistan and India (Hodgson et al., 2008), China (Wang et al., 2009; Wu and Zhang, 2009), Sri Lanka (Prishanthini and Laxmi, 2009) and Australia (Admin, 2010). In India, P. solenopsis was predominant mealy bug infested cotton in Andhra Pradesh, Gujarat, Haryana, Maharastra, Rajsthan and Tamil Nadu (Nagrare et al., 2009). With the introduction of a fortuitously parasitoids, Aenasius bambawali Hayat (Hymenoptera : Encyrtidae) in India (Gautam et al., 2009; Hayat, 2009; Pala and Saini, 2010) and Aneasius sp. Longiscapus (Hymenoptera: Encyrtidae) in Pakistan (Bodlah et al., 2010) in combination with the already established IPM tactics, seemed to manage the mealy bug population.

Despite this biological and chemical control, the risk of spreading P. solenopsis remains high due to its polyphagous nature and high fecundity. Some other crops that are most susceptible to attack of P. solenopsis are cotton, Trianthema monogyna L. (Shanti or itsit), Xanthium strumarium L. (gutputna), Achyranthus aspera Linn. (puthkanda), ornamental plant (Hibiscus rosa-sinensis L.), okra, tomato, brinjal and chilli (Tanwar et al., 2008). Scientists at the National Centre for Integrated Pest Management, New Delhi have reported the Spatio-temporal distribution of host plant of P. solenopsis in India. Based on the report, the P. solenopsis had spread all over India (South > Centre >North region) (Anonymous, 2010).

P. solenopsis feed on 154 species of field vegetables, ornamental plants and weeds. Four parasitoid species and four predator species have reported against P. solenopsis. Some of these natural enemies such as Aenasius sp and Cryptoleamus montrozieiri (Coleoptera: coccinellidae), Coccinella septumpuntata have reported, identified and used to control P. solenopsis. Because of the polyphagous nature of P. solenopsis, researchers used different host plants for the study of biology (David et al., 2009; Abbas et al., 2010; Sana-Ullah et al., 2011). The variations in the methodologies complicated the efforts in estimating the life table parameter of P. solenopsis. Certain chemicals belongs to organophosphates (Profenophos, Acephate, etc), Carbaryl group (Sevin, Hexavin, etc.) and IGR's (Buperofezin) gave effective control of this pest (Anonymous, 2011).

Although number of chemicals and biological control agents are effective against cotton mealy bug (P. solenopsis). Parthenogentic reproduction of this pest can give birth to some young ones which may act as biotypes of this pest and may lead to resistance against these insecticides and biological control agents. Therefore, there is need to investigate the life table parameters of P. solenopsis and identification of the ecological factors (including both biotic and abiotic factors like predators, parasitoids, rainfall, relative humidity, etc.) associated with this pest as no such factors are reported prior to this study. This investigation presents concise information on the life table of P. solenopsis conducted on cotton (Gossypium hirsutus L.) at three stages (vegetative, flowering and maturation stage). We determined the life table parameters, age-schedule of survival (lx), gross reproductive rate (GRR) ($\sum$mx), net reproductive rate (R_o), mean length of generation (T), intrinsic rate of increase (rm), finite rate of increase/ day (λ) and doubling time (DT) of P. solenopsis under natural conditions. The main objective of this study is to provide a better understanding of life table of P. solenopsis, and to provide information about the favorable phenological stage for development and prediction of P. solenopsis distribution.

MATERIALS AND METHODS

Maintenance of insect colonies

Nymphs and adults of P. solenopsis were collected from different host and non-host plants in cotton fields of Southern region of Punjab State and from the University fields. These stages of P. solenopsis were maintained and mass multiplied on the various hosts (Congress grass, Gutputna, Hibiscus sp. and cotton) in earthen pots under screen house (4 × 3× 2 m) of 20 cm square mesh size at Entomological Farm, Punjab Agricultural University (PAU), Ludhiana. These hosts were selected for mass multiplications due to their preference by P. solenopsis. From these colonies, gravid females of P. solenopsis were transferred to cotton (RCH 134) plants for conditioning (3 to 4 days) in another screen house of same dimensions, so that of P. solenopsis adapt to the cotton plant and then produce the offspring /ovisacs on cotton plant. Colonies of P. solenopsis on cotton plants were used in experiments.

In winter, the colonies of P. solenopsis were reared on cuttings of tender parts of Hibiscus sp. in battery jars in the laboratory. The old cuttings replaced with newer when required. Temperature was control by using electrical heater to provide better condition for development and survival of P. solenopsis. Prior to the oncoming of main season (April to December), P. solenopsis were transferred to the cotton plant for conducting the experiments.

Methodology adopted

Seeds of cotton hybrid RCH 134 were procured from Rasi Seed Pvt. Ltd. Madhya Pradesh, India, and were sown in the 10 × 10 m^2 plots with three replication in randomized design at Entomological Research Farm, PAU, Ludhiana during 2008 and 2009 by following recommended guidelines for cotton cultivation (Anonymous, 2008). The experiments were conducted at three different phenophases that is, vegetative, flowering and maturation stage during June- mid July, mid-July - August and September – October, respectively, in both the years. P. solenopsis feed on the apical portion of cotton plant, thus the crawlers released on the upper canopy. Ten plants have tagged in each plot; hence thirty plants were used for one phenophase of the cotton crop to record the data. Ten newly hatched crawlers were collected from the insect colonies and transferred on the tagged cotton plants. Crawlers on each plant represented a cohort and each plant was act as replicate in the experiment under field conditions. Each plant and cohort has observed after 24 h and data on the life table parameters recorded. The presence of exuviae indicated the successful development from one instar to another. Various biotic (predators and

Table 1. Effect of different phenophases on the life table parameters of *P. solenopsis* on cotton (RCH 134).

Parameter	Vegetative stage		Flowering stage		Maturation stage	
	2008	2009	2008	2009	2008	2009
Gross reproductive rate(GRR) ($\sum$mx)	329.83	349.10	328.65	327.81	231.17	246.54
Net reproductive rate ($R_o = \sum$lxmx)	65.29	278.46	18.69	28.68	79.20	72.76
Mean length of generation (T=$\sum$ l$_x$.m$_x$/R$_o$)	24.51	23.23	27.00	28.86	30.05	29.94
Intrinsic rate of increase (r$_m$ = log$_e$ R$_O$/T)	0.169	0.242	0.108	0.116	0.145	0.143
Precise value of r$_m$ ($\sum$e^{7-rmx}. lxmx = 1097)	0.183	0.246	0.111	0.117	0.148	0.145
Finite rate of increase/ day (λ=e^{rm})	1.184	1.274	1.114	1.123	1.156	0.154
Doubling time (DT= loge 2 /rm)	4.10	2.86	6.42	5.98	4.78	4.84

parasitoids) and abotic (rainfall, temperature and R.H.) factors were recorded to know that which factors are responsible for the mortality of *P. solenopsis* under field conditions.

Life table analysis

The effect of three phenophases (vegetative, flowering and full bloom) on the population growth and age structure of *P. solenopsis* was determined by assessing different life table parameters. Data measured on the survival and reproduction were used to estimate l$_x$ (age-schedule of survival) and m$_x$ (age-schedule of female birth). Age-specific survival is the fraction of initial cohort alive at age x, while age–schedule of female birth was the mean number of females produced by each female at a pivotal age in days (x). The formulae for estimating life table parameters of female *P. solenopsis* at each phenophase were (Atwal and Bains, 1974).

Gross reproductive rate (GRR) = $\sum$mx, Net reproduction rate (R$_0$) = $\sum$ lx mx, Mean length of generation (T) = $\sum$ x lx mx / R$_o$, Doubling time (DT) = log$_e$ 2 / rm, Intrinsic rate of increase (rm) = log$_e$ R$_0$ / T, Finite rate of increase (λ) = e^{rm}

To calculate the precise value of rm, we followed Southwood (1978) graphical method by using the equation mentioned below:

$$\sum e^{(7-rmx)}. (lxmx) = 1097$$

Along with life table parameters, data on biotic (natural enemies) and abiotic (weather conditions) were also recorded. In the beginning, 300 nymphs used and then with the passage of time the mortality assessed along with the cause of death. The different observations recorded were (Atwal and Bains, 1974) pivotal age (x), number of individuals in the beginning (lx), number of individuals died (dx), factors responsible for death (dxf). Based on these observations, apparent mortality (100qx) and survival rate (Sx) were computed by using formula given below:

Apparent mortality (100qx) = dx/lx × 100

Survival rate (Sx) = lx of subsequent stage / lx of particular stage

RESULTS

The life table parameters of *P. solenopsis* show clear cut differences at three phenophases viz. vegetative, flowering and maturation stage (Table 1). Gross reproductive rate (GRR) was maximum (329.83 and 349.10) at vegetative stage followed by flowering stage (328.65 and 327.80) and minimum (231.17 and 246.54)

at maturation stage during 2008 and 2009, respectively. Same trend was observed in finite rate of increase (λ) with corresponding values 1.184 and 1.274, 1.114 and 1.123, 1.156 and 0.154 at vegetative, flowering and maturation stage during both seasons, respectively. Maximum (65.29 and 278.46) net reproductive rate (R$_o$) and intrinsic rate of increase (rm) (0.169 and 0.242) was recorded at vegetative stage followed by maturation stage and minimum (18.69 and 28.68) net reproductive rate was recorded at flowering stage. Similar results were recorded in precise value of rm as in net reproductive rate. Mean length of generation (T) increased as the cotton phases goes towards maturation stage. Maximum mean length of generation (30.05 and 29.94 days) was observed at maturation stage followed by flowering stage (27.00 and 28.86 days) and minimum (24.51 and 23.23 days) at vegetative stage in 2008 and 2009, respectively. The minimum (4.10 and 2.86) doubling time (DT) was recorded at vegetative stage and maximum (6.42 and 5.98 days) at flowering stage during respective years of study. Age specific survival (lx) of *P. solenopsis* is presented in Figures 1 and 2. Quite similar results were obtained during both years. Pivotal age (x in days) to complete the life cycle was least (approx. 32.5 days) at vegetative stage, while, maturation stage took maximum (approx. 39.5 days) time to complete one generation. This difference in the pivotal age may be due to hardening of the plant parts as it reaches to maturity. Precise value of rm was calculated only at vegetative stage as the intrinsic rate of increase has maximum value in this stage. The corresponding values for precise rm were 0.183 and 0.246 during 2008 and 2009, respectively (Figure 3).

During 2008, heavy rainfall (199 mm) responsible for 74.33% mortality in second instar nymphs developing on cotton plants in vegetative stage, while no substantial mortality in other developmental stages (Table 2). At flowering stage, heavy rainfall (161.8 mm), Aenasius sp. and Coccinella sp contributed 77.31% mortality in second instar nymphs, while it was 71.21% in the third instar. At maturation stage, the highest percent mortality (40.64%) observed in the second instar nymphs due to heavy rainfall (64.2 mm) coupled with high relative humidity

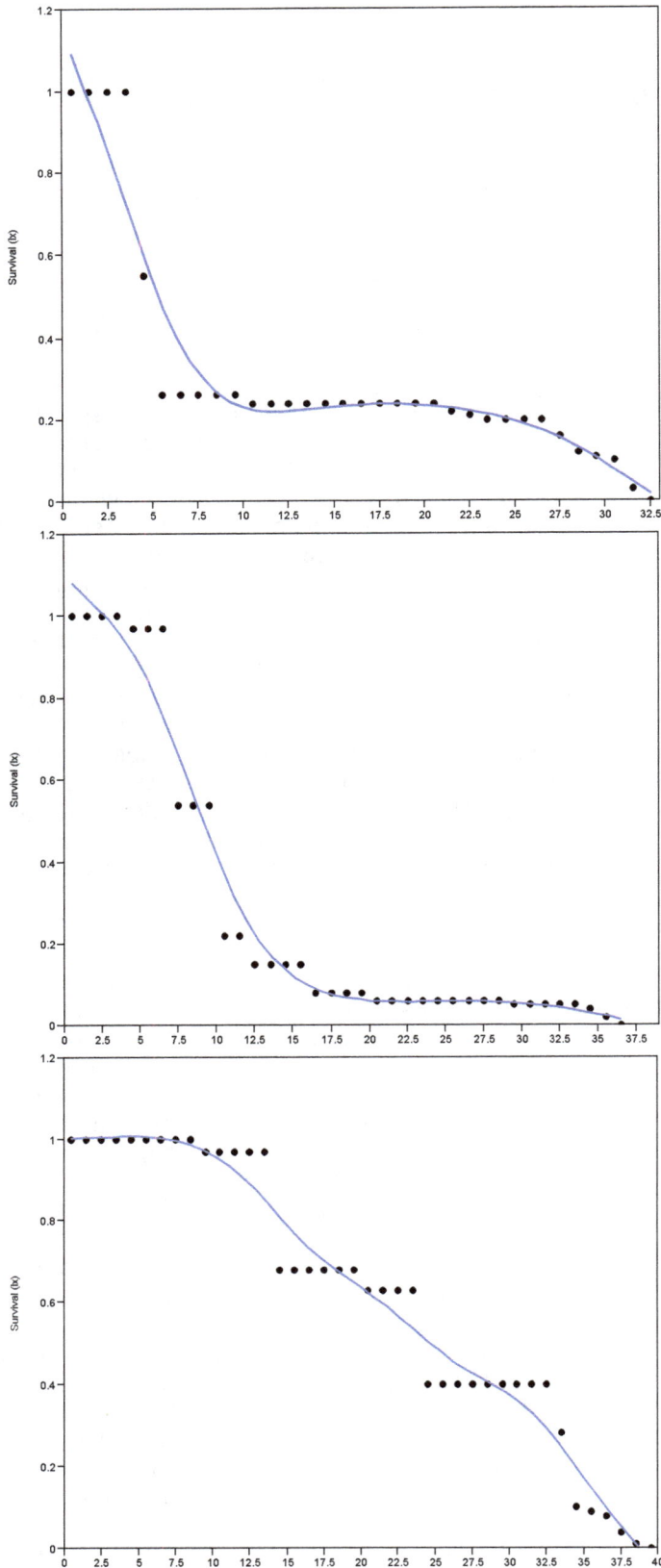

Figure 1. Age- specific survival (lx) of *P. solenopsis* at different phenophases of cotton during 2008.

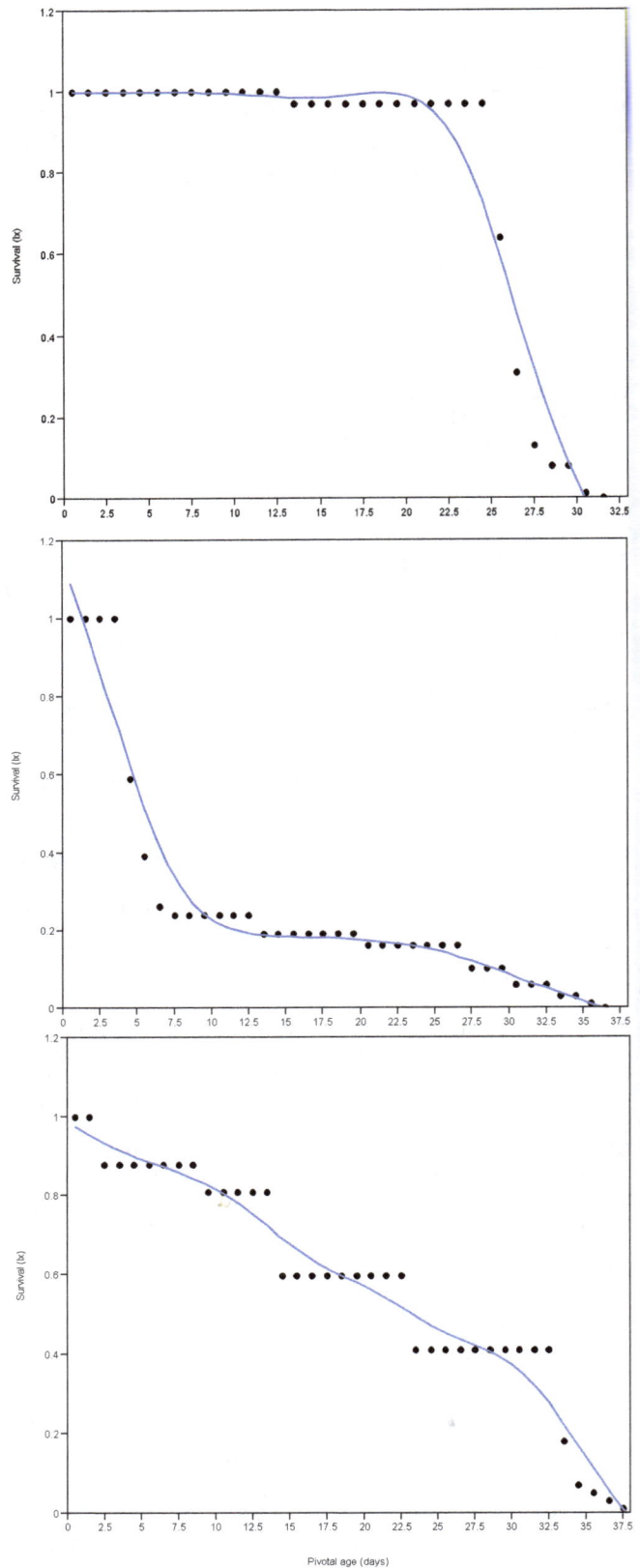

Figure 2. Age- specific survival (lx) of *P. solenopsis* at different phenophases of cotton during 2009.

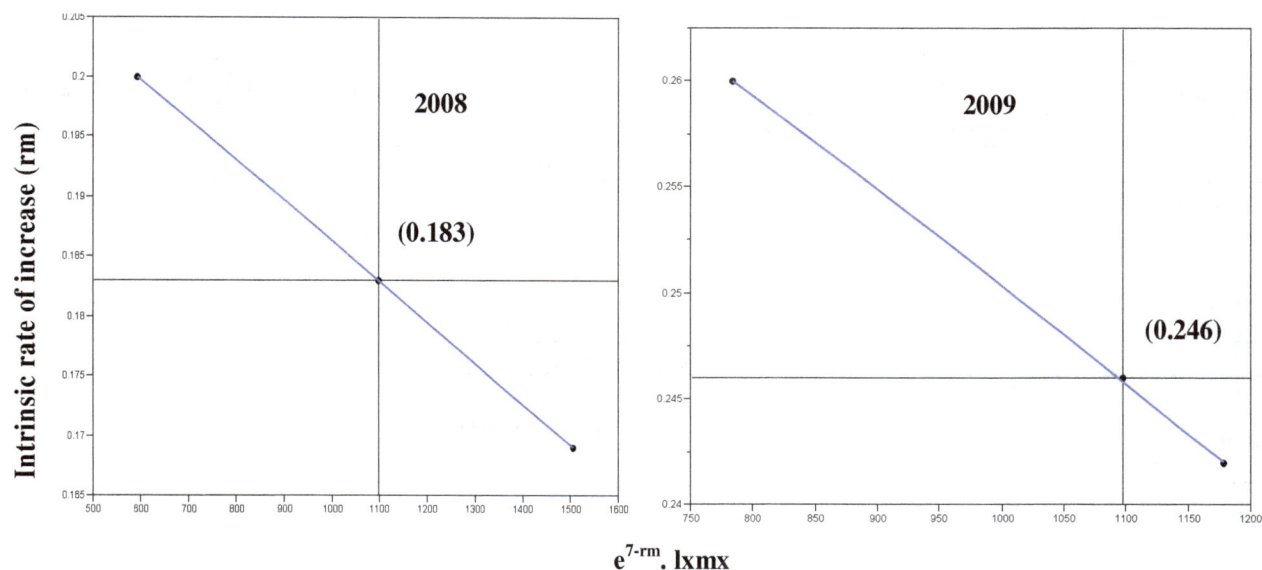

Figure 3. Precise value of r_m of *P. solenopsis* at vegetative stage on cotton under natural conditions.

Table 2. Life table of *P. solenopsis* at different phenophases under natural conditions in field during 2008.

Pivotal age in days (x)	Number of individuals in the beginning (l_x)	Number of individuals died (d_x)	Factors responsible for death (d_{xf})	Apparent mortality (100qx)	survival rate (s_x)
Vegetative stage					
Ist instar	300	-	-	-	1.00
2nd instar	300	223	Heavy rain fall	74.33	0.26
		4	*Aenasius* sp. and *Coccinella* sp, Rainfall in traces	5.19	0.94
3rd instar	73	8	Rain , high R H and *Aenasius* sp.	10.95	0.89
	65	6	*Aenasius* sp.	9.23	0.91
Adult stage		2	High RH	3.22	0.97
		57	Due to aging Life-cycle completed		
Flowering stage					
Ist instar	300	9	Probably > RH	3.00	0.97
2nd instar	291	225	Heavy Rain fall and *Coccinella* sp	77.31	0.23
3rd instar	66	47	Rain and *Aenasius* sp.	71.21	0.29
Adult	19	3	Probably R H >87%	15.79	0.84
	16	5	Rain	31.25	0.69
	11	11	Due to aging Life-cycle completed		
Maturation stage					
Ist instar	300	-	-	-	1.00
2nd instar	300	96	Probably > RH, Low temp and Heavy Rain fall	32.00	0.68
3rd instar	204	83	RH 90%, Low temp	40.69	0.59
Adult stage	121	37	Low temp	30.58	0.69
		84	Due to aging Life-cycle completed		

Table 3. Life table of *P. solenopsis* at different phenophases under natural conditions in field during 2009.

Pivotal age in days (X)	Number of individuals in the beginning (l_x)	Number of individuals died (d_x)	Factors responsible for death (d_{xf})	Apparent mortality (100qx)	Survival rate (s_x)
Vegetative stage					
1st and 2rd instar	300	-	-	-	1.00
3rd instar	300	97	*Aenasius* sp. and high rainfall	32.33	0.68
	203	155	Heavy rain	76.35	0.24
Adult stage		48	Due to aging Life-cycle completed		
Flowering stage					
1st instar	300	124	Heavy rain fall	41.33	0.59
2nd instar	176	104	Rain and *Aenasius* sp. and *Coccinella* sp.	59.09	0.41
3rd instar	72	23	Rain	31.94	0.68
Adult stage	49	19	Rain	38.78	0.61
	30	13	*Aenasius* sp.	43.33	0.57
		17	Due to aging Life-cycle completed		
Maturation stage					
1st instar	300	35	Rain fall	11.67	0.88
2nd instar	265	136	Probably > R H 85% and rainfall	51.32	0.49
3rd instar	129	6	Probably low temp	4.65	0.95
Adult stages	123	70	High rain fall	56.91	0.43
		53	due to aging Life-cycle completed		

(90%) and low temperature. During 2009, no mortality in 1st and 2nd instar nymphs was observed when reared on vegetative stage (Table 3). The mortality in 3rd instar was due to Aenasius sp. and heavy rain fall (79.3 mm). At flowering stage, the apparent mortality in all the nymphal instars was due to the same biotic and abiotic factors as in vegetative stage. The main factors responsible for substantial mortality at maturation stage were rainfall (111 mm) and high relative humidity (>85%) in first and second instars, respectively. Mortality of third instar nymphs due to low temperature was negligible and 56.91% mortality observed at third instar and adult stage due to high rainfall and aging factors. The survival rate (Sx) was decreased with increase in the development stage during 2008 and 2009.

DISCUSSION

Results of this study indicated that different phenophases of cotton, as well as biotic and abiotic factors, had direct correlation with the life history of *P. solenopsis*. Previous studies on the table of mealy bugs were conducted under controlled and laboratory conditions; therefore, it is difficult to directly compare the results. Different scientists investigated the life table parameters of *Phenacoccus*

species on different host plants. In this study, we found that the net reproductive rate was 65.29 and 278.46 at vegetative stage of cotton during 2008 and 2009. While the study conducted by Marohasy (1997) on age specific life tables of *Phenacoccus parvus* on seven plant species revealed that the net reproductive rate was maximum on lantana (263) followed by tomato (249) and eggplant (231), where as it was least in *Clerodendron*. However, the intrinsic rate of increase after adjustment was highest on eggplant while vegetative stage of cotton have maximum intrinsic rate of increase among the three phenophases studied in this research. In this study, heavy rain fall and high relative humidity are the major abiotic factors responsible for the mortality of *P. solenopsis* on cotton. Our study is in line with the findings of Suresh and Kavitha (2008) in which they concluded that high rainfall reduced the population of *P. solenopsis* on *Parthenium hysterophorus*. They further reported that for every unit increase in maximum temperature, evening relative humidity, rainfall caused significant population reduction (8.9, 0.46 and 0.96 units, respectively). Similarly, Akintola and Ande (2009) concluded that the mealy bugs are abundant on host plant during the dry season and more or less absent in rainy season. During the rainy season, crawlers were dislodging by the effect of rainfall and wind.

Survivorship and fertility of *Helopeltis antonii* measured under laboratory and field condition. The highest mortality occurred in the immature stages (first and second instar). The females lived for a maximum of 24 days. The intrinsic rate of increase (rm) was 0.092 per female per day and daily finite rate of increase (λ) was 1.097 females per female per day, with mean generation time of 27.70 days.

The net reproductive rate (R_o) of the population was 12.84 (Siswanto et al., 2001). Joshi et al. (2010) reported the bionomics, natural enemies and host range of *P. solenopsis* under controlled and laboratory conditions. The total life cycle completed in 16 to 38 days, whereas, in this study, 32.5, 37.0 and 39.5 days (approx.) at vegetative, flowering and maturation stage, respectively for both the seasons.

At the vegetative stage of cotton crop, the average temperature was varied between 23.7 to 40.3°C during 2008 to 2009. So this study is also compare with the studies conducted at control temperature too, although, R_o, T and λ were (65.29 and 278.46, 0.169 and 0.242, 1.184 and 1.274, respectively) at vegetative stage for 2008 and 2009 (in this study). However, in results on age specific cohort life table on *P. solenopsis* at 27 ± 2°C and 65 ± 5% RH (Fand et al., 2010) these were 123.41 females/female/generation, 28.34 days and 1.185 females/female/day, respectively. The mortality rate was high for the first instar crawlers, which declined sharply in subsequent instars. Study also revealed that R_o and T were higher than (15.51 females/female/generation and 34.23 days, respectively) reported in Persad and Khan (2002). The estimated intrinsic rate of increase (rm) at vegetative stage was 0.169 and 0.246 females/female/generation which was more than 1.4 to 2 times than those achieved at 25 and 27°C by *M. hirsutus* (Chong et al., 2008). This difference with the results of Chong et al. (2008) could be because of controlled conditions (at constant temperatures) and qualitative differences between cotton and hibiscus plant.

An understanding of the life table of *P. solenopsis* has important implications in management. Vegetative stage is most suitable for the production of its parasitoids and predators (e.g. *Aenasius* sp. and *Coccinella* sp.) to take advantage of high development rate, more fecundity and less mortality. A comparison between the life table parameters of *P. solenopsis* and its natural enemies, which not investigated in this study, is also helpful in selecting the most appropriate biological control agents especially these having high rm relative to this pest. Therefore, this study provides a pathway to the worker for comparing the life tale of natural enemies and *P. solenopsis*, to achieve efficient biological control. This study also provides farmer with information about the biotic and abiotic factor especially, rainfall and low temperature, which are detrimental for the growth of *P. solenopsis*. Such information used for the successful management of this pest under the Indian conditions.

ACKNOWLEDGEMENTS

Financial assistance provided by University Grant Commission (UGC) to carry out the present studies is gratefully acknowledged. Corresponding author is also grateful to Dr G. S. Dhaliwal for critically reviewing the manuscript, and Dr B. K. Kang for providing all-necessary inputs for conducting various experiments.

REFERENCES

Abbas G, Arif MJ, Ashfaq M, Aslam M, Saeed S (2010).The impact of some environmental factors on the fecundity of *Phenacoccus solenopsis* Tinsley (Hemiptera: Pseudococcidae): A serious pest of cotton and other crops. Pak. J. Agric. Sci. 47:321-325.

Admin (2010). Exotic mealybug species-a major pest in cotton. Published February 12, 2010 http:/ /thebeatsheet.com.au/mealy bugs/exotic mealy bug species a major new pest in cotton/ Accessed on 25th May 2010.

Akintola AJ, Ande AT (2009). Life history and behavior of *Rastrococcus invandens* Williams on *Ficus thonningii* in Nigeria. Australian J. Crop Sci. 3:1-5.

Anonymous (2008). Package and practices for kharif crop. Ludhiana, India, Punjab Agricultural University.

Anonymous (2011). Package and practices for kharif crop. Ludhiana, India, Punjab Agricultural University.

Anonymous (2010). Spatio-temporal distribution of host plants of cotton mealybug, Phenacoccus solenopsis. Technical Bull. 26:1-48.

Atwal AS, Bains SS (1974). Applied Animal Ecology. Kalyani Publisher, Ludhiana. pp. 44-46.

Ben-Dov Y (1994). A systematic catalogue of the mealy bugs of the world (Insecta: Homoptera: Coccoidea: Pseudococcidae and Putoidae) with data on geographical distribution, host plants, biology and economic importance. Intercept Limited, Anover, UK. P. 686.

Bodlah I, Ahmad M, Nasir MF, Maeem M (2010). Record of Aenasius bambawalei Hayat, 2009 (Hymenoptera: Encyrtidae), a parasitoid of *Phenacoccus solenopsis* (Sternorrhyncha: Pseudococcidae) from Punjab, Pakistan. Pak. J. Zool. 42:533-536.

Chong JH, Roda AL, Mannion CM (2008). Life history of the mealybug, Maconellicoccus hirsutus (Hemiptera: Pseudococcidae), at Constant Temperatures. Environ. Entomol. 37:323-332.

Culik MP, Gullan PJ (2005). A new pest of tomato and other records of mealy bugs (Hemiptera: Pseudococcidae) from Espirito Santo, Brazil. Zootaxa 964:1-8.

David PM, Elanchezhyan K, Rajkumar K, Razak TA, Nelson SJ, Suresha S (2009). Simple petrifigure bracket cage and host plants to culture cotton mealybug, *Phenacoccus solenopsis* (Tinsley) and its predator, *Harmonia octomaculata* (Fab.). Karnataka J. Agric. Sci. 22(3):676-677.

Dhawan AK, Singh K, Saini S, Mohindru B, Kaur A, Singh G, Singh S (2007). Incidence and damage potential of mealybug, Phenacoccus solenopsis Tinsley, on cotton in Punjab. Indian J. Ecol. 34:110-116.

Fand BB, Gautam RD, Chander Subhash S, Sachin S (2010). Life table analysis of the mealybug, *Phenacoccus solenopsis* Tinsley (Hemiptera: Pseudococcidae) under laboratory conditions. J. Entomol. Res. 34:175-179.

Gautam RD, Suroshe SS, Gautam S, Saxena U, Fand BB, Gupta T (2009). Fortuitous Biological control of exotic mealy bug, *Phenacoccus solenopsis*- A boon for Indian growers. Ann. Plant Protec. Sci. 17:459-526.

Granara de Willink MC (2003). New records and host plants of *Phenacoccus* sp. for Argentina (Hemiptera: Pseudococcidae). Revta Sociedad Entomologia Argentina. 62:80-82.

Hayat M (2009). Description of a new species of Aenasius Walker (Hymenoptera: Encyrtidae), parasitoid of the mealybug, *Phenacoccus solenopsis* Tinsley (Homoptera: Pseudococcidae) in India. Biosystematica 3:21-26.

Hodgson C, Abbas G, Muhammad JA, Saeed S, Karar H (2008). *Phenacoccus solenopsis* Tinsley (Sternorrhycha: Coccoidea:

Pseudococcidae), an invasive mealy bug damaging cotton in Pakistan and India, with a discussion on seasonal morphological variation. Zootaxa 1913:1-35.

Joshi MD, Butani PG, Patel VN, Jeyakumar P (2010). Cotton mealy bug *Phenacoccu solenopsis* Tinsley- A Review. Agric. Rev. 31:113-119.

Larraín P (2002). Incidencia de insectos y ácaros plagas en pepino dulce (Solanum muricatum Ait.) cultivado en la IV Región, Chile. Agricultura Técnica. 62:15-26.

Marohasy J (1997). Acceptability and suitability of seven plant species for the mealy bug Phenacoccus parvus. Entomol. Exp. Appl. 84:239-246.

Nagrare VS, Kranthi S, Biradar VK, Zade NN, Sangode V, Kakde G, Shukla RM, Shivare D, Khadi BM, Kranthi KR (2009). Widespread infestation of the exotic mealy bug species, *Phenacoccus solenopsis* (Tinsley) (Hemiptera: Pseudococcidae), on cotton in India. Bull. Entomol. Res. 99:537-541.

Pala R, Saini RK (2010). Biological control solenopsis mealy bug *Phenacoccus solenopsis* Tinsley on cotton: A typical example of fortuitous biological control. J. Biol. Contr. 24:104-109.

Persad A, Khan A (2002). Comparison of life table parameters for *Maconellicoccus hirsutus*, *Anagyrus* kamali, Cryptolaemus montrouzieri and Scymnus coccivora. Biol. Contr. 47:137-149.

Prishanthini M, Laxmi VM (2009). The *Phenococcus solenopsis*. Department of Zoology, Eastern University, Sri Lanka. Available online: http:// www.dailynews.lk/2009/07/01/fea30.asp.

PWQCP (2005). *Weekly pest scouting reports*. Directorate of General. Pest warning and quality control of pesticides (PWQCP) Punjab, Lahore, Pakistan.

Sana-Ullah M, Arif MJ, Gogi MD, Shahid MR, Adid MA, Raza A, Ali A (2011). Influence of different plant genotypes on some biological parameters of cotton mealybug, *Phenacoccus solenopsis* and its predator, *Coccinella septempunctata* under laboratory conditions. Int. J. Agric. Biol. 12:125-129.

Siswanto M R, Dzolkhifli O, Elna Karmawati (2001). Life tables and population parameters of *Helopeltis antonii* (hemiptera: miridae) reared on cashew (anacardium occidentale l.) J. Biol. Sci. 19(1):9 – 101.

Southwood TRE (1978). Ecological Methods. Methwen and Co. Ltd, London.

Suresh S, Kavitha CP (2008). Seasonal incidence of economically important coccoid pests in Tamil Nadu. Proceedings of XI International Symposium on Scale *Insect Studies*. Oeiras, Portugal. pp. 285-289.

Tanwar RK, Bhamare VK, Ramamurthy VV, Hayat M, Jeyakumar P, Bambawale OM (2008). Record of new parasitoids on mealybug, *Phenacoccus solenopsis*. Indian J. Entomol. 70:404-405.

Tinsley JD (1898). Notes on Coccidae, with descriptions of new species. *Can. Entomol.* 30:317-320.

Wang YP, Wu SA, Zhang RZ (2009). Pest risk analysis of a new invasive pest *Phenacoccus solenopsis* to China. (in Chinese; Summary in English). China Bull. Entomol. 46:101-106.

Wu SA, Zhang RZ (2009). A new invasive pest, *Phenacoccus solenopsis* threatening seriously to cotton production. China Bull. Entomol. 46:159-162.

7

Effect of *in ovo* injection of cadmium on chicken embryo heart

Krzysztof Pawlak[1], Małgorzta Dżugan[2], Dorota Wojtysiak[3], Marcin Lis[1] and Jerzy Niedziółka[1]

[1]Department of Poultry and Fur Animal Breeding and Animal Hygiene, University of Agriculture in Krakow, Aleja Adama Mickiewicza 21, Kraków, Poland.
[2]Department of Food Chemistry and Toxicology, University of Rzeszów, Aleja Tadeusza Rejtana 16, Rzeszów, Poland.
[3]Department of Animal Reproduction and Anatomy, University of Agriculture in Kraków, Aleja Adama Mickiewicza 21, Kraków, Poland.

Due to the ease of absorption, accumulation in tissues, and extremely long biological half-life in the body, cadmium is considered one of the most hazardous heavy metals. The aim of the study was to investigate the effect of *in ovo* injection of cadmium on chicken embryo heart. A total of 160 chicken hatching eggs were used in the study. On day 4 of incubation, eggs from the experimental groups were injected with cadmium at a dose of 1 and 5 μg/egg and the incubation was prolonged to 21 day until hatching. Cadmium was found to slow the heart rate and reduce heart weight. In embryos exposed to 5 μg of cadmium, the histological analysis and aminotransferases concentration confirmed the occurrence of inflammatory processes in the heart muscle.

Key words: Cadmium, heart, chicken embryo, cardiotoxicity.

INTRODUCTION

Cadmium (Cd) is a heavy metal whose divalent cations are easily absorbed and accumulated in plant and animal tissues (Satarug et al., 2003). Exposure to cadmium occurs as a result of atmospheric emission during Cd production and processing, from combustion of fossil energy sources, waste and sludge, phosphate fertilizers, and deposition of waste and slag in at disposal sites (Satarug et al., 2003).

The main source of cadmium exposure for non-occupational population is food and tobacco smoking (EFSA, 2009). Meat, fish, and fruits generally contain up to 50 μg Cd/kg on fresh weight basis, whereas vegetables, potatoes, and grain products may contain up to 150 μg Cd/kg fresh weight. Higher concentrations are found in the kidneys of animals slaughtered for food, in wild mushrooms, and in seafood such as mussels and oysters (EFSA, 2009). The average cadmium intake from food generally varies between 8 and 25 μg per day (Jarüp and Äkesson, 2009).

Due to the exceptionally long biological half-life (10-30 years) in the body, the toxicity of cadmium ions increases with advancing age and may persist after the end of exposure (Nordberg, 2009). In animals exposed to cadmium compounds, the ions of this metal are particularly abundant in kidneys, heart and liver, and, to a lesser degree, in pancreas and brain (Jarüp and Äkesson, 2009). This can induce irreversible damage to these organs in addition to anemia, osteoporosis and carcinogenic (for kidney, lung, prostate, testis and breast) (Jarüp and Äkesson, 2009) and teratogenic lesions (Järup et al., 1998). The effects at the cellular level are increased oxidative stress, damage to mitochondria, disturbances in trace element (e.g. calcium and zinc) and vitamin metabolism, and disruption of cell signaling pathways (Czeczot and Skrzycki, 2010). Cadmium also destabilizes lysosomal membranes, which is possibly associated with the secretion of lysosomal enzymes into cellular fluids, blood and urine (Fotakis et al., 2005; Nordberg, 2009). Induction of an inflammatory reaction is probably one of the mechanisms for the toxic effect of cadmium.

T0he proinflammatory activity of cadmium was observed in kidneys, liver and the respiratory system (Mlynek and Skoczyńska, 2005). Research also shows that cadmium may cause an inflammatory reaction in the cardiovascular system (Yazihan et al., 2011). In experimental chronic cadmium poisoning, a greater incidence of lipid infiltration in the aorta as well as arteriosclerosis and arteriolosclerosis in rat heart, lungs, kidneys and adrenals were observed (Mlynek and Skoczyńska, 2005).

It is known that juvenile organisms are particularly sensitive to the adverse effects of heavy metals (Wonga et al., 1980; Yasuda et al., 2012). What is more, research concerning the effect of Cd on embryos is all the more important because cadmium inhaled by females crosses the placenta and accumulates in the fetal body (Trottier et al., 2002).

Chick embryo provides a useful model for studying the effect of various factors on the developing embryos. Development outside the mother's body, the high rate and peculiar characteristics of embryogenesis, and the relatively well known process of organogenesis make chick embryos a frequent subject of such studies (Tadahiro et al., 1998; Davison, 2003; Rajendra et al., 2004; Mitta and Weber, 2005; Lahijani et al., 2009; Dżugan et al., 2011; Pawlak et al., 2011). Treating avian embryos with Cd has always resulted in dorsally growing upper limbs, malposition of predominantly the lower limbs, and primary ventral body wall defect (Thompson and Bannigan, 2008; Cullinane et al., 2009). The heart is one of the earliest developing embryo organs, the activity of which can be recorded. Therefore, it is apparent that studies concerning the effect of cadmium on the developing embryo heart may be of great importance, especially since cadmium is cardiotoxic in small doses that have no negative effect on other organs (Limaye and Shaikh, 1999).

The aim of the study was to investigate the effect of *in ovo* injection of cadmium on the heart rate, histological appearance and morphometric parameters of chick embryo heart.

MATERIALS AND METHODS

Hatching eggs (62.0 ± 5.4 g) from a *Ross 308* broiler breeder flock were incubated in a Masalles 65 DIGIT incubator under standard conditions (1to 18 day of incubation: T = 37.8°C, RH = 50%; 19–21 day: T = 37.2 ± 0.1°C, RH = 70%). On day 4 of incubation the eggs were candled, and eggs with live embryos were randomly divided into 4 groups (n=40 eggs per group). Embryos in Group I were incubated without interference (control group). On day 4 of incubation in Groups II, III and IV, a hole (approximately 5 mm in diameter) was made in the air cell end of each egg to inject 50 µl of 0.7% natrium chloride solution (NaCl; Sigma USA) containing Cd ions (as cadmium chloride; $CdCl_2$, Sigma USA) into ovalbumin, just under the chorioallantoic membrane. The used cadmium dose were: Group II - 0 µg Cd/egg - (sham); Group III - 1 µg Cd/egg ; and Group IV- 5 µg Cd/egg. The dose was based on our earlier study, where under the study conditions; the lethal dose (LD_{50}) was 3.9 µg Cd/egg (Dżugan et al., 2011). After injection, the hole in the shell was sealed with Parafilm® (area approximately 150 mm^2) and the incubation was continued until hatching. On days 6 and 18 eggs were recandled to remove eggs with dead embryos.

The experiment was approved by the First Local Ethics Committee at the Medical University in Lublin, Poland (No. 9/2011).

Heart rate measurements

Cardiac work of chick embryo was measured from day 9 to 21 day of incubation at the same hour, using contactless ballistocardiography (Pawlak and Niedziółka, 2004). In the contactless ballistocardiography of the chick embryo, an egg shell bearing electric charges is one capacitor plate, the other being a receiving antenna of the measuring equipment. The heart work of embryo induces micro-movements of the whole egg, resulting in changes of the distances between the plates and thus in the difference of potentials between the shell and the receiving antenna, which is registered by the measuring equipment (Szymański et al., 2002).

Once the measurements were completed the signal from the embryo heart work was analysed by a computer. In the computer analysis performed to determine the heart rate, envelope signal was calculated using a moving average (11 points) from the modulus of the recorded signal in accordance with the following formula:

$$Y_i = \frac{1}{11} \cdot \sum_{k=i-5}^{i+5} |y_k|$$

where: yk is kth sample of the recorded signal, and Yi is ith sample of the newly formed time series (envelope).

Next, the signal was analysed by dividing the measurement series into 15 second fragments. The signal power spectrum with a resolution of 0.06 Hz was calculated from each fragment of measurement data. Power spectra were calculated using fast Fourier transform (Brigham, 2002). These spectra were summed to determine mean power spectrum of the whole measurement, which finally made it possible to determine the heart rate of the developing embryo on successive days of growth.

Morphometric measurements

Immediately after hatching (21 day of incubation) 10 chicks were randomly selected from each group. Chicks were weighed and decapitated. Bleeding out was followed by collection of their hearts, which were weighed and fixed in 4% paraformaldehyde (PFA), dehydrated in increased concentrations of alcohol, and embedded in paraplast. Paraffin blocks were cut using a Leica RM2145 rotary microtome into sections 5 µm thick, which were stained with haematoxylin and eosin. So prepared histological preparations were measured for thickness of left and right heart ventricles using Multi Scan Base 98 software. Fifty measurements of left and right ventricular wall thickness was taken in each preparation, starting from auriculoventricular valves toward the apex of the heart.

Table 1. Effect of *in ovo* injection of cadmium ions on the cardiac work of chick embryos on successive days of incubation; (n=40); means ± SD.

Injection day of incubation	Group I non-injected (control)	Group II 50 µl of 0.7% NaCl (sham)	Group III 1 µg Cd in 50 µl of 0.7% NaCl /egg	Group IV 5 µg Cd in 50 µl of 0.7% NaCl /egg
10	258±11.2ᵃ	258±14.3ᵇ	253±12.9	248±17.5ᵃᵇ
11	246±10.8ᵃ	250±15.1ᵇ	244±14.7	236±14.8ᵃᵇ
12	237±7.6ᵃ	237±9.6ᵇ	232±12.4	223±16.7ᵃᵇ
13	231±8.3ᵃ	230±10.0ᵇ	225±15.6	216±16.0ᵃᵇ
14	232±9.1ᵃ	229±18.4ᵇ	224±10.0	217±12.7ᵃᵇ
15	231±11.2ᵃ	230±12.6ᵇ	225±15.1	217±11.0ᵃᵇ
16	231±7.9ᴬ	230±8.9ᴮ	224±18.2ᵉ	206±18.6ᴬᴮᵉ
17	220±12.3ᴬᶜ	220±11.4ᴮᵈ	207±11.0ᶜᵈᵉ	190±20.5ᴬᴮᵉ
18	214±11.7ᴬᶜ	213±14.2ᴮᵈ	200±15.7ᶜᵈᵉ	183±20.2ᴬᴮᵉ
19	205±14.6ᴬᶜ	202±15.1ᴮᵈ	190ᵉ±14.2ᶜᵈᵉ	170±19.5ᴬᴮᵉ
20	197±13.9ᴬᶜ	193±19.5ᴮᵈ	183±20.6ᶜᵈᵉ	163±18.5ᴬᴮᵉ
21	179±22.1ᴬᶜ	181±24.9ᴮᵈ	166±22.9ᶜᵈᵉ	149±24.1ᴬᴮᵉ

Values in rows with the same small letters differ significantly at p ≤ 0.05; those with the same capital letters differ significantly at p ≤ 0.01.

Aminotransferase measurement

Concentrations of alanine aminotransferase (ALT; AlaAT) and aspartate transferase (AST; AspAT) were measured in plasma samples collected from chick embryos on day 20 of incubation. The enzymes were determined by means of the automatic kinetic method recommended by IFCC (1986) using kits produced by Alpha Diagnostics (Poland).

Statistical analysis

The results were analysed statistically by two-way analysis of variance for repeated measures, and significant differences between the means were evaluated using Student's t-test. Prior to the analysis, normality of distribution was assessed using Shapiro-Wilk's test. Probability of 0.05 was considered an important indicator of statistical differences between the means.

RESULTS AND DISCUSSION

The heart rate of avian embryos is one of the most often reported parameters of cardiac work (Ono et al., 1997; Tazawa et al., 2001). In our research, comparison of the heart rate in different groups showed that it was always lower in embryos exposed to cadmium than in embryos from the control and sham groups. Differences between the group injected with 5 µg of cadmium and Groups I and II were statistically significant from 10 to 15 days (p≤0.05) and from 16 to 21 days of incubation (p≤0.01). Differences in the heart rate between the group exposed to 1 µg of Cd and Groups I and II were statistically significant (p≤0.05) between 17 and 21 days of incubation (Table 1).

Starting from day 16 of incubation, the heart rate in Group IV was significantly lower than in Group III (p ≤ 0.01) (Table 1). The slowing of heart rate in the final stage of incubation was probably due to the embryo absorbing and accumulating cadmium injected into ovalbumin. Similar findings were obtained by Nishiyama et al. (1990), who observed a slowed heart rate in rats receiving cadmium-supplemented feed. Also, the experiments with fish embryos support the thesis that cadmium has a slowing effect on heart rate (Westernhagen et al., 1975; Jezierska et al., 2002).

Our experiment also revealed that cadmium has a negative effect on morphometric elements of the heart. Measurements of heart weight showed a significant (p≤0.05) decrease in this parameter in embryos exposed to 5 µg of cadmium (Group IV) compared to the control and sham groups (Table 2). When comparing the relative heart weight, it was found that in chickens originating from cadmium-injected eggs, the value of this parameter was always lower than in embryos unexposed to cadmium. Statistical analysis revealed that differences between the group of embryos exposed to 1 µg of cadmium and the control and sham groups was significant at 0.05, while differences between the control and sham groups and the group injected with 5 µg of cadmium were significant at 0.01 (Table 2). Similar changes in heart weight in female Pekin ducks exposed to cadmium were observed by Hughes et al. (2000). A decrease in relative heart weight in broiler chickens receiving dietary cadmium was also described by Bokori et al. (1996). Mikhaleva et al. (1991) report that chronic oral administration of cadmium chloride in rats decreased the weight of heart ventricles. On the other hand, Jamall and Smith (1985) demonstrated increased heart weight in cadmium-exposed rats.

Our study, for the first time, showed that in embryos exposed to cadmium, the thickness of the right ventricle wall decreased considerably. Differences between the

Table 2. Effect of *in ovo* injection of cadmium ions on morphometric parameters of the heart of newly-hatched chicks (n=10); means ± SD.

Item	Group I non-injected	Group II 50 µl of 0.7% NaCl (sham)	Group III 1 µg Cd in 50 µl of 0.7% NaCl /egg	Group IV 5 µg Cd in 50 µl of 0.7% NaCl /egg
Heart weight (g)	0.258 ± 0.06^a	0.262 ± 0.04^b	0.252 ± 0.06	0.247 ± 0.07^{ab}
Heart weight/body weight	0.588 ± 0.04^{aB}	0.590 ± 0.06^{cD}	0.574 ± 0.08^{ac}	0.561 ± 0.05^{BD}
Thickness of the right ventricle (µm)	682.7 ± 50.1^{aB}	685.0 ± 56.2^{cD}	670.4 ± 55.3^{ac}	665.3 ± 66.2^{BD}
Thickness of the left ventricle (µm)	861 ± 48.2	863.2 ± 52.5	858.3 ± 47.2	853.2 ± 42.4

Values in rows with the same small letters differ significantly at p ≤ 0.05, those with the same capital letters differ significantly at p ≤ 0.01.

Table 3. Effect of *in ovo* injection of cadmium ($CdCl_2$) on aspartate (AST) and alanine aminotransferase (ALT) levels in the blood of 20-day-old chick embryos (n=10); means ± SD.

Enzyme	Group I Non-injected	Group II 50 µl of 0.7% NaCl (sham)	Group III 1 µg Cd in 50 µl of 0.7% NaCl /egg	Group IV 5 µg Cd in 50 µl of 0.7% NaCl /egg
AST (U/L]	6.72 ± 0.25^A	7.43 ± 0.46^B	6.99 ± 0.53^C	24.7 ± 0.51^{ABC}
ALT (U/L)	14.04 ± 1.75^a	13.60 ± 4.83^b	16.58 ± 7.12^c	29.6 ± 9.47^{abc}

Values in rows with the same small letters differ significantly at p ≤ 0.05; those with the same capital letters differ significantly at p ≤ 0.01.

control (Group I) and sham groups (Group II) and the group exposed to 1 µg of cadmium were statistically significant at 0.05, and those between the group injected with a higher dose of cadmium ions and Group I and II were significant at P<0.01 (Table 2). Measurements of the left ventricle wall did not show any significant differences in the value of this parameter (Table 2). The available literature provides no data regarding this issue. So, we suggest that observed decrease of ventricle wall thickness could be associated with cadmium-induced apoptosis (Yazihan et al., 2011).

Our study also showed a negative effect of a higher cadmium dose (5 µg) on cardiac muscle structure. In embryos from this group, the microscopic images for the first time showed inflammatory changes indicative of interstitial myocarditis. Muscle fibres were spaced apart, with a large inflammatory infiltration of mononuclear cells (mainly lymphocytes) in between. Lesions of this type were observed in all birds exposed to the higher cadmium dose, both in the muscle fibres of the ventricles and in the interventricular septum (Figure 1d). In embryos from the control and sham groups and in those exposed to 1 µg cadmium, no pathological changes in heart muscle structure were observed (Figure 1a to c). Meanwhile, Mikhaleva et al. (1991) showed the effect of Cd on left ventricular cardiomyocyte hypertrophy. Kolakowski et al. (1983) observed distinct alterations of the intercalated disc structure. The damage to intercalated discs varied from the enlargement of the fissure between membranes (within unspecialized segments) to disruption of the complex junctions.

The results of histological examination correspond with the results of biochemical analysis. The injection of 5 µg of Cd/egg caused a significant increase in the

concentrations of aspartate (p ≤ 0.01) and alanine aminotransferase (p ≤ 0.05) in relation to both the control and sham groups (Table 3). The injection of 1 µg of Cd/egg caused no significant changes in the concentration of these enzymes (Table 3).

Aspartate and alanine aminotransferases are two of the enzymes most frequently measured by the clinical laboratory (Rej, 1989). They are most commonly used in the differential diagnosis of various liver diseases where the ratio of the two enzymes provides additional clinical insight. AST is much less liver-specific, therefore is also useful in many cases for diagnosis, or estimating severity of myocardial infarction (Rej, 1989). In plasma of chicks treated with cadmium during embryogenesis the most intensive increase in AST versus ALT levels was observed. In control and sham group the relation between AST and ALT was 1:2, whereas in Cd-group this relation was 1:1. Although used Cd dose caused the chick liver damage, observed the more intensive increase in AST level (by 4-times as compared to control) than in ALT level may indicate other non-liver cadmium toxic action. This observation is confirmed by the study of the effect of local heart irradiation in a rat model which showed, among other myocardial enzymes, an increase in plasma AST activity whereas plasma ALT levels remained unchanged (Franken et al., 2000). Chronic Cd administration (15 ppm per 8 weeks) induces inflammation and apoptosis in rat hearts, which was evaluated using much more specific biomarkers TNF-α and IL-6 (Yazihan et al., 2011).

In conclusion, the present study showed a clear effect of cadmium on the heart rate, and on the morphometric and histological results of chick embryo hearts, with the intensity of these changes being positively correlated.

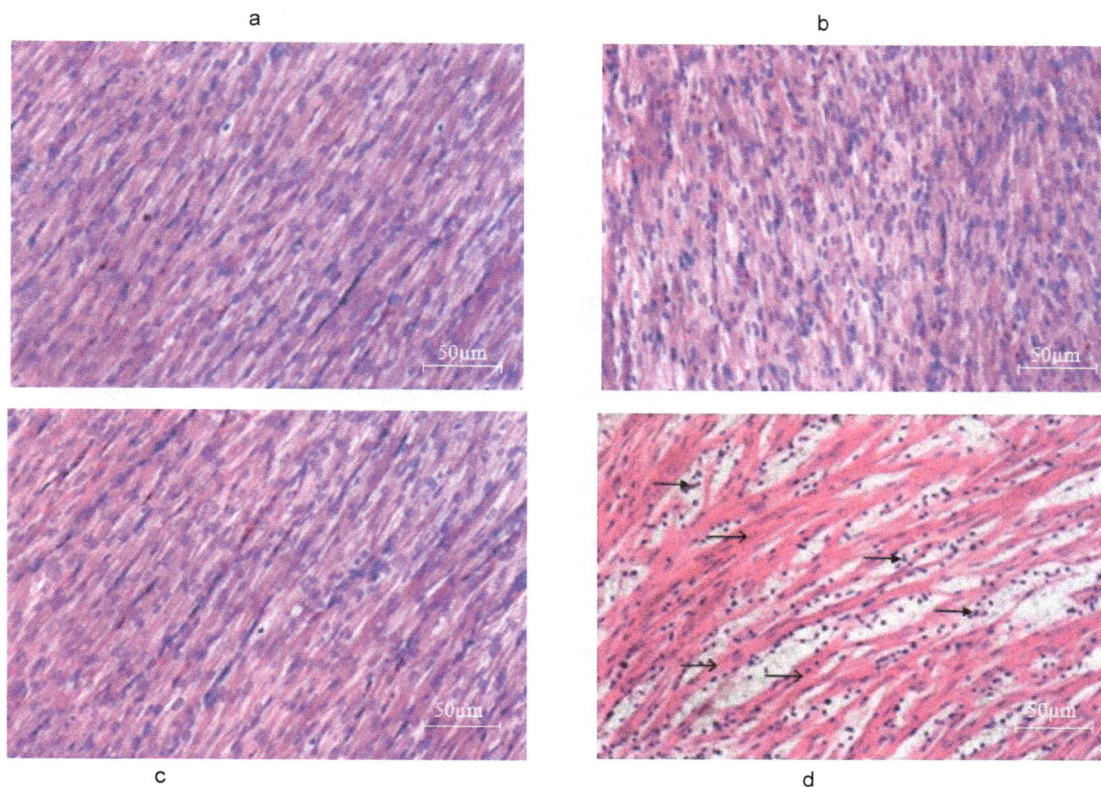

Figure 1. Cardiac muscle microstructure (a) Group I non-injected; (b) Group II sham; (c) Group III injected 1 µg of Cd/egg; (d) Group IV injected 5 µg of Cd/egg. A dense mononuclear and predominantly lymphocytic infiltrate (⟶) among cardiomyocytes (→).

The mechanism of cadmium cardiotoxicity needs explanation and probably is connected with calcium, one of the elements responsible for normal heart work (Dhalla et al., 1982). It is known that cadmium may interfere with calcium metabolism by removing it from the body (Czeczot and Skrzycki, 2010). In addition, cadmium is a blocker of calcium channels (Bridges and Zalups, 2005; Martelli et al., 2006). By acting through various metabolic pathways, cadmium affects the activity of the renin-angiotensin system (Martynowicz and Skoczyńska, 2003). Cadmium may reduce concentration of angiotensin II, which is produced as a result of the action of dipeptide carboxylase on angiotensin. The lower concentration of this compound inhibits stimulation of the AT-1 receptor, which is responsible for increased influx of calcium into the heart muscle cells and releases calcium from the sarcoplasmic reticulum (Martynowicz and Skoczyńska, 2003).

It is now a real challenge to understand the precise nature of processes associated with the cadmium exposure and intracellular calcium deficiency in order to achieve proper management of cardiac disorders. It is generally known that cadmium from cigarette smoke, penetrating through the placenta may reach the embryos and accumulates in the fetal body (Trottier et al., 2002).

Assumption than can be made, based on our research is that the cadmium contained in the cigarette smoke can damage the heart of the human embryo. Thus further study of cadmium influence on heart with particular regard to its effects on calcium intracellular balance using in ovo model are planned.

ACKNOWLEDGEMENTS

This work was financed by University of Agriculture in Krakow (DS 3210/KHDZFiZ/2012).

REFERENCES

Bokori J, Fekete S, Glávits R, Kádár I, Koncz J, Kövári L (1996). Complex study of the physiological role of cadmium. IV. Effects of prolonged dietary exposure of broiler chickens to cadmium. Acta. Vet. Hung. 44(1):57-74.

Bridges CC, Zalups RK (2005). Molecular and ionic mimicry and the transport of toxic metals. Toxicol. Appl. Pharmacol. 204:274-308.

Brigham EO (2002). Fast fourier transform and its applications. Ed Prentice-Hall New York.

Cullinane J, Bannigan J, Thompson J (2009). Cadmium teratogenesis in the chick: Period of vulnerability using the early chick culture method, and prevention by divalent cations. Reprod. Toxicol. 28:335-341.

Czeczot H, Skrzycki M (2010). Cadmium – element completely unnecessary for the organism (in Polish). Postepy. Hig. Med. Dosw. 64: 38-49.

Dhalla NS, Pierce GN, Panagia V, Singal PK, Beamish RE (1982). Calcium movements in relation to heart function. Basic Res. Cardiol. 77(2): 117-139.

Davison TF (2003). The immunologists' debt to the chicken. Br. Poult. Sci. 44(1):6-21.

Dżugan M, Lis M, Droba M, Niedziółka WJ (2011). Effect of cadmium injected in ovo on hatching results and the activity of plasma hydrolytic enzymes in newly hatched chicks. Acta. Vet. Hung. 59(3):337-347.

EFSA (2009). Scientific opinion of European Food Safety Authority: Cadmium in food. The EFSA Journal, 980, 1-139, http://www.efsa.europa.eu/EFSA/efsa_locale-1178620753812_1211902396263.htm.

Fotakis G, Cemeli E, Anderson D, Timbrell JA (2005). Cadmium chloride-induced DNA and lysosomal damage in hepatoma cell line. Toxicol. *In vitro* 19:481-489.

Franken NA, Strootman E, Hollaar L, van der Laarse A, Wondergem J. (2000). Myocardial enzyme activities in plasma after whole-heart irradiation in rats. J. Cancer Res. Clin. Oncol. 126(1):27-32.

Hughes MR, Smits JE, Elliott JE, Bennetta DC (2000). Morphological and pathological effects of cadmium ingestion on pekin ducks exposed to saline. J. Toxicol. Environ. Health, Part A 66(7):591-608.

IFCC (1986). Method for the measurement of ALP. J. Clin. Chem. Clin. Biochem. 24:481-595.

Jamall IS, Smith JC (1985). Effects of cadmium treatment on selenium-dependent and selenium-independent glutathione peroxidase activities and lipid peroxidation in the kidney and liver of rats maintained on various levels of dietary selenium. Arch. Toxicol. 58(2): 102-105.

Järup L, Berglund M, Elinder CG, Nordberg G,Vahter M (1998). Health effects of cadmium exposure: A review of the literature and risk estimate. Skand. J. Work Environ. Health 24:1-51.

Jarüp L, Äkesson A (2009). Current status of cadmium as an environmental health problem. Toxicol. Appl. Pharm. 238(3):201-208.

Jezierska B, Ługowska K and Witeska M (2002). The effect of temperature and heavy metals on heart rate changes in common carp Cyprinus carpio L. and grass carp Ctenopharyngodon idella (Val.) during embryonic development. Arch. Pol. Fish. 10:153-165.

Kolakowski J, Baranski B, Opalska B (1983). Effect of long-term inhalation exposure to cadmium oxide fumes on cardiac muscle ultrastructure in rats. Toxicol. Lett. 19(3):273-278.

Lahijani MS, Tehrani DM, Sabouri E (2009). Histopathological and ultrastructural studies on the effects of electromagnetic fields on the liver of preincubated white Leghorn chicken embryo. Electromagn. Biol. Med. 28(4):391-413.

Limaye DA, Shaikh ZA (1999). Cytotoxicity of cadmium and characteristics of its transport in cardiomyocytes. Toxicol. Appl. Pharmacol. 154:59-66.

Martelli A, Rousselet E, Dycke C, Bouron A, Moulis JM (2006). Cadmium toxicity in animal cells by interference with essential metals. Biochimie 88:1807-1814.

Martynowicz H, Skoczyńska A (2003). Effects of cadmium on the vascular endothelium function (in Polish). Med. Pr. 54(4):383-388.

Mikhaleva LM, Zhavoronkov AA, Cherniaev AL, Koshelev VB (1991). Morphofunctional characteristics of cadmium-induced arterial hypertension. Bull. Exp. Biol. Med. 111(4):420-423.

Mitta B, Weber CC, Fussenegger M (2005). *In vivo* transduction of HIV-1-derived lentiviral particles engineered for macrolide-adjustable transgene expression. J. Gene. Med. 7:1400-1408.

Mlynek V, Skoczyńska A (2005). Proinflammatory effects of cadmium (in Polish). Postepy. Hig. Med. Dosw. 59:1-8.

Nishiyama S, Saito N, Konishi Y, Abe Y, Kusumi K (1990). Cardiotoxicity in magnesium-deficient rats fed cadmium. J. Nutr. Sci. Vitaminol. (Tokyo) 36(1):33-44.

Nordberg GF (2009). Historical perspectives on cadmium toxicology. Toxicol. Appl. Pharmacol. 238:192-200.

Ono H, Akiyama R, Sakamoto Y, Pearson JT, Tazawa H (1997). Ballistocardiogram of avian eggs determined by an electromagnetic induction coil. Med. Biol. Eng. Comput. 35: 431-435.

Pawlak K, Niedziółka J (2004). Non-invasive measurement of chick embryo cardiac work. Czech. J. Anim. Sc. 49(1):8–15.

Pawlak K, Lis MW, Sechman A, Tombarkiewicz B, Niedziółka JW (2011). Effect of in ovo injection of acetylsalicylic acid on morphotic parameters and heart work in chicken embryos exposed to hyperthermia. Bull. Vet. Inst. Pulawy. 55:95-100.

Rajendra P, Sujatha H, Devendranath D, Gunasekaran B, Sashidhar R, Subramanyam C, Channakeshava L (2004). Biological effects of power frequency magnetic fields: Neurochemical and toxicological changes in developing chick embryos. Biomagn. Res. Technol. 312(1):1.

Rej R (1989). Aminotransferases in disease. Clin Lab Med. 9(4):667-687.

Satarug S, Baker JR, Urbenjapol S, Haswell-Elkins M, Reilly EB, Wiliams DJ, Moore MR (2003). A global perspective on cadmium pollution and toxicity in nonoccupationally exposed population Toxicol. Lett. 137:65-83.

Szymański JA, Pawlak K, Mościcki JM (2002). A device for reception of slow changing electric field produced by developing hen embryo Rev. Sci. Instrum. 73:3313-3317.

Tadahiro K, Kiyoshi K, Yoshio E, Tetsuya I, Tadao U, Takuma S, Mikic N, Eur A (1998). Chick Embryo Model for Metastatic. Hum. Urol 34:154-160.

Tazawa H, Person J, Komoro T, Ar A (2001). Allometric relationship between embryonic heart rate and egg mass in birds. J. Exp. Biol. 204:165-174.

Thompson J, Bannigan J (2008). Cadmium: toxic effects on the reproductive system and the embryo. Rev. Reprod. Toxicol. 25:304-315.

Trottier B, Athot J, Ricard AC, Lafond J (2002). Maternal-fetal distribution of cadmium in the guinea pig following a low dose inhalation exposure. Toxicol. Lett. 28,129(3):189-197.

Westernhagen V, Dethlefsen and Rosenthal H (1975). Combined effects of cadmium and salinity on development and survival of garpike eggs. Helgolander Wiss. Meer. 27:268-282.

Wonga KL, Cachiaa R, Klaassena CD (1980). Comparison of the toxicity and tissue distribution of cadmium in newborn and adult rats after repeated administration. Toxicol. Appl. Pharmacol. 56, 3:317-325.

Yasuda HA, Yoshida KA, Yasuda, YB, Tsutsui TA (2012) Two age-related accumulation profiles of toxic metals. Curr. Aging Sci. 5(2):105-111.

Yazihan N, Kocak MK, Akcil E, Erdem O, Sayal A, Guven C, Akyurek N (2011). Involvement of galectin-3 in cadmium-induced cardiac toxicity. Anadolu Kardiyol Derg 11(6):479-484.

8

Effect of different feeding system on body weight, testicular size developments, and testosterone level in pre-pubertal male camel (*Camelus dromedarius*)

Al-Saiady, M. Y.[1], Mogawer H. H.[1], Al-Mutairi S. E.[2], Bengoumi M.[2], Musaad A.[2], Gar-Elnaby A.[3], and Faye B.[2]

[1]ARASCO R & D Department, P. O. Box 53845, Riyadh 11593, Kingdom of Saudi Arabia.
[2]Camel Breeding, Range Protection and Improvement Center in Al-Jouf area, Kingdom of Saudi Arabia.
[3]Animal Production Department, College of Food and Agricultural Sciences, P. O. Box 2460, King Saud University, Riyadh 11451, Kingdom of Saudi Arabia.

Eighteen dromedary males (*Camelus dromedarius*) were used to investigate the effect of nutrition on body weight, size of the testes, and blood testosterone concentrations in pre-pubertal male camels. Animals were divided into two groups of 9 animals each. Group A received a diet with 13% crude protein (CP) and 2.9 MCal (ME), whereas Group B received the traditional diet of the farm, and each animals' feed intake was calculated after allowing a 14 day adaptation period. Diets contain 25:75 (roughage: concentrate, respectively). Blood samples were taken from the same five animals from each group, every 15 days during the whole experimental period and plasma testosterone concentrations were measured. There was no significant difference in total body weight gain over the whole experimental period between Group A and B, although Group A showed a significant increase in body weight over the last 6 months compared with Group B. Group A consumed less feed and were more efficient at converting feed to body weight than Group (B), as shown by the (FCR) over the whole period which was 9.25 for Group A and 13.03 for Group B. There was no significant difference in testicle size between Groups A and B at the start of the experimental period, blood testosterone levels were significantly higher in Group A compared with Group B, but although there was an increase in testicle size over the experimental period, there was no significant increase in blood testosterone levels.

Key words: *Camelus dromedarius*, puberty, body weight gain, testicular size, testosterone.

INTRODUCTION

The one humped camel (*Camelus dromedarius*) has the capacity of being a better provider of food in the desertareas of the world than the cow which can be severely affected by heat and scarcity of feed and water. One of the most important factors affecting productivity, other than nutrition and disease, is the low reproductive performance of the camel, short breeding season and long gestation period of 13 months (Aboul-Ela, 1991).

This low reproductive performance has remained a major obstacle to the growth of populations of dromedaries over the years (Tibary and Anouassi, 1997). High fertility levels in the camel are essential, not only for profitable production, but also to provide opportunities for selection and genetic improvements. The breeding season of camel in India extends from December to March, that is, the period of short day length (Matharu, 1966) and similar short day breeding seasons have been reported in Sudan (Musa and Abusineina, 1978). However, as stated by Musa et al. (1993), no organized attempts to manipulate the onset of the breeding season, or to extend it, have been reported. Thompson and Johnson (1995), Ott (1991), Osman et al. (1979), and Abdel-Raouf et al. (1975) reported that size and weight of testes are affected by the age of the camel, and the season of the year. In addition, nutrition effects the age at which male camels reach puberty as animals on a good plane of nutrition come to puberty earlier, in fact body weight has more influence on puberty than age. Nolan et al. (1991), Abdel-Rahim et al. (1994) and Abdel-Rahim (1997) however, reported that there was a highly negative correlation between testicular dimensions and the age at which spermatogenesis starts.

Testosterone is the main sex hormone controlling most of the reproductive functions including libido, later stages of spermatogenesis, and the activity of accessory sex glands in male animals (Hafez and Hafez, 2000). Azouz et al. (1992) reported a significant decrease in testosterone levels in male camels in the non-breeding season as compared with the levels during the rut. This agreed with results of El-Bahrawy and El-Hassanein (2011) who reported that basal concentrations of testosterone during the non-rutting season were 2.89±0.26 ng/ml, which significantly increased to 5.8±0.74 ng/ml during the pre-rut. This increase continued to maximum concentrations of 7.95±1.85 ng/ml during the rutting season, then finally decreased during the post rut to 3.15±0.38 ng/ml before finally reaching basal concentrations again. Also, Rateb et al. (2011) reported that the average values of blood serum testosterone were significantly lower in sub-fertile camels (1.7±0.2 ng/ml) comparing with fertile animals (3.7±0.2 ng/ml) and immature camels had overall significantly lower levels of testosterone (Al Qarawi and ElMougy, 2008).

In view of these results and considering that diet could have an effect on the age at which male camels reach puberty, the objective of this study was to evaluate the effect of diet on live body weight gain, testes size development, and blood testosterone concentrations, as reproductive parameters for pre-pubertal male camels.

MATERIALS AND METHODS

Animals and diets

This experiment was conducted in October at Camel Breeding, Range Protection and Improvement Center in Al-Jouf area, K. S. A. Eighteen dromedary males (*C. dromedarius*) were used to investigate the nutritional effect on body weight and testes size development, as well as blood testosterone level in pre-puberty camel males. Animals were divided into two equal groups according to body weight and age at the start of the trial (265 kg and 17 month, respectively). Group (A) received diet with 13% crude protein (CP) and 2.9 Mcal (ME). Group (B) received the traditional diet of the center (Table 1).

Animals' individual feed intake was calculated after allowing a 14 day period to adapt to the feed. The feed offered and orts were recorded daily for the entire experimental period of 12 months. Animals were fed with diets containing 25:75 (roughage: concentrate, respectively) and in Diet A, roughage and concentrates were in one pellet. Fresh water was available all time. Jugular vein blood samples were collected into anticoagulant, evacuated tubes from the same five animals from each group every 15 days during the whole experimental period. The plasma was then separated and frozen at -20°C until further analyses.

Lab analysis and measurements

The following parameters were measured or calculated: (i) body weight every 15 days in kg, the animals were weighed after 10 h of fasting; (ii) body weight gain in kg, (iii) daily weight gain in Kg/day.

Jugular vein blood samples were collected in ethylenediaminetetraacetic acid (EDTA) vials twice a month always at 8.00 Am, samples centrifuged at 5000 rpm for 20 min, plasma were separated and stored at -20°C until used to measure plasma testosterone concentrations using commercial Elisa kits (Diagnostic Automation inc. CA. USA).

The axes of the testicles were measured using calipers and the size calculated by using the equation for ellipsoid volume: '4/3 π a*b*c' a.b.c. = axes of ellipsoid (http://en.wikipedia.org/wiki/volume).

Statistical analysis

Data had been subjected to statistical analysis using the SAS program (SAS, 2000). Data for changes in body weight were analyzed according to the following model:

$$Y_{ij} = \mu + T_i + e_{ij}$$

Where Y_{ij} is the observation of the dependent variable obtained from J^{th} animal of Ith treatment, μ is the overall mean; T_i is the effect of ith treatment (i = A or B); and e_{ij} is the residual term. For the testicular size and testosterone concentrations the model was:

$$Y_{ijk} = \mu + T_i + P_j + e_{ijk}$$

Where Y_{ijk} is the observation of the dependent variable obtained from K^{th} animal of Ith treatment, of Jth period, μ is the overall mean; T_i is the effect of ith treatment (i = A or B); P_j is effect of jth period (j = 1 to 4); and e_{ijk} is the residual term. The interaction between groups and months failed to be significant for that deleted from the model. The general linear model (GLM), least squares means (LSMEANS) procedures were used.

RESULTS AND DISCUSSION

Changes in live body weight

The total body weight gain over the whole period was not significantly higher in-Group (A) compared with Group (B), 146.6±6.9 vs 138.3±6.8, respectively, however Group (A) showed significantly (P<0.05) higher body weight gain over the last 6 months compared with group (B) 79.01±4.5 vs 62.3±4.5 kg, respectively (Tables 2 and 3). The difference between Group A and B in feed intake was significant (P<0.05) and interestingly Group A consumed less feed but showed higher body weight gain,

Table 1. Diet composition and chemical constituents (dry matter bases).

Item	Diets	
Raw materials (%)	**A**	**B**
Barley	60.22	62.23
Wheat bran	9.63	12.08
Soya meal 48%	4.25	-
Salt	0.47	-
Limestone	2.10	-
Acid buf	1.00	-
Molasses	3.00	-
Premix	0.30	-
Alfalfa	19.03	15.23
Wheat straw	-	10.46
Nutrient (%)		
Dry matter (DM)	90.20	92.52
Crude protein (CP)	13.08	12.43
ADF	16.00	18.5
NDF	33.8	37.7
Calcium	1.67	0.35
Phos.	0.42	0.27
Salt	0.78	1.38
ME Mcal/kg	2.9	2.7

ME = Metabolisable energy.

Table 2. Treatment effect on live body weight and body weight gain.

Treatment	LBW 6 m (kg)	LBW 12 m (kg)	BWG 6 (kg)	BWG 12 (kg)	Total BWG (kg)
A	349.6±12.2	398.6±15.4	67.6±3.6	79.0±4.5[a]	146.6±6.9
B	355.7±11.9	419.6±15.2	75.4±3.5	62.3±4.5[b]	138.3±6.8

Values with different superscripts within the same column are significantly different (P≤0.05). LBW6 = Live body weight at 6 months. LBW12 = Live Body Weight at 12 months. BWG6 = Body weight gain over first 6 months. BWG12 = Body weight gain over last 6 months.

Table 3. Treatment effect on feed conversion ratio over first, last 6 months and whole period.

Item	Groups	
First 6 months	**A**	**B**
FI (Kg)	4.36±0.08[b]	5.76±0.06[a]
DWG (Kg)	0.400±0.05	0.400±0.04
FCR	10.90	14.40
Last 6 months		
FI (Kg)	4.63±0.07[b]	5.68±0.07[a]
DWG (Kg)	0.560±0.09[a]	0.431±0.08[b]
FCR	8.26	13.17
Whole period		
FI (Kg)	4.35+0.03[b]	5.50±0.03[a]
DWG (Kg)	0.470±0.06	0.420±0.06
FCR	9.25	13.03

Values with different superscripts within the same row are significantly different (P≤0.05).
FI = Feed Intake. DWG = Daily body weight Gain. FCR = Feed conversion ratio.

Table 4. Treatment effect on testes size and blood testosterone level.

Treatment	R. Testicular /cm^3	L. Testicular /cm^3	Testosterone ng/ml
A	120.61±9.41	108.55±10.39	3.88±0.08[a]
B	114.44±13.95	90.39±13.83	3.65±0.08[b]
Month			
3	86.67±15.24[b]	67.17±16.20[b]	3.84±0.11
6	140.85±27.33[a]	102.86±29.49[a]	3.80±0.11
9	91.73±12.34[b]	99.55±15.57[b]	3.89±0.12
12	150.85±9096[a]	128.30±8.70[a]	3.54±0.12

Values with different superscripts within the same column are significantly different (P≤0.05). R = right L = left.

Table 5. Treatment effect on mean blood concentrations of 'total protein, albumin, globulin, glucose, and cholesterol'.

Items	Treatments	
	A	**B**
Total protein (TP) g/l	67.66±1.17	64.98±1.20
Albumin (Alb) g/l	44.28±0.66[a]	41.65±0.68[b]
Globulin (Glo) g/l	23.28±1.04	23.33±1.06
Alb/Glo	2.01±0.08	1.95±0.08
Glucose (Glu) mg/dl	140.29±4.73[b]	162.26±4.83[a]
Cholesterol (Chol) mg/dl	29.50±1.40	32.52±1.43

Different letters within row indicates significant difference (P<0.05).

thus indicating a better food conversion rate than Group B. This was shown in the results by the feed conversion ratio (FCR) which was 10.9 (Group A) *vs* 14.4 (Group B) over the first 6 months (Table 3); 8.26 in Group A *vs* 13.17 in Group B over the last 6 months (Table 3) and 9.25 (Group A) *vs* 13.03 (Group B) over the whole period (Table 3). These results agreed with those of Mohamed, (2007) who observed a clear variation in camel FCR when they were fed different types of rations. The daily body weight gain (DWG) in this study was higher than that obtained by Sahani et al. (1998) and Faye et al. (2001) who reported that the daily gain for male camels between 18 to 24 months of age ranged from 0.111±0.015 to 0.219±0.24 kg/day. This result could be due to the difference in the management system, the nutritive value of the diet distributed to the animals and the breed characteristics. Indeed, the conformation of adult camel breeds in Saudi Arabia (Waddah and Majaheem) which are used in the present trial) and consequently their adult weight was on average much higher than the Indian and Ethiopian breeds which were used by these authors, respectively. In Kenya, on Somali breed with similar conformation than Saudi breed, the DWG (387 g/day) was similar to our observations (Kaufmann, 1998). In very intensive systems for fattening young camels, it has been reported exceptional post-weaning DWG superior to 500 g (Faye, 1997).

Changes in testes size

The size of the testicles was not significantly different between Groups A and B and although the right testicle tended to be larger, it was not significantly bigger than the left testicle (Table 4). This difference between right and left testicle was already reported in adult camel by several authors, whatever the season or the age groups, but generally, it observed a higher size of the left one (Singh and Bharadwaj, 1980). Over the whole experimental period, there was a significant increase in size of both the right and left testicles which agrees with the results of Al-Asaad et al. (2007) and El-Harairy and Attia (2010) who reported that there was a considerable development in testicular dimensions with increasing age. Group A had significantly higher blood testosterone concentrations (*P*<0.05) compared with group B but although there was an increase in testicle size over the experimental period, there was no significant increase in blood testosterone levels (Table 5). This result indicates that the testes did not secrete testosterone yet. Testosterone levels in this study match those during the non-rutting season (2.89±0.26 ng/ml) reported by El-Bahrawy and El-Hassanein (2011), and Yagil and Etzion (1980). However, in prepubertal camel, El-Harairy and Attia (2010) reported lower values (0.31±0.05 ng/ml) than in our study. Reported testicular dimensions in the dromedary camel vary from one author to another. This variation is attributed to the age, to the sexual activity in the adultand probably to the breed (Tibary and Anouassi, 1997). These factors did not interfere in our case as breed composition was similar in the two groups, the mean ages were comparable and the animals were not yet adult. However, in spite of the juvenile status of the camel, a seasonal effect was observed, the testicle size being higher during the months 6 and 12 corresponding to the rutting time (Table 4). This seasonal variation is well known in adult camel (Tingari et al., 1984).

Physiological status

Regarding blood parameters as an indicator of the physiological status of the animals, only 2 parameters, albumin and glucose, showed significant differences due to treatment (Table 5). In Group A, the albumin level was significantly higher compared with Group B. Albumin provides the body with the protein needed to both maintain growth and repair tissues, which reflected in a higher body weight gain in Group A compared with Group B. The glucose level in Group A was lower than in Group B due to using more glucose as an energy source, to increase body weight. The values for albumin are in the normal range for camel in good conditions (Amin et al., 2007). For glucose, the values are higher than the normal range for adult (50 to 120 mg/100 ml) according to Faye and Mulato (1991). Indeed, it has been reported that the young camel has usually a higher glycemia than adult linked (Souilem et al., 1999).

Conclusion

In spite of only a slight difference in the protein contents of the two diets compared in this study, it appears that the nutrition had a slight effect on live body weight gain, and would thus affect the age at which they reach puberty. Pre-pubertal camels receiving a balanced diet with 13% crude protein, 2.9 Mcal ME and the required vitamins and minerals, improved body weight gain, testes size, and testosterone concentrations in the blood. Thus it would be expected that these animals would come into puberty earlier than those which received an unbalanced diet. However, more research is needed on the effect of nutrition on decreasing the age at which male camels reach puberty, especially by taking into account probable other limiting factors such as vitamins or minerals.

ACKNOWLEDGEMENTS

This study has been achieved within FAO project UTF/SAU/021/SAU with the support of Camel and range research Center (CRRC). We thank all the local staff for the monitoring of the camel in the farm. Special gratitude is extended to Dr. Moez Ayadi for editing and invaluable suggestion added to the manuscript. The authors are also grateful to thank ARASCO for sponsored feed for this trial and for their R&D team especially those of the data processing and statistical division for time and efforts invested and discussing of the results of this work.

REFERENCES

Abdel-Rahim S, Abdel-Rahman E, El-Nazier AE (1994). Production and reproduction of one-humped camel in Al-Gassim region, Saudi Arabia. J. Arid Environ. 26:53-59.

Abdel-Rahim SEA (1997). Studies on the age of puberty of male camels (Camelus dromedaries) in Saudi Arabia. Vet. J. 154:79-83.

Abdel-Raouf M, Fath El-Bab MR, Owaida MM (1975). Studies on reproduction in the camel (Camelus dromedaries). V. Morphology of the testes in relation to age and season. J. Reprod. Fert. 43:109-116.

Aboul-Ela MB (1991). Reproductive performance of camels (Camelus dromedarius) under field conditions in the United Arab Emirates. Proceedings of the International Conference on Camel Production and Improvement (Toburk). pp. 93-100.

Al Qarawi AA, ElMougy SA (2008). Seasonality and the melatonin signal in relation to age as correlated to the sexual cycle of the one-humped male camel (Camelus dromedarius). Biol. Rhythm Res. 39(2):131-142.

Al-Asaad A, Salhab SA, Al-Daker MB (2007). Development of testicular dimensions and relative puberty in Shami camel males. J. Damascus Univ. Agric. Sci. 23:233-250. (in Arabic).

Amin ASA, Abdoun KA, Abdelatif AM (2007). Seasonal Variation in Blood Constituents of One-humped Camel (Camelus dromedarius). Pakistan J. Biol. Sci. 10:1250-1256.

Azouz A, Ateia MZ, Shawky H, Zakaria AD, Farahat AA (1992). Hormonal changes during rutting and non-breeding season in male dromedary camels. Proceeding 1st International Camel Conference. pp 169-171.

El-Bahrawy KA, El-Hassanein EE (2011). Seasonal variation of some blood and seminal plasma biochemical parameter of male dromedary camels. Am. J. Agric. Environ. Sci. 10(3):354-360.

El-Harairy MA, Attia KA (2010). Effect of age, pubertal stage and season on testosterone concentration in male dromedary camel. Saudi J. Biol. Sci. 17:227-230

Faye B (1997). Guide de l'élevage du dromadaire. Ed. Sanofi, Libourne, France. P. 126.

Faye B, Bengoumi M, Cleradin A, Tabarani A Chilliard Y (2001). Body condition score in dromedary camel: A tool for management of reproduction. Proc. of the International twin conf. On reproduction/production in camelids Emir. J. Agric. Sci. 13:1-6.

Faye B, Mulato C (1991). Facteurs de variation des paramètres protéo-énergétiques, enzymatiques et minéraux dans le plasma chez le dromadaire de Djibouti. Rev. Elev. Med. Vét. Pays Trop. 44:325-334.

Hafez B, Hafez ESE (2000). Reproduction in farm animals, Seventh ed. Lippincott Williams and Wilkins, U.S.A.

Kaufmann BA (1998). Analysis of pastoral camel husbandry systems in Northern Kenya, Hohenheim Tropical Agricultural Series 5, Markgraf Verlag, Germany.

Matharu BS (1966) Management production and reproduction in dromedary. Indian Farm. 16:19-22.

Mohamed IM (2007). Growth performance of growing Maghraby camel fed on un-conventional feed. J. Agric. Biol. 9(1):18-21.

Musa B, Sieme H, Merkt, H, Hago BED, Cooper MJ, Allen WR Jochle W. (1993). Manipulation of reproductive functions in male and female camels. Anim. Reprod. Sci. 33:289-306.

Musa BE, Abusineina ME (1978). Clinical pregnancy diagnosis in the camel and a comparison with bovine pregnancy. Vet. Rec. 102:7-10.

Nolan CJ, Neuendorff DA, Godfrey RW, Harns PG, Mc Arthur NH, Randel RD (1991). Influence on dietary, intake on pre-pubertal development of Brahman bulls. J. Anim. Sci. 68:1087-1096.

Osman DI, Moniem KA, Tingari MD (1979). Histological observation on the testes of the camel with special emphasis on spermatogenesis. Acta. Anat. 104:164-171.

Ott RS (1991). Fertility potential of male animals in extensive breeding station. Contraception, Fertilities, Sexualite. 19:749-755.

Rateb SA, El-Hassanein EE, El-Koumy AG, El-Bahrawy KA, Zahraa R Abo El-Ezz (2011). Manipulation of reproductive hormones disorder in sub-fertile male Dromedary camels using exogenous gonadotropic-releasing hormone (GnRH). World J. Agric. Sci. 7(3):280-285.

Sahani MS, Bissa UK, Khanna ND (1998). Factors influencing pre and post weaning body weight and daily gain in indigenous breeds of camels under farm conditions. Proceeding of the third annual meeting for animal production under arid conditions, United Arab Emirates University. 1:59-64.

SAS (2000). User's guide. Statistical Analysis System, Cary, NC, USA.

Singh UB, Bharadwaj MB (1980). Histological studies on the testis pathway and epididymis of camels (Camelus dromedarius). Acta Anat. 101:274-279.

Souilem O, Chine O, Alguemi C, Gogny M (1999). Glycemia in the camel (Camelus dromedarius) in Tunisia preliminary results, Rev. Elev. Med. Vet. 150:537-542.

Tibary A, Anouassi A (1997). Theriogenology in camelidae. Anatomy, physiology, pathology and artificial breeding. Actes Editions Publ., IAV HassanII, Rabat, Maroc. P. 489.

Tingari MD, Ramos AS, Gaili ESE, Rahma BA, Saad AH (1984). Morphology of the testis of the one-humped camel in relation to reproductive activity. J. Anat. 139:133-143.

Thompson DM, Johnson WH (1995). Scrotal size of yearling sire and early calving in beef herds: epidemiological investigation of possible causal pathways. Theriogenology. 43:1279-1287.

Yagil R, Etzion Z (1980). Hormonal and behavioural patterns in the male camel (Camelus dromedarius). J. Reprod. Fert. 58:61-65.

9

Propolis as a natural antibiotic to control American foulbrood disease in honey bee colonies

Kamel, A. A.[1], Moustafa, A. A.[2] and Nafea, E. A.[1]

[1]Beekeeping Research Department, Plant Protection Research Institute (PPRI), Agricultural Research Center, Giza, Egypt.
[2]Department of Biochemistry, Faculty of Agriculture, Cairo University, Egypt.

American FoulBrood (AFB) is one of the most severe bacterial diseases that affect larvae of honey bee *Apis mellifera*, causing a decrease of bee population and colony production and due to the serious effects associated with AFB disease and the problems related to the use of antibiotics, it is necessary to develop alternative strategies for the control of the disease. The aim of this study was to determine, under field conditions, the effectiveness of tylosin and three kind of ethanolic extract propolis (Chinese, Egyptian and old wax comb extract propolis) for controlling AFB disease in honey bee colonies. Identification of Phenolic composition of the ethanolic extract samples were investigated by high performance liquid chromatography (HPLC) instrument. Laboratory studies were conducted to determine the LC_{50} (half lethal concentration) and LT_{50} (half lethal time) values were determine. In field trials the colonies were inoculated with AFB disease for three weeks before initiation of trial or treated with tylosin, 0.1, 0.05 and 0.025% ethanolic extract propolis (EEP) which extracted from Chinese propolis, Egyptian propolis, old wax comb extract and fed with sugar syrup only for three weeks at one week intervals. Field assays showed that the treatment of beehives affected with AFB disease by tylosin 1% and Egyptian EEP in concentration of 0.1 and 0.05% had elimination of clinical symptoms at 100% of reduction rate.

Key words: American foulbrood, ethanolic extract propolis (EEP), high performance liquid chromatography (HPLC), honey bee disease, natural treatments.

INTRODUCTION

The causative agent of American Foulbrood (AFB) disease is *Paenibacillus larvae*, a Gram positive and spore-forming bacterium that is distributed worldwide (Genersch et al., 2006). Lodesani et al. (2005) reported that AFB is a virulent brood disease and is caused by *P. larvae*, which has a long-lived, resistant spore that can remain dormant for many years in combs and honey. AFB is spread by the exchange of infected honey and combs among colonies, either by the beekeeper tools or by robber bees. If no measures are taken by the beekeeper the colony is very likely to be destroyed by the infection, thus becoming a source of contagion for the whole apiary.

Macrocyclic lactone tylosin, (which has current US Food and Drug Administration approvals for agricultural uses) may soon gain approval for the control of AFB. This antibiotic inhibits ribosomal protein synthesis and has recently shown a good efficacy, while other substances failed to cure AFB in colonies with a high level of spores. The β- lactams (penicillins and cephalosporins), while active *in vitro*, are apparently not effective in the field. No antibiotic is capable of acting through the thickened wall of the bacillus spore and for this reason antibiotics are said to 'mask' the infection for the whole duration of their use; usually the disease reappears when the treatment is interrupted because the spores remain viable for several decades or longer (Marco et al., 2006).

A common strategy for the prevention and treatment of affected colonies is the use of antibiotics, particularly oxytetracycline hydrochloride (Hansen and Brodsgaard, 1999). However, several problems may be associated with its extended use. Chemical residues can persist in honey affecting its quality for human consumption while application of antibiotics may reduce the lifetime of bees and raise the risk of resistant strains emergency (Shuel and Dixon, 1960; Martel et al., 2006). Al Zen et al. (2002) reported that tylosin applied in a confectioner's sugar dust was effective in reducing and eliminating symptoms of OTC-resistant AFB disease in the apiary of the study and treated hives with tylosin was significantly reduced to no diseased hives. Resistance to this and other macrolides together with lincosamides and streptogramin B occurs in Gram-positive bacteria and was first shown, in *Staphylococcus aureus* (Lai et al., 1973). The presence of *P. larvae* OTC- resistant strains has been reported so far in Argentina, the United States, Italy, New Zealand and United Kingdom (Alippi, 1996; Miyagi et al., 2000; Evans, 2003).

Propolis is a natural product derived from plant resins and collected by honey bees to seal the walls and entrance of the hive and contributes to protect the colony against different pathogens (Ghisalberti, 1979), propolis has several biological properties such as antibiotic, antifungal, antiviral, anti-inflammatory activities (Manolova et al., 1985; Marquee, 1995; Drago et al., 2000; Tichy and Novak, 2000; Santos et al., 2003).

Furthermore, different compositions and amounts of the active substances are detected in separate samples of propolis (Bankova et al., 2002). The findings of various studies confirm that chemical composition of propolis depends on trees and plants available to the bees, on the season in which it is collected, on the geographical area, and other factors (Kartal et al., 2002; Abd El Hady and Hegazi, 2002). The volatile substances (aromatic oils) determine the flavor of propolis, and the variety of flavor depends on the geographical area and assortment of plants (Bankova, 1994). Flavonoids comprise the major part of biologically active substances in propolis (Havsteen, 2002). Furthermore, propolis has been found to contain phenolic acids (for example, prenilic derivates of cinnamic and coumaric acids), characterized by very potent antimicrobial activity (Hegazi et al., 2000). The antimicrobial and anti-inflammatory activity of European propolis is associated with the presence of flavonoids, flavones, and phenolic acids and their derivates

(Bankova, 2005). Due to the serious effects associated with AFB and the problems related to the use of antibiotics, the aim of the present work was to evaluates the biological activities of the EEP as a natural antibiotic, as well as its chemical composition for develop a new control strategy of this disease.

MATERIALS AND METHODS

Propolis samples

Three propolis samples were used, the first sample was Egyptian propolis (E.) which collected by glass slides (plaques of transparent glass close to the internal and lateral walls, near the little boxes 1 and 10 (Breyer, 1995) from honey bee colonies located in the apiary of Beekeeping Research Department, Plant Protection Research Institute, Agriculture Research Center at Dokki, Giza governorates, Egypt, through two years (2006-2007) and the second sample was Chinese propolis (C.) which imported from China and purchased commercially in Egyptian market and the third sample was old wax combs (W.) which collected from experimental apiary.

Preparation of EEP solution

Extraction procedures

Finely ground propolis was extracted by maceration at room temperature, with occasional shaking, in the proportion of 10 g of (C, E and W) propolis to 100 ml of solvent (ethanol 80%v/v), extracts were obtained after 7 days of maceration and the ethanolic extracts were then filtered in Whatmann N° 1 filter paper and incubated at room temperature until ethanol evaporated and the product obtained a honey-like consistence are referred to as EEP according to the method reported by Ildenize et al. (2004). This extract was diluted in sugar syrup 1:1 (1 kg of sugar in 1 L of water) at a final concentration of 2, 1, 0.5, 0.1, 0.05 and 0.025% EEP (w/v).

Identification of phenolic compounds in EEP by HPLC instrument

Identification of individual phenolic compounds of the three kind of EEP was performed on a HPLC instrument, 1 g sample was soaked in 20 ml of ethanol (80%v/v) and filtered through 0.45 μm filter membrane prior to HPLC analysis. High performance liquid chromatography, analytical HPLC was run on HPLC (JASCO, Japan), equipped with a pump (model PU-980) and a UV detector (UV-970). Separation was achieved on a hypersil BDS C18 (Thermo Hypersil-keystone, Germany) reversed-phase column (RP-18, 250 × 4.6 mm) with 5 μm particle size, a constant flow rate of 0.7 ml min^{-1} was used with two mobile phases: (A) 0.5% acetic acid in distilled water at pH 2.65 and solvent (B) 0.5% acetic acid in 99.5% acetonitrile, the system was run with a gradient program: 100% A (0 min); 0% B (0 min);100 to 50% A (50 min); 0 to 50% B (50 min), using an UV detector set at wavelength 254 nm. Phenolic compounds of each sample were identified by comparing their retention times with those of the standards mixture chromatogram. The concentration of an individual compound was calibrated and calculated on the basis of peak area measurements, and then converted to g phenolic /100 g fresh weight. All chemicals and solvents used were HPLC spectral grade and obtained from sigma (st. Louis, USA) and Merck - (Munich, Germany chemical companies), 24 components which presented the identical UV spectrum as standards compounds.

Detection of the half lethal concentration and half lethal time, (LC_{50} and LT_{50}) of EEP on worker honey bees

Susceptibility of worker honey bees to EEP was detected using a technique developed by (Maggi et al., 2010). Hybrid carniolan race (F1) bees were collected from frames in healthy colonies from the experimental apiary through July and August, 2007. Tests were conducted using 100 workers of honey bee 1 day old removed from the emergence boxes and placed in special wooden cages (16 × 12 × 6 cm) with wire screened side and glass fronts. The workers fed with 10 ml of different (C, E and W) EEP concentrations 2, 1, 0.5, 0.1, 0.05 and 0.025% in sugar syrup (1:1) were placed into each box and a negative control was performed using sugar syrup without EEP and the assay was carried out by 4 replicates then boxes were incubated at 32°C and 65% RH. Along the experiment period, the feeding solution had been changed daily and dead bees were counted and discarded. At the end of the experiment, bees were sacrificed and mortality percentages were corrected according to natural mortality (Abbott, 1925), and subjected to probit analysis according to the method of Finney (1952).

Determination of diagnosis of American foulbrood disease in honey bee colonies

The AFB infection was determined by number of infected larvae per colony according to diagnosis reported by Shimanuki and Knox (2000). Infected colonies spotty brood have been found, capping tend to be darker, concave larvae colored and extended length wise in the cell and contents of the cell rope out forming fine elastic thread up to 30 mm (Nikola, 2001). Larvae that have died of American foul brood disease exhibit a "ropy" condition that can be demonstrated by inserting a matchstick or similar implement into the dead and mass and drawing out the material into a threadlike projection longer than 2.5 cm (Morse and Nowogrodzki, 1990).

Field experiment

The efficiency of EEP to control the AFB on *P. larvae* artificially infected colonies was evaluated on hybrid carniolan race (F1) colonies which located in the experimental apiary through year 2008, Forty-four apparently healthy colonies (without clinical symptoms of AFB) and untreated with any antibiotics before were used. Colonies consisted of three brood, two honey and pollen combs were present in each hive and all hives were inoculated two weeks before initiation of trial. The inoculation process consisted of removing 100 cells of actively diseased brood from a local commercial apiary and agitating them in sucrose solution 50%. All hives were then fed, with 500 ml of this syrup/slurry mixture until all was consumed. At initiation, AFB disease evaluation was determined by removing brood frames from each individual hive and categorizing (Hitchcock et al., 1970) infected larvae (diseased cells) per hive were count every week. After 3 weeks we have thirty-three colonies had approximately 100 diseased cells/colony (sever degree). The thirty-three Colonies were divided into five groups in a randomized design, group one, Tylosin (T.) as positive control consisted of a confectioner's sugar dust, which made by combining 200 mg of tylosin tartrate with 20 g confectioner's sugar (a dose found efficacious in a previous study). The full 20 g of this dust were applied on 3 colonies by sprinkling over end of top bars for

once a week for three weeks, for a total dose of 600 mg tylosin tartrate over 3 week. Group two, Chinese propolis (C.) feeding with 500 ml of 0.1, 0.05 and 0.025% C. EEP solution, 3 colonies for each concentration for three weeks at one week intervals. Group three, Egyptian propolis (E.) and Group four, old wax comb extract propolis (W.) were used the same methodology and doses of group two. Group five, (Con.) as a control, 500 ml of sugar syrup 1:1 were performed once a week, during 3 consecutive weeks. The all treatment groups were reassessed from June to August, 2008. All these colonies were recorded with regard to their disease rating prior to the all treatments and subsequently evaluated 30 days after the third treatment, AFB disease re-examined by removing brood frames from each individual hive and the infected larvae (diseased cells) per hive were count and colonies with no visible signs of AFB disease at this time were considered recovered. The reduction percentage (rate) of infection was calculated according to the equation given by Henderson and Tilton (1955).

Reduction percentage of infection

$$= 1 - \frac{\text{n in Control before treatment } \mathbf{x} \text{ n in treatment after treatment}}{\text{n in Control after treatment } \mathbf{x} \text{ n in treatment before treatment}} \times 100$$

Where: n, number of diseased cells/colony.

Statistical analysis

For each evaluation data were compared by analysis of variance (ANOVA) and means were separated by least significance test at L.S.D0.05 on the other hand the data in Table 4 were transformed by Arcosin (angular transformed) according to Sokal and Rohef (1995).

RESULTS

Separation of phenolic compounds in three kind of EEP by HPLC

Phenolic compounds might be responsible for the biological activity in the three kind of EEP (Table 1). HPLC analysis was used to give information about the chemical composition of E.EEP, C.EEP and W.EEP. The phenolic compounds content found in E.EEP were salicylic acid, caffeic acid, ferulic acid, quercetin ,pinocembrin, pinostrobin, genistein and daiazein higher than that in C.EEP and W.EEP, in addition the phenolic compounds found in C.EEP were phenol, para hydroxy benzoic acid, p. coumaric acid, 3,5 dimethoxy benzyl alcohol, trans – cinnamic acid, chrysin, galangin, daidzin, acacetin higher than that in E.EEP and W.EEP, on the other hand in W.EEP were pyrogallic acid, protocatechuic acid, catechines, higher than that in E.EEP and C.EEP. It is evident from (Table 1) that composition of phenolic constituents were different in the three kinds of EEP and E.PEE were contained more phenolic compounds than in the C.PEE and W.PEE.

Table 1. Phenolic compounds concentration of three kinds of ethanolic extract propolis (E.EEP, C.EEP and W.EEP) determinate by HPLC.

Phenolic compound.		mg/100 g		
		C.EEP	E.EEP	W.EEP
Phenol: *phenol	C_6H_6O	3.757	15.968	0.000
Pyrogallic acid: *benzene-1,2,3-triol	$C_6H_6O_3$	0.00	0.00	31.710
Resorcinol: *benzene-1,3-diol	$C_6H_6O_2$	1.11	0.00	0.00
Salicylic acid: *2-hydroxybenzoic acid	$C_7H_6O_3$	15.72	716.80	15.13
para hydroxy benzoic: *4-hydroxybenzoic acid	$C_6H_6O_3$	9.18	11.60	0.000
Protocatechuic acid: *3,4-dihydroxybenzoic acid	$C_7H_6O_4$	29.66	54.60	25.45
Vanillin: *4-hydroxy-3-methoxy-benzaldehyde	$C_8H_8O_3$	0.00	0.00	13.60
p-Coumaric acid: * 3-(4-hydroxyphenyl)-2-proponic acid	$C_9H_8O_3$	1.25	0.00	0.00
Coumarine: * chromen-2-one	$C_9H_6O_2$	5.88	38.64	0.000
Caffeic acid: *3-(3,4-dihydroxyphenyl)prop-2-enoic acid	$C_9H_8O_4$	7.119	10.77	7.97
Trans-Cinnamic acid: * (E)-3-phenylprop-2-enoic acid	$C_9H_8O_2$	32.58	38.64	2.04
Ferulic acid: *3-(4-hydroxy-3-methoxy-phenyl)prop-2- enoic acid	$C_{10}H_{10}O_4$	1.56	193.55	0.00
Quercetin: *2-(3,4-dihydroxyphenyl)-3,5,7-trihydroxy- chromen-4-one	$C_{15}H_{10}O_7$	0.00	98.11	0.00
Pinocembrin: *2S)-5,7-dihydroxy-2-phenyl-chroman-4- one	$C_{15}H_{12}O_4$	0.00	73.70	0.00
Chrysin: *5,7-dihydroxy-2-phenyl-chromen-4-one	$C_{15}H_{10}O_4$	67.03	53.29	1.73
Galangin: *3,5,7-trihydroxy-2-phenyl-chromen-4-one	$C_{15}H_{10}O_5$	70.13	63.51	1.95
3.5 dihydroxy isoflavone: *3.5-Dihydroxy-3-(4- hydroxyphenyl)chromen-4-one	$C_{15}H_{10}O_5$	0.546	16.79	0.0393
Pinostrobin: *5,7-dihydroxy-2-phenyl-chroman-4-one	$C_{15}H_{12}O_4$	0.00	76.79	0.00
Daidzin: *7-hydroxy-3-(4-hydroxyphenyl)chromen-4-one	$C_{15}H_{10}O_4$	0.199	25.09	0.269
Genistein: *5,7-dihydroxy-3-(4- hydroxyphenyl)chromen-4-one	$C_{15}H_{10}O_5$	9.90	87.40	0.00
Catechines: *(2R,3S)-2-(3,4-dihydroxyphenyl)chroman-3,5,7-triol	$C_{15}H_{14}O_6$	9.10	12.13	29.60
Acacetin: *5,7-dihydroxy-2-(4- methoxyphenyl)chromen-4-one	$C_{16}H_{12}O_5$	48.32	11.60	1.93
Phenolphthalein: *2-[(4-hydroxyphenyl)-(4-oxo-1-cyclohexa-2,5-dienylidene)methyl]benzoic acid	$C_{20}H_{14}O_4$	10.44	14.85	10.44
Daidzein: *7-(-D-Glucopyranosyloxy)-3-(4-hydroxyphenyl)-4H-1-benzopyran-4-one	$C_{21}H_{20}O_9$	42.97	50.97	0.00
Total Peak Area		42533437	112470140	11014825

*IUPAC name: E.EEP: Egyptian ethanolic extract propolis, C.EEP: Chinese ethanolic extract propolis, W.EEP: Wax comb extract ethanolic extract propolis.

Median lethal concentration and time of using EEP on honeybee workers

The objectives of the present study are to determine the acute oral toxicity, expressed as half lethal time and concentration (LT_{50} and LC_{50}) of three kind of EEP (C., E. and W.) on honeybee workers and to evaluate the safe concentration of them to be applied on colonies infected with AFB, about the LT_{50} data. Table 2 demonstrated that there were significant differences among the concentration 2, 1 and 0.5% of three tested kind of EEP and there was no significant difference in 0.1, 0.05 and 0.025% of three kind of EEP in comparison to control (0.00% of EEP), so the high concentration of C., E. and W.EEP (2, 1 and 0.5%) had effected toxically on honeybee worker (oral administration), on the other hand the low concentration of C., E. and W.EEP (0.1, 0.05 and 0.025%) had a safely effect on honeybee workers.

Data in Table 3 demonstrated that the W.EEP was more toxic (LC_{50} = 1.404) than C.EEP (LC_{50} = 15.047) and E.EEP (LC_{50} = 8.223), in addition there are a significant deference among the three kind of EEP in LC_{50}, lower and upper limit of LC_{50} were reported in the table.

The reduction percentage (rate) of infection

The effect of EEP on the counts of infected larvae per hive was assessed by feeding, result obtained are summarized in Table 4, it is clear that tylosin and 0.1 and 0.05% E.EEP had a high significant positive influence on controlling the growth of *Paenibacillus larvae* with 100% reduction rate. The C.EEP and W.EEP group had a significant deference when compared with untreated (Con.) in three concentrations 0.1, 0.05 and 0.025%, with the mean rate of reduction 69.13, 64.98 and 40.66, for C.EEP group, respectively. In addition the reduction rates

Table 2. Median lethal time at least at 50% (LT_{50}) of C., E. and W.EEP on honey bee workers.

Concentration of EEP. Soluble in sugar solution 50% (% (w/v))	C.EEP	E.EEP	W.EEP
		LT_{50} (day)	
2.000	$12.40^b \pm 0.11$	$10.60^b \pm 0.49$	$9.10^d \pm 0.72$
1.000	$12.10^b \pm 0.23$	$11.20^b \pm 0.26$	$8.70^d \pm 0.46$
0.500	$13.80^b \pm 0.46$	$11.10^b \pm 0.89$	$10.80^c \pm 0.14$
0.100	$19.90^a \pm 0.86$	$18.50^a \pm 0.40$	$16.00^b \pm 0.26$
0.050	$19.70^a \pm 0.69$	$18.70^a \pm 0.75$	$21.06^a \pm 0.80$
0.025	$20.30^a \pm 0.40$	$19.30^a \pm 0.66$	$20.60^a \pm 0.40$
0.000	$19.80^a \pm 0.63$	$19.80^a \pm 0.17$	$19.80^a \pm 0.63$

E.EEP: Egyptian ethanolic extract propolis, C.EEP: Chinese ethanolic extract propolis, W.EEP: Wax comb extract ethanolic extract propolis.

Table 3. Median lethal concentration (LC_{50}) of C., E. and W.EEP on honey bee workers.

	C.EEP	E.EEP	W.EEP	F	P	$L.S.D_{0.05}$
LC_{50} (%)*	$15.04^a \pm 0.16$	$8.223^b \pm 0.20$	$1.404^c \pm 0.19$	139.598	0.000	1.998
Upper limit %	131.07 ± 0.08	31.63 ± 0.16	4.41 ± 0.15			
Lower limit %	5.04 ± 0.06	3.73 ± 0.18	0.72 ± 0.10			

*percentage of ethanolic extract propolis in sugar syrup (50% w/v), E.EEP: Egyptian ethanolic extract propolis, C.EEP: Chinese ethanolic extract propolis, W.EEP: old wax comb extract ethanolic extract propolis.

in W.EEP group were 87.95, 57.29 and 60.67%, respectively. Therefore, from mentioned results it could be concluded that the two investigated concentration (0.1 and 0.05% E.EEP) had inhibitory effect on viability and growth of *Paenibacillus larvae* under filed conditions.

DISCUSSION

The present investigation is a systematic study to evaluated using the ethanolic extract propolis, EEP for controlling AFB disease infected colonies. The differences observed in the propolis composition in the three kinds of EEP, it may be due to the deferent in vegetal source available in the collecting area (Egypt and China) (Table 1), the chemical composition of propolis is dependent on its geographical location; as a result, its biological activity is closely related to the vegetation native to the site of collection (Park et al., 2002; Christov et al., 2005). The antibacterial activity of EEP could be related to the chemical composition of propolis, which includes phenolic compounds (flavonoids and aromatic acids), terpenes and essential oils among others (Forcing, 2007). The antibacterial and antifungal activities of European and Uruguayan propolis are mainly due to flavonones, flavones, phenolic acids and their esters while in the case of Brazilian propolis such activities are due to prenylated o-coumaric acids and diterpenes (Ghisalberti, 1979; Kujmgiev et al., 1993; Marquee, 1995; Kanazawa et al., 2002; Bankova, 2005).

Results in Tables 2 and 3 indicated that the high concentration of EEP (2, 1 and 0.5%) affected toxically on honeybees that may be due to the anti nutritive compound like phenolic compound which occurs in propolis in high concentration so we cannot use it in field experiment. The ANFs (Anti-nutritive factors) which have been implicated in limiting the utilization of shrub and tree forages include non-protein amino acids, glycosides, phytohemagglutinins, polyphenolics, alkaloids, triterpenes and oxalic acid, ANFs may be regarded as a class of these compounds which are generally not lethal and they diminish animal productivity but may also cause toxicity during periods of scarcity or confinement when the feed rich in these substances is consumed by animals in large quantities (Agenda and Tshwenyane, 2003).

The present work reports the systematic study about the use of the ethanolic extract of propolis for the treatment of *P. larvae*-affected bee colonies. These results (Table 4) indicate that EEP has a direct *in vivo* antibacterial activity against *P. larvae* vegetative cells and that very low concentrations of propolis were required to inhibit its growth and these results are based on the compounds soluble in organic solvents (phenolic compounds) these compounds are responsible for the main part of the biological activity of propolis, on other hand it is important to note that the concentration of ethanolic extract propolis were significantly different, especially in regards to the active components. These results are in accordance with previous works that reported the antibacterial activity of EEP against

Table 4. Evaluation of the effect of EEP administered by feeding on the mean number of infected larvae per hive and reduction rate.

EEP.	Concentration of EEP. soluble in sugar solution 50% (w/v)	Number of Infected larvae (diseased cells) per hive		Reduction rate (%)
		Before	After	
	0.100	117	171	$69.13^c \pm 4.32$
C. EEP	0.050	134	205	$64.98^{cd} \pm 4.22$
	0.025	87	245	$40.66^e \pm 2.43$
	0.100	115	0	$100.0^a \pm 0.00$
E. EEP	0.050	111	0	$100.0^a \pm 0.00$
	0.025	111	41	$91.47^b \pm 0.56$
	0.100	113	65	$87.95^b \pm 0.86$
W. EEP	0.050	110	222	$57.29^d \pm 0.63$
	0.025	106	200	$60.67^d \pm 5.083$
T.	1.000*	105	0	$100.00^a \pm 0.00$
Con.	0.000**	92	444	$0.00^f \pm 0.0000$
F				113.544
P				0.000
L.S.D$_{0.05}$				7.388

* 200 mg of tylosin tartrate with 20 g confectioner's sugar, ** sugar syrup 1:1 without EEP, T.: Tylosin tartrate, E.EEP: Egyptian ethanolic extract propolis, C.EEP: Chinese ethanolic extract propolis, W.EEP: old wax comb extract ethanolic extract propolis, T: Tylosin. Con: Control.

diverse pathogens (Drago et al., 2000; Garedcw et al., 2004). Antibacterial effect of propolis was also demonstrated, since a significant decrease in the number of *P. larvae* spores/g of honey was found in naturally infected beehives treated with EEP. The proposed mechanism of action, includes the oral ingestion of EEP by adult honeybees and its delivery to larvae with feeding, facilitating the interaction and direct antibacterial effect on *P. larvae* vegetative cells, the addition of honey to the larval diet is around the third day of the larval stadium, coinciding with germination and multiplication of vegetative cells of *P. larvae* (Shuel and Dixon, 1960; Hansen and Brodsgaard, 1999).

The site(s) and number of hydroxyl groups on the phenol ring are thought to be related to their relative toxicity to microorganisms, with evidence that increased hydroxylation results in increased toxicity (Weissman, 1963). The mechanisms thought to be responsible for phenolic toxicity to microorganisms include enzyme inhibition by the oxidized compounds, possibly through reaction with sulfhydryl groups or through more nonspecific interactions with the proteins (Mason and Wasserman, 1987). Flavones are phenolic structures containing one carbonyl group their activity is probably due to their ability to complex with extracellular and soluble proteins and to complex with bacterial cell walls (Tsuchiya et al., 1996). Simuth et al. (1986) reported that the mechanism of propolis action on microorganisms

seems to be complex with respect to those components which are presently known. The inhibition of cell division and of cross wall separation of daughter cells by EEP led to the formation of pseudo-multicultural streptococci. This effect could be due to the blockage of the so-called splitting system of the cross wall as was demonstrated by *S. aureus* during treatment with trimethoprim (Nishino et al., 1987). The inhibition of cell division observed in the presence of EEP suggested that this natural drug would act like nalidixic acid which is known to inhibit DNA replication and, indirectly, cell division and propolis inhibited the synthesis and secretion of proteins from the bacterial cells (Nintendo et al., 1994).

Karina et al. (2008) propose that this mechanism cannot prevent the infection of new larvae with *P. larvae* spores, but can inhibit the replication of vegetative cells in the larval gut. Moreover, we cannot rule out a possible indirect effect of the propolis due to the stimulation of the bee immune system. Several authors have reported the stimulating effect of propolis in the innate and adaptive immune response of mouse, bovines and humans. *In vitro* and *in vivo* assays demonstrated that propolis activates macrophages, increasing their microbecide activity, enhances the lytic activity of natural killer cells and stimulates antibody production (Forcing, 2007). Enhancement of the defense response of honeybees by propolis could also be important for the control of other honeybee diseases (Evans et al., 2006). The mixture and combined effects of its different components decrease the chance of propolis-resistant bacterial strains emergency, due to the several target sites probably present in a bacterial cell (Rios et al., 1988; Denyer and Stewart, 1998). The present findings indicate that the antibacterial activity, and perhaps other biological properties of propolis, could not be correlated with their propolis concentration but mostly to their chemical composition which can be variable according to the collection site and vegetal source.

Conclusion

The aim of this study develop a new strategy for controlling American FoulBrood (AFB) disease by using a natural antibiotic collected by honeybees from plant resins which is called propolis, to avoid using a common antibiotic (tylosin and oxytetrecycline) for its several problems, chemical residues, reduce the life time of bees and the risk of resistant strains emergency. Thirty-three colonies had a sever degree of AFB disease which is located in experimental apiary of beekeeping research department, plant production research institute, Egypt were treated by several concentration of three kind of propolis ethanolic extract (Chinese, Egyptian and old wax comb propolis) soluble in sugar solution 50%. Result indicated that tylosin, 0.1 and 0.05% of Egyptian propolis ethanolic extract eliminating of AFB clinical symptoms at 100% of reduction rate. This result could be related to the chemical composition of propolis which includes a high active phenolic compounds.

ACKNOWLEDGEMENTS

The authors wish to express their sincere feeling gratitude to Dr. Farag, A. M., Plant Protection Research Institute, Dokki, Giza, for his great help, and fruitful remarks and suggestions that enriched this work.

REFERENCES

Abbott WS (1925). A method for computing the effectiveness of an insecticide. J. Econ. Entomol. 18:265-267.

Abd El Hady F, Hegazi AG (2002). Egyptian propolis: 2. Chemical Composition, antiviral and antimicrobial activity of East Nile Delta propolis. Z Naturforsch. 57(3-4):386-394.

Agenda AA, Tshwenyane SO (2003). Feeding Values and Anti - Nutritive Factors of Forage Tree Legumes. Pak. J. Nutr. 2(3):170-177.

Alippi A (1996). Characterization of isolates of Paenibacillus larvae with biochemical type and oxytetracycline resistance. Rev. Argent. Microbiol. 28:197-205.

Al Zen PJ, Westervelt D, Causey ID, Ellis IJ, Hepburn HR, Neumann P (2002). Method of Application of Tylosin, an Antibiotic for American Foulbrood Control, with effects on Small Hive Beetle (Coleoptera Nitidulidae) Populations. J. Econ. Entomol. 95(6):1119-1122.

Bankova V (2005). Recent trends and important developments in propolis research. Evid. Based Complcm. Altern. Med. 2:29-32.

Bankova V, Christov R, Popov S, Pureb O, Bocari G (1994). Volatile constituents of propolis. Z Naturforsch Sect C. 49(1-2):6-10.

Bankova V, Popova M, Bogdanov S, Sabatini AG (2002) Chemical composition of European propolis: Expected and unexpected results. Z Naturforsch C. 5:530-533.

Breyer HFE (1995). Aspectos de produção, coleta,limpeza, classificaçãoe acondicionamento de própolis bruta de abelhas Apis mellifera L. in: x simpósio estadual de apicultura do paranáe VII exposição de equipamentos e materiais apicolas. Prudentópolis Anais Prudentópolis P. 143.

Christov R, Trusheva B, Popva, M, Bankova V, Bertrand M (2005). Chemical composition of propolis from Canada, its antiradical activity and plant origin. Nat. Prod. Res. 19:673-678.

Denyer SP, Stewart SA (1998). Mechanisms of action of disinfectants. Int. Biodeterior. Biodegrad. 41:261-268.

Drago L, Momhclli B, Vecchio F, Fascino M, Tocalli M, Gismondo M (2000). In vitro antomicrobial activity of propolis dry extract. J. Chemother. 12:390-395.

Evans E (2003). Diverse origins of tetracycline resistance in the honey bee bacterial pathogen Paenibacillus larvae. J. Invertebr. Pathol. 83:50-56.

Evans JD, Aronstein K, Chen YP, Hetru C, Imler JL, Jiang H, Kanost M, Thompson GJ, Zou Z, Hultmark D (2006). Immune pathways and defense mechanisms in honey bees Apis mellifera. Insect Mol. Biol. 15:645-656.

Finney DJ (1952). Probit Analysis. A statistical treatment of the sigmoid response curve. Cambridge Univ. Press, London.

Forcing JM (2007). Propolis and the immune system: A review. J. Ethnopharmacol. 113:1-14.

Garedcw A, Schmolz F, Lamprcchlb I (2004). Microbiological and electric investigations on the antimicrobial actions of different propolis extracts: an in vitro approach. Thermo Chim. Ada. 422:115-124.

Genersch E, Forsgran E, Pentikainen J, Ashiralieva A, Rauch S, Kilwinski FI (2006). Reclassification of Paenibacilluslarvae subsp. Pulvifaciens and Paenibacillus larvae subsp. larvae as Paenibacillus larvae, without subspecies differentiation. Int. J. Syst. Eval. Microbiol. 56:501-511.

Ghisalberti E (1979). Propolis: A review, Bee World 60:59-84.

Hansen H, Brodsgaard C (1999). American Foulbrood: A review of its biology, diagnosis and bee control, Bee World. 80:5-23.

Havsteen BH (2002). The biochemistry and medical significance of the flavonoids. Pharmacol. Ther. 96:67-202.

Hegazi A, Abd El Hady F, Abd Allah F (2000). Chemical composition and antimicrobial activity of European propolis. Z Naturforsch. 55(1-2):70-75.

Henderson CF, Tilton EW (1955). Tests with acaricides against the brow wheat mite. J. Econ. Entomol. 48:157-161.

Hitchcock JD, Moffett JO, Lackett JJ, Elliott JR (1970). Tylosin for control of American foulbrood disease in honey bees. J. Econ. Entomol. 63:104-207.

Ildenize BSC, Alexandra CHFS, Fabio MC, Mario TS, Maria CM, Flavia TD, Giovanna SP, Patricia DOC (2004). Factors that Influence the Yield and Composition of Brazilian Propolis Extracts. J. Braz. Chem. Soc. 15(6):964-970.

Kanazawa S, Hayashi K, Kajiya K, Ishii T, Hamanaka T, Nnkayama T (2002). Studies of the constituents of Uruguayan propolis. J. Agric. Food Chem. 50:4777-4782.

Karina A, Jorge H, Liesel G, Matias M, Martin E, Pablo Z (2008). Efficacy of natural propolis extract in the control of American Foulbrood. Veter. Micro. 131:324-331.

Kartal M, Kaya S, Kurucu S (2002). GC-MS analysis of propolis samples from two different regions of Turkey. Z Naturforsch. 57:905-909.

Kujmgiev A, Bankova V, Ignatov A, Popov S (1993). Antibacterial activity of propolis, some of its components and their analogs. Pharmazic 48:785-786.

Lai CJ, Weisblum B, Fahnestock SR, Nomura M (1973). Alteration of 23S ribosomal ribonucleic acid and erythro mycin-induced resistance to lincomycin and spiramycin in Staphylococcus aureus. J. Mol. Biol. 74:67-72.

Lodesani M, Costa C, Calin MM (2005). Limits of chemotherapy in beekeeping: Development of resistance and the problem of residues. Bee World 86(4):102-109.

Maggi M, Rufrinengo S, Gende L, Sarlo G, Bailac P, Ponzi M, Eguaras M (2010). Laboratory evaluations of *Syzygium aromaticum* (L.) Merr. et perry essential oil against *Varroa destructor*. J. Essent. Oil Res. 22(2):119-122.

Manolova N, Maximova V, Gegova G, Sekedjieva Y, Uzunov S, Marekov Bankova V (1985). On the anti influenza action of fractions from propolis. Comptes rendus de l'Academie bulgare de Sciences. 38:735-738.

Marco L, Cecilia C, Man MC (2006). limits of chemotherapy in beekeeping:development of resistance and the problem of residues. Bulletin University Agricultural Sciences and Veterinary Medicine Cluj-Napoca. Anim. Sci. Biotechnol. P. 62.

Marquee MC (1995). Propolis: Chemical composition, biological properties and therapeutic activity. Apidologie 26:83-99.

Martel AC, Zeggane S, Rijndael P, Faucon JP, Aubert M (2006). Tetracycline residues in honey after hive treatment. Food Addil. Contam. 23:265-273.

Mason TL, Wasserman BP (1987). Inactivation of red beet betaglucansynthase by native and oxidized phenolic compounds. Phytochemistry 26:2197-2202.

Miyagi T, Ping CYST, Chuang RY, Mussen EC, Spivak MS, Doi RH (2000). Verification of oxytctracycline-resistant American Foulbrood pathogen *Paenibacillus larvae* in the United States. J. Invertebr. Pathol. 75:95-96.

Morse RA, Nowogrodzki R (1990). Honey bee pests, predators and diseases, second edition. Cornell Univ., press Itnaca and London. pp. 29-47.

Nikola EZ (2001). Studies on brood diseases in honeybees, Ph.D. Thesis Fac. Agric.Moshtohor. Zagazig. Univ.

Nintendo B, Takaishi-kikuni, Heinz S (1994). Electron microscopic and microcalorimetric Investigations of the possible mechanism of the antibacterial action of a defined propolis provenance. Planta Med. 60:222-227.

Nishino T, Wacke J, Krüger D, Giesbreucht P (1987). Trimethoprim-induced structural alterations in *Staphylococcus aureus* and the recovery of bacteria in drug- freemedium. J. Antimicrob. Chemother. 19:147-159.

Park YK, Alecar SM, Aguiar CL (2002). Botanical origin and chemical composition of brazilin propolis. J. Agric. Food Chem. 50:2502-2506.

Rios JL, Recio MC, Villar A (1988). Screening methods for natural products with antimicrobial activity: A review of the literature. J. Eflnopharm. 23:127-140.

Santos F, Bastos E, Maia A, Used M, Carvalho M, Farias I, Morcira E (2003). Brazilian propolis: Pliysicochemical properties, plant origin and antibacterial activity on pcriodontopatlio gens. Phytother. Res. 17:285-289.

Shimanuki H, Knox DA (2000). Diagnosis of honey bee diseases. U.S. Dep. Agric. Handbook No. AH-690.

Shuel RW, Dixon SE (1960). The early establishment of dimorphism in the female honeybee. *Apis mellifera* L. Insect Sociaux 7:265-282.

Simuth J, Trnovsky J, Jelokova J (1986). Inhibition of RNA polymerases by UV- absorbent components from propolis. Pharmazie 41:131-132.

Sokal RR, Rohlf FJ (1995). Biometry: the principles and practice of statistics in biological research. 3rd edition. W. H. Freeman and Co.: New York. P. 887.

Tichy J, Novak J (2000). Detection of Antimicrobials in bee Products with Activity Against *Viridans Streptococci*. J. Altern. Complem. Med. 6(5):383-389.

Tsuchiya H, Sato M, Miyazaki T, Fujiwara S, Tanigaki S, Ohyama M, Tanaka T, Iinuma M (1996). Comparative study on the antibacterial activity of phytochemical flavanonesagainst methicillin resistant *Staphylococcus aureus*. J. Ethnopharmacol. 50:27-34.

Weissman TA (1963). Flavonoid compounds, tannins, lignans and related compounds, In M. Florkin and E. H. Stotz (ed.), Pyrrole pigments, isoprenoid compounds and phenolic plant constituents. Elsevier, New York, N. Y. 9:265.

10

Evaluation of pure and crossbred progenies of Red Sokoto and West African Dwarf goats in the Rainforest Zone of South Eastern Nigeria

E. N. Nwachukwu[1], K. U. Amaefule[2], F. O. Ahamefule[3], S. C. Akomas[4], T. U. Nwabueze[5], U. A. U Onyebinama[6] and O. O. Ekumankama[6]

[1]Department of Animal Breeding and Physiology, Okpara University of Agriculture, Umudike, Umuahia, Abia State, Nigeria.
[2]Department of Animal Nutrition and Forage Sciences, Okpara University of Agriculture, Umudike, Umuahia, Abia State, Nigeria.
[3]Department of Animal Production and Management, Okpara University of Agriculture, Umudike, Umuahia, Abia State, Nigeria.
[4]College of Veterinary Medicine. Okpara University of Agriculture, Umudike, Umuahia, Abia State, Nigeria.
[5]Department of Food Processing and Analysis, Okpara University of Agriculture, Umudike, Umuahia, Abia State, Nigeria.
[6]College of Agricultural Economics and Rural Sociology, Michael Okpara University of Agriculture, Umudike, Umuahia, Abia State, Nigeria.

A total of 56 kids produced from four breeding goat units consisting of pure Red Sokoto (RS ×RS), pure West African Dwarf (WAD × WAD), main cross (RS × WAD), and reciprocal cross (WAD × RS) were used to investigate performance and cost benefit of producing progenies of these indigenous goat breeds in a rain forest zone of South Eastern Nigeria. The experimental design was a completely randomized design with genetic group as the factor of interest. Data obtained showed that at birth, the male RS × RS kids weighed significantly (P<0.05) highest (1840.0 ± 230.98 g), followed by WAD × RS (1430 ± 144.36 g), RS × WAD (1371.00 ± 56.54 g) while the WAD x WAD kids had the lowest birth weight (1150.00 ± 39.36 g). However, the RS × AD female kids, at birth weighed significantly highest (1500 ± 54.10 g), followed by the RS × RS (1328.50 ± 98.69g), WAD × RS (1312.50 ± 17.81 g) and WAD × WAD (1087.50 ± 106.80 g). The male and female kids of the RS × WAD had improved body weight and the linear body measurements such as body length, height-at-withers and heart girth. They also had a higher average daily gain (46.03 ± 1.41 g/day) and better feed conversion ratio (5.38 ± 0.27) than the RS x WAD and WAD x WAD goats. This genetic group had the lowest cost of production (₦953.40 ± 10.21) and as such the highest gross margin (₦2111.06 ± 21.7). They appear more promising hybrid goats for commercial meat goat production in the rainforest zone of South Eastern Nigeria.

Key words: Red Sokoto goats, West African Dwarf (WAD) goats, crossbreeding, conformation traits, cost benefit, rainforest zone.

INTRODUCTION

Goats constitute a very important part of the rural economy in Nigeria, with more than 95% of the rural households keeping goats (Ukpabi et al., 2000). As a multipurpose animal, goats provide meat, milk, hides and

Table 1. Percentage Composition of concentrate ration fed to breeding goats

Feed ingredients	
Maize offal	50.00
Wheat offal	6.50
Palm kernel cake	39.00
Bone meal	2.00
Periwinkle	1.00
Minerals – vitamin premix	0.25
Salt	1.25
Total	**100**
% Crude protein (%)	13.9
ME (kcal/kg)	2980

skin and manure.

They rank next to cattle in income generation and their meat (chevon) is quite popular and well relished (Ladele et al., 1996).

The need to develop productive and adaptive goat breed for the rainforest zone is desirable. Multiplication and distribution of such high quality hybrid goat definitely would increase small ruminant animal production and animal protein supply in South East and South-South Nigeria; where the level of livestock production is quite low. These geopolitical zones correspond to the agro-ecological area described as the rainforest zone where tse-tse fly infestation and typanosomiasis infection are serious menace to livestock production. The use of well adapted West African Dwarf (WAD) and highly productive indigenous Red Sokoto (RS) goat in 'new breed' formation is an appropriate breeding plan especially from the view point of utilizing local animal genetic resources (AnGR) in realizing local needs (Nwosu, 2005).

It is a common knowledge that not every mating scheme yields satisfactory result. To overcome some suspected growth and reproductive problems associated with crossbreeding in goats e.g. low birth weight, poor kids survival rate, insufficient milk supply to the young and dystocia (Malik et al., 1980); mating of unproductive local males to improved productive females has been suggested in sheep (Dickerson, 1992). Mating of WAD goats with Red Sokoto breed has been carried out to determine growth and reproductive potentials of the offsprings (Taiwo et al., 2005). Result of that study revealed that present breeding plan would be feasible and beneficial.

The objective of this study was therefore, to evaluate growth performance and cost benefit of producing pure and crossbred lines of meat goats in a rainforest zone of Nigeria.

MATERIALS AND METHODS

Location of study

The investigation was carried out in the Goat Unit of the Teaching and Research Farm of Michael Okpara University of Agriculture, Umudike, Abia State, Nigeria. The study area lies within the rainforest zone of South East Nigeria with a bimodal rainfall pattern. Total annual rainfall ranges from 1700 to 2100 mm with temperature range of 27 to 38°C during the dry season (November to April) and 18 to 26°C during rainy season (May to October). This agro-ecological zone has been described as warm-wet humid tropics.

Management of breeding stock

A total of 34 mature goats with an average age of 10 months and consisting of 16 West African Dwarf (WAD) and 18 Red Sokoto (RS) goats were the breeding stock. They were housed in a conventional dwarf walled pen structured house. The sex ratio was 1:8 for the WAD and 3:6 for the RS. The animals were quarantined for 28 days during which vaccination against pest des petit ruminant (PPR) disease was administered.

The breeding stock was maintained on a daily feed allowance of 25% concentrate ration and 75% fresh fodder made up of *Panicum maximum* and a mixture of other browse plants. The composition of the concentrate ration fed to breeding WAD and RS goats is shown in Table 1.

Mating scheme

Mating schemes adopted to generate progenies for the study were pure line, main cross and reciprocal crossing as shown below:

Pure line mating   - RS × RS
                                - WAD × WAD
Main crossing   - RS × WAD
Reciprocal crossing - WAD × RS

Management of pure and crossbred kids

A total of 56 kids were produced by the breeding herd. Kids were allowed with their dams in nursing pens for 4 months before weaning. Apart from a feeding allowance of 300 g/doe/day, additional 150 g/kid/day was added for the number of kids each doe nursed. Protection against ecto- and endo-parasites among kids was achieved using Ivomectin, while vaccination against *PPR* disease was carried out at 3 months of age.

Data collection

Data collected on pure and crossbred kids were weekly body weight changes, body length, heart girth and height-at-withers in both sexes as well as scrotal length and scrotal circumference in the buck kids only. The body length was measured as the distance from the pole of the animal to the base of the tail. Heart girth was taken as the circumference of the chest close to the forelegs, while height-at-withers was measured as the distance from the withers to the base of the hoof while the animal stood erect on a platform. The

Table 2. Body weight changes and linear body dimensions of male progenies of pure and crossbred lines of Red Sokoto and West African Dwarf goats.

Parameter	Genetic group			
	RS X RS	RS X WAD	WAD X RS	WAD X WAD
BWT at week 1 (g)	1840[a].00 ± 230.98	1371.43[b] ± 56.54	1430[b].00 ± 144.56	1150[b].00 ± 35.36
BWT at week 18 (g)	9820.00[a] ± 794.45	8340.00[ab] ± 852.08	7610.00[b] ± 126.51	7510.00[b] ± 606.28
BL at week 18 (cm)	67.60[a] ± 3.07	59.86[b] ± 0.10	57.16[b] ± 1.19	57.02[b] ± 0.75
HW at week 18(cm)	52.72[a] ± 2.28	46.16[b] ± 0.94	42.86[c] ± 0.76	40.24[c] ± 0.76
HG at week 18 (cm)	53.20[a] ± 1.45	47.26[ab] ± 1.91	47.30[b] ± 1.62	46.98[b] ± 2.16
SL at week 18 (cm)	8.82[a] ± 0.29	7.53[a] ± 0.31	6.97[ab] ± 0.43	5.85[b] ± 0.27
SΘ at week 18 (cm)	17.08[a] ± 1.74	16.05[a] ± 1.52	13.83[b] ± 1.62	12.55[c] ± 1.02

[abc] Means on the same row bearing different superscripts are significantly different (P<0.05); BWT = Body weight; BL = body length; HW = height at withers; HG = heart girth; SL = scrotal length; SΘ = scrotal circumference.

Table 3. Body weight changes and some linear body dimensions of female progenies of pure and crossbred lines of Red Sokoto and West African Dwarf Goats.

Parameter	Genetic group			
	RS X RS	RS X WAD	WAD X RS	WAD X WAD
BWT at week1 (g)	1328.50[b] ± 98.69	1500.00[a] ± 54.01	1312.50[b] ± 71.81	1087.50[c] ± 106.80
BWT at week 18 (g)	6875.00[a] ± 852.08	7637.50[a] ± 521.37	5242.50[b] ± 368.47	5112.00[b] ± 308.47
BL at week 18 (cm)	58.85[a] ± 3.00	56.35[a] ± 1.00	53.65[ab] ± 0.25	50.10[b] ± 1.14
HW at week 18 (cm)	51.38[a] ± 2.78	44.55[b] ± 0.36	42.38[b] ± 0.75	42.2[b] ± 0.60
HG at week 18 (cm)	44.78[a] ± 2.02	42.45[ab] ± 0.34	41.33[b] ± 0.81	39.28[b] ± 0.36

[abc] Means on the same row bearing different superscripts are significantly different (P<0.05); BWT = Body weight; BL = body length; HW = height at withers; HG = heart girth.

scrotal length was taken as the distance from the base of the scrotum to the tip of the scrotal sac, while the scrotal circumference was measured as the region of largest scrotal expansion. All linear body measurements were taken with a tailor's tape in centimeters. Cost-benefit of raising each genetic group was computed based on feed consumed, other variable costs and prevailing market price of life goats in the environment. Average daily gain (ADG) and feed conversion ratio (FCR) of the various breeding groups were also computed.

Experimental design and data analysis

The experimental design was completely randomized design with genetic group as the factor of interest.

$$Y_{ij} = u + G_i + e_{ij}$$

Where, Y_{ij} = j[th] individual in the ith genetic group; U = Overall mean G_i = Effect of the ith genetic group (I = 1,--,4); e_{ij} = Random error assumed to be independently identical and normally distributed with zero mean and constant variance.

Means and their associated standard deviations were computed for the measured parameters. Significant means were separated using Duncan's New Multiple Range Test (Duncan, 1955).

RESULTS AND DISCUSSION

The performance of male and female progenies of pure and crossbred Red Sokoto and West African dwarf goats

are presented in Tables 2 and 3, respectively. Birth weight of the male kids was significantly (P< 0.05) highest with the RS × RS (1840.0 ± 230.98g), followed by WAD × RS kids (1430.0 ± 144.36 g); RS × WAD (1371.0 ± 56.54 g) while the WAD × WAD kids had the least birth especially between the two pure lines (RS × RS and WAD × WAD kids) since such difference in body weight is common knowledge and underscores the need for the upgrading programme. The RS × RS kids maintained their superiority in body weight and were only equaled by the RS × WAD kids. This indicates obvious improvement in body weight following crossbreeding for the RS × WAD individuals. Significant improvement in body weight and in some linear body parameters in half bred RS ×WAD goats have been reported by Ozoje and Herbert (1997) and in crossbred sheep by Wiener and Hayter (1974).

The body measurements namely body length, height-at-withers, heart girth, scrotal length and scrotal circumference followed similar pattern as body weight in the various genetic groups. This indicated strong influence of body weight on other body components. Indeed, strong and positive associations between body weight and most conformation traits in farm animal species have been well reported for goats (Ozoje and Herbert, 1997), Ayrshire cattle (Russel, 1975) and humpless indigenous cattle (Ibe and Ezekwe, 1994).

The performance of the female kids (Table 3) revealed

Table 4. Average daily gain (ADG), feed conversion ratio (FCR), cost of production (CP), revenue (R) and gross margin (GM) realized from pure and crossbred progenies of Red Sokoto and West African Dwarf goats.

Parameter	Genetic group			
	RS × RS	RS × WAD	WAD ×RS	WAD × WAD
ADG (g/d)	45.37^b ± 1.34	46.03^a± 1.41	43.45^b ± 3.34	38.62^c ± 3.04
FCR	7.97^a ± 0.15	5.38^b ± 0.27	6.64^b ± 0.18	7.02^a ± 0.21
CP (N)	1392.16^a± 13.64	953.40^c ± 10.21	1110.76^{ab} ± 12.32	1032.00^b± 10.25
R (N)	3280.00^a ± 24.01	3064.00^b ± 20.43	2930.00^b± 26.01	2556.00^c ± 19.06
GM (N)	1888.00^b ± 13.22	2111.06^a ± 21.71	1819.24^b ± 15.01	1523.57^c ± 18.74

[a, b, c] Means on the same row bearing different superscripts are significantly different (P<0.05).

that the main crossbred (RS × WAD) doe- kids had significantly higher birth weight (1500 ± 54.01g) than the RS × RS (1328.50 ± 98.69), WAD × RS (1312.50 ± 71.81) and WAD × WAD (1087.50 ± 106.50).   Birth weight of a kid is a reproductive trait that could be influenced by such factors as body weight and condition score of dam, nutritional status, type of birth, sex and season of the year (Cassard et al., 1956). The WAD dams were well adapted to the rearing environment and may have utilized the concentrate – fodder based ration better. It could also be that the Red Sokoto bucks mated to the WAD does were able to stamp their superior genetic worth on their female progenies, thus resulting in the RS × WAD doe kids showing obvious improvement in body weight even above the RS × RS pure line. This superior-parent improvement in body weight in the females is desirable and is indeed, the essence of crossbreeding (Shrestha and Fahmy, 2007). Body weight attained by the pure and crossbred lines at 18 weeks (Table 3) showed that the RS × RS and RS × WAD female goats were not significantly different. However, the RS × WAD goats had numerically higher final body weight (7,637.50 ± 521.37 g) than even the RS × RS females (6,875 ± 852.08 g). The WAD goat is a known meat animal (Jean, 1993) and with improvement in its body weight, both male and female hybrid WAD goats could grow fast and mature early. This observation holds a good promise for the development of the RS × WAD hybrids as possible candidate meat goats especially in the rainforest zone which is a natural habitat for the West African Dwarf goats.

The performance of the WAD × RS hybrid males (Table 2) and females (Table 3) is noteworthy. This genetic group did not differ significantly in final body weight from the WAD × WAD individuals. This result indicated that there was no obvious improvement in body weight for both male and female WAD × RS kids. The practical implication of this finding is that the use of Red Sokoto was mated to WAD bucks to improve a goat herd as some uniformed goat keepers do would give disappointing results.

The production and economic performance indices of raising the pure and crossbred progenies of Red Sokoto

and West African dwarf goats are presented in Table 4. Average daily gain of 46.03 ± 1.41 g/day at 126 days (18 weeks) achieved by the RS × WAD was higher than that of the RS × RS individuals (45.37 ± 1.34g/day) and this, perhaps must have precipitated the higher final body weight attained especially by the female kids of this genotype (Table 3). This result indicated that the WAD dams and their RS crossbreds (RS × WAD) were well adapted to the nutrition and rearing environment they were subjected to, when compared to their supposedly superior RS × RS counterparts. The ADG value of 37g/day recorded for half bred WAD × RS goats fed legumes and fodders with concentrate supplementation for 150 days at Ibadan (Ebozoje, 1992) was even poorer than the value achieved by the RS × WAD hybrid individuals in this study. This seems to confirm further that main crossbreeding and the main crossbred goats are the most preferred mating system and most desirable genotype, respectively.

The feed conversion ratio (FCR) was significantly low and more efficient for the two crossbreds (RS × WAD and WAD × RS) with values of 5.38 ± 0.27 and 6.64 ± 0.18, respectively when compared to the RS × RS and WAD × WAD pure lines which had FCRs of 7.97 ± 0.15 and 7.02 ± 0.21 respectively at 18 weeks of age. Efficient feed conversion of the RS × WAD showed that the upgrading exercise was effective and, of course with improved weight gain, appetite and feed intake were expected to increase within this genotype. This finding also showed that the RS × WAD goat was able to utilize the local feed resources more efficiently than the parental RS and WAD lines. This potential of the RS × WAD individuals is desirable since good growth rate and efficient feed utilization are common attributes required for meat animals (Roge, 1992).

The RS × WAD hybrids which recorded the highest ADG and a better FCR also had significantly (P <0.05) least cost of production (₦953.40 ± 10.21) at 18 weeks of age. This observation is understandable, since these animals converted much of their feed resulting in higher final body weight. Gross margin which is an index of efficient production was also highest (₦2111.06 ± 21.71) for the main crossbred individuals. The performance of this genetic group suggests that it is a promising genotype for the development of hybrid meat goat especially in the rainforest zone of Nigeria.

Conclusion

It is evident from the results of this study that the progenies of the main crossbred (RS × WAD) - males and females showed superiority in body weight and in most of the linear body measurements studied. Average daily gain, feed conversion ratio, revenue and gross margin were also better for the RS × WAD individuals compared to their reciprocal (WAD × RS) and pure (WAD × WAD) counterparts. The performance of these main crossbreds suggests that they are preferred goat genotype for the production of hybrid meat goat in the rainforest zone of South Eastern Nigeria.

ACKNOWLEDGEMENTS

The authors are grateful to the Directorate of Research and Development, Michael Okpara University of Agriculture, Umudike for supporting this research work through Research Grant No. API/07/05 of 2007. We are also grateful to the staff of Goat Unit who assisted in the daily care of the experimental animals.

REFERENCES

Cassard DNPW, Weir WC, Wilson JF (1956). Environmental factors affecting body dimension in yearling Hampshire ewes. J. Anim. Sci. 15:922–929.

Dickerson GE (1992). Manual for evaluating breeds and crosses of domestic animals publications Div., FAO, Rome, Italy.

Duncan CB (1955). Multiple range and multiple F-test. Biometrics International Biometric Society 11(1):1-42.

Ebozoje MO (1992). Preweaning performance of West African Dwarf and West African Dwarf X Maradi Halfbred goat in Ibadan, Ph.D Thesis. University of Ibadan, Ibadan, Nigeria, P. 329.

Ibe SN, Ezekwe AG (1994). Quantifying size and shape difference between Muturu and N'dama breeds of cattle Nig. J. Anim. Prod. 21:51-58.

Jean P (1993). Anim. production in the tropics and subtopics. Published by the Macmillan Press Ltd. London and Basingstoke.

Ladele AA, Joseph K, Omotesho OA, Ijaiya TO (1996). Sensory quality ratings, consumption pattern in Nigeria. Brit. J. Food Sci. Nut. 47:141-145.

Malik RC, Singh RN, Acharya RM, Dutta OP (1980). Factors affecting lamb survival in crossbred sheep. Trop. Anim. Health Prod. 12:217-223.

Nwosu CC (2005). Strategies for improvement of animal genetic resources in a developing economy. Invited paper. Proc. 30th Ann. Conf. genetic Soc. Of Nigeria Nsukka, 5-8 Sept, 2005.

Ozoje MO, Herbert U (1997). Linear measurements in West African Dwarf (WAD) and WAD X Red Sokoto goats. Nig, J. Anim. Prod. 24(1):13-19.

Roge JEO (1992). Indigenous African small ruminants: A case for characterization and improvement. Proc. Of 2nd Biennial conf. of the African Small Ruminant Res. Network, Arusha Tanzania. pp. 205-211.

Russel WS (1975). The growth of Ayrshire cattle: an analysis of linear body measurements. Anim. Prod. 20:217-226.

Shrestha JNB, Fahmy MA (2007).Breeding goats for meat production: 3 Selection and breeding strategies. Small Rum. Res. 67(2):113-126.

Taiwo BBA, Buvanendran V, Adu IF (2005). Effects of body condition on the reproductive performance of Red Sokoto goats. Nig. J. Anim. Prod. 32(1):1-6.

Ukpabi UH, Emerole CO, Ezeh CI (2000). Comparative study of strategic goat marketing in umuahia regional market in Nigeria. Proc. 25th Annual conf. NSAP. pp. 364-365.

Wiener G, Hayter S (1974). Body size and conformation in sheep from birth to maturity as affected by crossbreeding, maternal and other factors. Anim. Prod. 19:47-65.

11

The perception of climate and environmental change on the performance and availability of the edible land snails; a need for conservation

Ivo Ngundu WOOGENG[1,2], J. Paul GROBLER[2], Kingsely Agbor ETCHU[3] and Kenneth Jacob N. NDAMUKONG[1]

[1]Department of Zoology and Animal Physiology, University of Buea, P. O. Box 63 Buea, Cameroon.
[2]Department of Genetics, University of the Free State, P. O. Box 339, Bloemfontein, 9300, South Africa.
[3]Institute of Agricultural Research for Development (IRAD) Ekona, P. M. B. 25 Buea, Cameroon.

Gastropods are vulnerable to extremes of climate and could be useful indicators of climate change and environmental manipulations. The aim of this study was to determine the perceptions of producers and harvesters of snails on climate change and habitat destruction, their effects on snails and the availability of the snail's feed. Questionnaires (402) were distributed in the Buea, Tiko, Muyuka, Kumba, Tombel, Ekondo Titi and Mamfe areas of Cameroon. Most respondents confirmed that they have heard about climate change (91.2%) and habitat destruction (87.3%). High proportions of 90.4 and 77.6% have noticed recent negative changes in the population of snails and their production respectively. Asked if these changes affect snail feeds as well, respondents said yes (53.5%) or partially (17.3%), explaining that it brings about a reduction in their feed mostly in the wild (showing a need for more domestication). Respondents (62.6%) also indicated that these changes expose the snails to pests, parasites and diseases, with 56.9% saying that these changes affect the performance and availability of snails. It can be deduced from the responses that climate change and environmental manipulations have an influence on the performance and availability of snails in captivity and in the wild, which necessitates conservation measures.

Key words: Snails, climate change, environmental manipulations, feed, performance, South west region - Cameroon.

INTRODUCTION

Snail farming/gathering and consumption is becoming increasingly popular in Cameroon. It is an important source of livelihood for rural dwellers and a source of protein for both rural and urban dwellers in Tropical Africa in particular but also the rest of the world (FAO, 1986; Ngenwi et al., 2010). Most of the land snails of West African origin (*Archachatina* spp., *Achatina* spp.,

Limcolaria spp.) are forest dwellers found mostly along the coastal zone (Imevbore, 1990). The problem of malnutrition in humans, which has its roots from shortage of animal protein, has encouraged researchers to look for alternative sources of animal protein which can be gotten with little or no capital investment. Micro-livestock or non-conventional livestock such as snails, grass-cutters etc have been domesticated and a significant amount of research is ongoing, aimed at increasing their availability at an affordable price. Snails are an important source of animal protein in many parts of West and Central Africa (Blay et al., 2004), having good quality protein (69% dry weight) and being rich in potassium, phosphorus, essential amino acids and vitamins C and B complex (Baba and Adeleke, 2006, Okpeze et al., 2007). Snail farming is environmentally friendly and can be done with minimal skill (NRC, 1991; Akinnusi, 1998) and low capital requirement for establishment and running. Snails and snail farming are also characterized by very high fecundity and low mortality, low labor requirement, lack of noise and with readily available markets (Cobbinah, 1993; Baba and Adeleke, 2006).

Recently, the production of snails has not kept pace with demand (Etchu et al., 2008), with different environmental and technical factors implicated. Environmental manipulations (urbanization, deforestation, burning of biomass, the use of harmful chemicals) and climate change as well as lack of training on intensive snail rearing have been identified as impediments to increased snail supply from the wild and in captivity (Ngenwi et al., 2010). Climatic variables are among the determining factors in the survival, growth and sustenance of any organism in its niche, including Mollusks. The performance of an organism is directly or indirectly dependent upon the innate ability of the organism as well as the totality of its surrounding. Climate change is already having a negative impact on Africa through extremely high temperatures (IPCC, 2007), which are not favorable to snail growth and development. The scale of domestication and intensive management of the edible land snail is bound to increase and conservation is consequently needed. To date, research on climate change in the tropics has focused mostly on crops, less on livestock and little on biodiversity conservation. Change in species abundance can provide useful insights on climate change and drivers of this change. However, there is dearth of information on this in most parts of the tropics.

Also, to understand the ecology and evolutionary biology of any given population, we need to understand how these populations respond to changes to an ideal environment (Maynard, 1989; Hoffman and Parsons, 1991; Peters and Lovejoy, 1992). The risk of extinction is eminent in populations that face the risk of, or are subjected to substantial environmental stress such as seen during colonization attempts, human-influenced introductions and reintroductions or the repercussions of the global climate change. Climate change is likely to have major impacts on poor livestock keepers and on the ecosystem services on which they depend.

Seasonal variations in land snail physiology and biochemical composition have been linked to annual cycles of photoperiod, temperature, humidity, water availability and reproduction (Machin, 1975; Riddle, 1983; Cook, 2001; Storey, 2002). Seasonal physiological data

for land snails may be useful in understanding species-specific habitat requirements and in predicting their response to environmental changes. Of special interest are responses and adaptations of land snails to the climatically unpredictable ecosystems (Blondel and Aronson, 1999). However, there are relatively few informative, long-term field studies on aspects of the physiological ecology of land snails (Wieser and Wright, 1979; Riddle, 1983; Arad, 1993; Rees and Hand, 1993; Pedler et al., 1996; Arad and Avivi, 1998; Arad et al., 1998; Sinos et al., 2007).

While climate change is a global phenomenon, its negative impacts are more severely felt by poor people in developing countries who rely heavily on the natural resource base for their livelihoods. Agriculture and livestock keeping are amongst the most climate sensitive economic sectors and rural poor communities are more exposed to the effects of climate change (FAO, 2008, 2009). There are global health implications related to biodiversity loss and many of the anticipated health risks brought about by climate change will be caused by loss of genetic diversity (Cohen et al., 2002; ARD, 2008).

This study was therefore carried out to assess the human perception (snail famers and gatherers) about the effects of both climate change and man-made environmental changes (anthropogenic stressors) on the performance and availability of edible land snails (*Archachatina* spp. and *Achatina* spp.) both in the wild and under conditions of artificial propagation. We sampled the opinion of the snail farmers and gatherers because, firstly, they have been dealing with the snails for quite a while and secondly because they literally deal with the snails on a daily bases.

MATERIALS AND METHODS

This study was carried out in 5 {Fako (4° 9'N, 9° 13'E), Kupe Muanenguba (4° 50'N, 9° 40'E), Manyu (5° 45'N, 9° 18'E), Meme (4° 37'N, 9° 26'E) and Ndian (4° 36'N, 9° 5'E)} divisions out of the 6 divisions of the South west region of Cameroon between June and October, 2011. The South west region is a cosmopolitan region with many and different ethnic groups with different cultural values and it is exclusively a humid forest ecosystem characterized by a monomodal rainfall pattern. A total of 52 questionnaires were distributed to snail farmers and those who harvest the snails from the wild in each of six localities (Buea, Tiko, Muyuka, Kumba, Tombel, and Ekondo Titi). Mamfe, the chief town of snail consumption received 90 questionnaires.

The questionnaire carried both open and closed questions, with the open questions given tags during analysis. The questions were asked in order to obtain the candid opinion and experience of the respondents about climate change (CC) and environmental manipulations (EM), and the effects of the climate change and environmental manipulations on both the snails (their availability and performance), their feeds (generally), feed types (individual or the different feed types), and number of snails etc.

Data collected was analyzed using statistical package for social sciences (SPSS) software, standard version, release 17.0 (SPSS Inc. 2008). Descriptive statistics provided insights into the snail farmers' and gatherers' perceptions of climate change and anthropogenic stressors on the edible land snails. Measures of

Table 1. Climate change and habitat/environmental destruction awareness of snail farmers and gatherers in the different Divisions of the South West Region of Cameroon.

Division of residence		Have you heard about climate change?				Do you know what habitat destruction is?			
		Yes	No	Partially	Total	Yes	No	Partially	Total
Fako	n	106	7	0	113	110	5	1	116
	%	93.8	6.2	0.0	100.0	94.8	4.3	0.9	100.0
Kupe Muanenguba	n	44	0	1	45	34	6	3	43
	%	97.8	0.0	2.2	100.0	79.1	14.0	7.0	100.0
Manyu	n	85	11	2	98	71	13	13	97
	%	86.7	11.2	2.0	100.0	73.2	13.4	13.4	100.0
Meme	n	37	10	0	47	43	4	0	47
	%	78.7	21.3	0.0	100.0	91.5	8.5	0.0	100.0
Ndian	n	50	0	0	50	50	0	0	50
	%	100.0	0.0	0.0	100.0	100.0	0.0	0.0	100.0
Total	n	322	28	3	353	308	28	17	353
	%	91.2	7.9	0.8	100.0	87.3	7.9	4.8	100.0

association between variables were carried out using Chi-Square test of independence or of equality of proportions for nominal Vs nominal and nominal Vs ordinal variables and Sommers'd for Ordinal Vs Ordinal variables. Multiple response analysis for multiple-responses question (with possibility for more than one response to a single question).

RESULTS

From a total of 402 questionnaires sent to the field, 393 were returned and validated for data. Out of the 393 received, 358 questionnaires were effectively used for data analysis following exploratory statistics, with 35 questionnaires discarded due to lack of relevant information (because either the respondent did not complete the questionnaire or responses to questions were scanty) after exploratory statistics.

Table 1 gives a summary of the awareness about climate change by snail farmers and gatherers while Table 2 gives data on awareness of habitat or environmental destruction. For climate change awareness, the majority of the farmers or gatherers answered in the affirmative to having a knowledge on climate change (91.2%, n = 322) and partial knowledge (0.8%, n = 3) as opposed to those professing no knowledge on climate change (7.9%, n = 28) (Table 1). The knowledge on climate change also varied significantly between the different Divisions (P <

0.05). A majority (87.3% ± 4.8%) of the snail farmers and gat-herers have knowledge of habitat destruction (Table 1).

Also, a majority of the farmers and gatherers (90.4%, n = 321) have noticed a decline in the snail population recently and there is a significantdifference between those who responded "Yes" from those who said "No" (χ^2 = 38.351, P < 0.05) (Table 2).

Slightly more than 85.8% of those interviewed have noticed changes in the wild (Table 3) and 56.9% believe it is due to the combined effects of climate change and habitat destruction (Table 3). Data in Table 4 reveals that 77.6% believe these climatic and environmental anomalies have greatly affected their production of snails and some 56.6% thinks that these changes also affect the performance of these animals significantly (Table 4). Asked if these changes affect the feed of the snails, 53.5% said that the changes do affect the snail feeds greatly and in a negative way (Figure 1). A total of 62.6% affirmed that these changes in the environment also expose their snails to pests, parasites and diseases, as the environment created is favorable for these pathogens to strive (Figure 2). Some negative effects of climate and environmental changes on the availability and performance of the edible land snails as seen by the farmers or gatherers, with more mortalities of snails (56.96%), harsh environment and reduction in

Table 2. Perceived changes in snail Population in recent times.

Division of residence		Have you noticed any changes on the snail population recently?		
		Yes	No	Partially
Fako	n	104	9	2
	%	90.4%	7.8%	1.7%
Kupe Muanenguba	n	36	1	8
	%	80.0%	2.2%	17.8%
Manyu	n	88	9	1
	%	89.8%	9.2%	1.0%
Meme	n	47	0	0
	%	100.0%	.0%	.0%
Ndian	n	46	0	4
	%	92.0%	.0%	8.0%
Total	n	321	19	15
	%	90.4%	5.4%	4.2%

water availability (47.34%), lack of food for the snails (43.68%), low production of snails and reduction in population (Av - 37.28%) and deforestation (which exposes snails to too much sunlight) at 36.10% (Table 5). The majority of snail farmers or gatherers have been involved in this activity for more than a year (Figure 3).

DISCUSSION AND CONCLUSION

Climate change is well known to farmers or gatherers of edible land snails, who show considerable awareness of the phenomenon. Habitat or environmental destruction which is a man-made phenomenon has also been noticed by farmers or gatherers of snails. Both phenomena influence to a great extent the performance and availability of the edible land snails as has been observed by farmers and gatherers of the snails. Naturally, snail production and availability is higher in the rainy season than in the dry season, as there is readily available and more feed in an environment conducive for habitation which ensures effective performance of the snails. This is confirmed by studies by Marcelo (2000), who reported that snails consistently respond to changes in their physical environment. Environmental and climatic factors play an important role in determining the survival (production and availability) and performance of snails, as these not only affect the snails directly but also indirectly via the introduction of new pests, parasites and diseases (Ngenwi et al., 2010) as well as influencing their feed (Clarke, 1998; Sternberg et al., 1999).

Environmental factors such as deforestation, increased temperatures with low rainfall, slash and burning (bush fires), uncontrollable collection from the wild, high use of agrochemicals and lack of training on intensive snail farming are all impediments to increase snail supply both from the wild and in captivity (Ngenwi et al., 2010), in response to increasing demand. All these factors can cause a decline in the population of the edible land snails (Amusan and Omidiji, 1998). Climatic factors are among the determining factors of growth and survival of mollusks. Land snails prefer humid environments for optimum performance in the presence of their choice feed (Ejidike et al., 2004).

Decreasing rainfall and increasing temperatures would have an adverse effect on the conditions necessary for adequate development and reproduction. Temperatures in West Africa have changed faster than the Global trend, with increases in the range of 0.2 to 0.8°C from the late 1970s (ECOWAS-SWAC/OECD, 2008). There is therefore a need for conservation of snails via the improvement of feed or alternative feed (Agbogidi et al., 2008; Ngenwi et al., 2010), promotion of aforestation, moderation of collection and the creation of awareness among people about the need for conservation of natural resources (snails).

The present study focused on local observations of snail farmers or gatherers on the consequences of climate change and environmental perturbation. This study is the first step towards understanding the complexity of the interactions of snails and their environments under conditions of global climate change.

Table 3. Observed changes in the wild and reasons for observed changes in the wild; distribution by division of residence.

Division of residence		In the wild, have you noticed such changes?				Are these changes in the wild related to climate change or habitat destruction?				
		Yes	No	Partially	Total	Climate change only	Habitat destruction only	Both climate change and habitat destruction	None	Total
Fako	n	98	4	3	105	10	33	58	8	109
	%	93.3	3.8	2.9	100.0	9.2	30.3	53.2	7.3	100.0
Kupe Muanenguba	n	19	6	14	39	1	3	39	0	43
	%	48.7	15.4	35.9	100.0	2.3	7.0	90.7	0.0	100.0
Manyu	n	77	16	3	96	8	26	40	18	92
	%	80.2	16.7	3.1	100.0	8.7	28.3	43.5	19.6	100.0
Meme	n	45	2	0	47	1	7	21	18	47
	%	95.7	4.3	0.0	100.0	2.1	14.9	44.7	38.3	100.0
Ndian	n	50	0	0	50	13	0	36	1	50
	%	100.0	0.0	0.0	100.0	26.0	0.0	72.0	2.0	100.0
Total	n	289	28	20	337	33	69	194	45	341
	%	85.8	8.3	5.9	100.0	9.7	20.2	56.9	13.2	100.0

This can then be extrapolated to other vulnerable organisms. Further experimental investigations of other invertebrates, with their intrinsic characteristics and interactions with other organisms and their respective environments, will provide useful tools to test predictions of the responses of snails to climatic and environmental changes. It can be deduced from the responses of the respondents (that climate change and manipulation of the environment may have an influence on the performance and availability of the edible land snails. The awareness of climate change and environmental changes match with the devastating effects of these phenomena on the edible land snails of the genus *Archachatina*. Both these processes create conditions that are unfavorable for the development and reproduction of the snails both in captivity and in the wild. As a result of these, there is a gradual but steady decline in the population of this highly priced non-conventional livestock, both in captivity and the wild, whereas its value and consumption rate are on a sturdy increase. Therefore, there is a need for conservation via the use of improved feed or alternative feed that can last throughout the year, training of farmers to intensify and improve management practices of their snail farms as well as the discouraging of practices that promote or enhance climate change and habitat destruction. The latter includes slash and burning (bush fires), deforestation, and the use of agrochemicals. This can only be achieved by creating awareness among the local population of farmers and collectors.

Table 4. Effects of climate change and environmental destruction on production and performance of snails both in captivity and the wild; distribution by divisions of the south west province of Cameroon.

Division of residence		Do these changes affect your production?				Do these changes affect their performance?			
		Yes	No	Partially	Total	Yes	No	Partially	Total
Fako	n	90	17	3	110	80	20	10	110
	%	81.8	15.5	2.7	100.0	72.7	18.2	9.1	100.0
Kupe Muanenguba	n	24	5	16	45	14	6	23	43
	%	53.3	11.1	35.6	100.0	32.6	14.0	53.5	100.0
Manyu	n	66	20	6	92	47	29	18	94
	%	71.7	21.7	6.5	100.0	50.0	30.9	19.1	100.0
Meme	n	42	4	1	47	42	3	1	46
	%	89.4	8.5	2.1	100.0	91.3	6.5	2.2	100.0
Ndian	n	45	2	3	50	11	6	33	50
	%	90.0	4.0	6.0	100.0	22.0	12.0	66.0	100.0
Total	n	267	48	29	344	194	64	85	343
	%	77.6	14.0	8.4	100.0	56.6	18.7	24.8	100.0

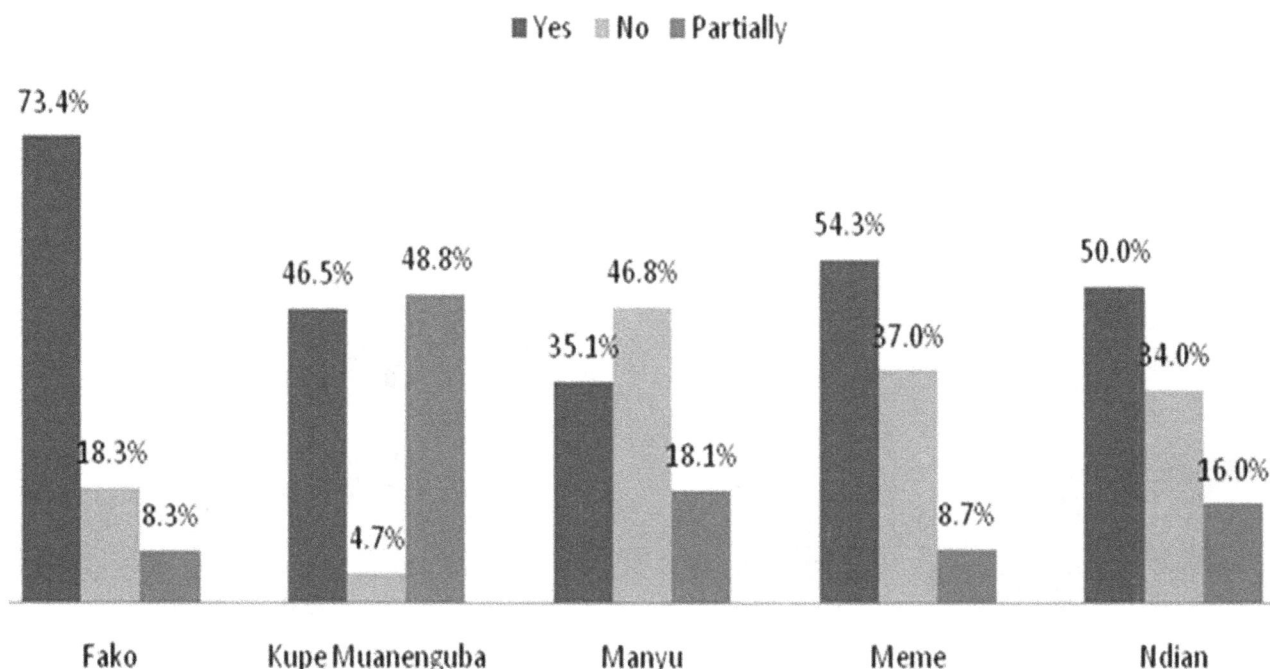

Figure 1. Effects of changes on feeds as perceived be farmers/gatherers of the edible land snails.

■ Yes ■ No ■ Partially

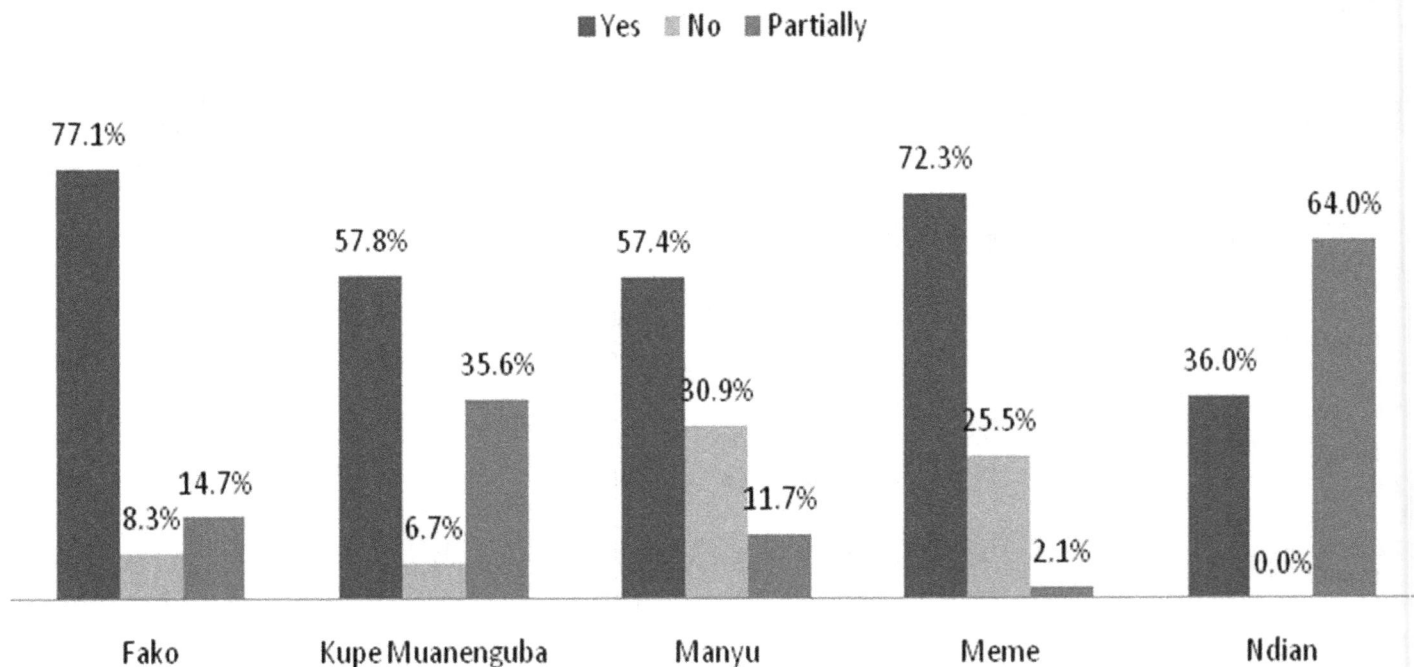

Figure 2. Exposure of the edible land snails to pests, parasites and diseases?

Table 5. Some major effects of climate and environmental changes on snails' availability and performance in the different study sites.

Effects		Division of residence					
		Fako	Kupe Muanenguba	Manyu	Meme	Ndian	Average percentage
Increase diseases	n	0	2	9	0	44	
	%	0.0	7.1	12.9	0.0	91.7	22.34
Low production of snails and reduction in population	n	20	15	18	0	40	
	%	23.8	53.6	25.7	0.0	83.3	37.28
Reduction in size	n	4	3	0	0	0	
	%	4.8	10.7	0.0	0.0	0.0	3.1
More dead of snails	n	52	15	49	39	2	
	%	61.9	53.6	70.0	95.1	4.2	56.96
Lack of food for the snails	n	52	9	13	34	11	
	%	61.9	32.1	18.6	82.9	22.9	43.68
Harsh environment makes habitation difficult and reduce water availability for snails	n	54	5	10	31	31	
	%	64.3	17.9	14.3	75.6	64.6	47.34
Deforestation exposes snails to too much sunlight	n	49	6	4	39	0	
	%	58.3	21.4	5.7	95.1	0.0	36.10

Table 5. Contd.

Inadequate soil texture	n	2	6	3	0	0	
	%	2.4	21.4	4.3	0.0	0.0	5.62
Toxic waste disposal that kill snails	n	2	0	0	0	0	
	%	2.4	0.0	0.0	0.0	0.0	0.48
Soil erosion	n	4	0	0	9	0	
	%	4.8%	0.0%	0.0%	22.0%	0.0%	5.36
Snails are burnt during slash and burning	n	2	0	2	0	0	
	%	2.4%	.0%	2.9%	0.0%	0.0%	1.06
Total	n	84	28	70	41	48	
	%	31.0%	10.3%	25.8%	15.1%	17.7%	

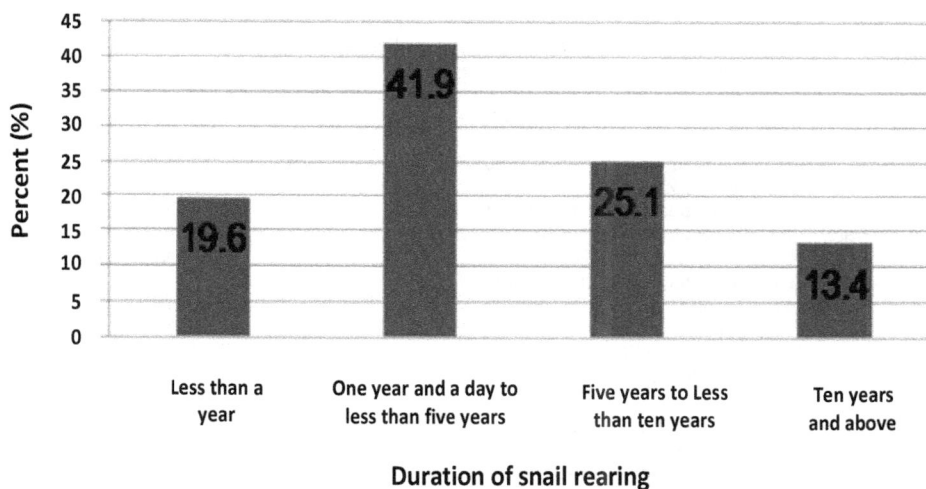

Figure 3. Duration of activity of the snail's farmers/gatherers.

ACKNOWLEDGEMENT

We acknowledge the enumerators who patiently distributed the questionnaires as well as a very big 'thank you' to the respondents who made this study possible.

REFERENCES

Agbogidi OM, Okonta BC, Ezeana EL (2008). Effects of two edible fruits on the growth performance of African giant land snail (*Archachatina marginata* Swainson). J. Agric. Biol. Sci. 3:26-29.

Akinnusi O (1998). Introduction to Snails and Snail Farming. Omega Science Publisher, Tinuoso House, Lagos, Nigeria.

Amusan JA, Omediji MO (1998). Edible land snails: A technical guide to snail farming in the tropics. Verity (ltd) Podo village, Ibadan, Nigeria.

Arad Z (1993). Water relations and resistance to desiccation in three Isr aeli desert snails, Eremina desertorum, Euchondrus desertorum and Euchondrus albulus. J. Arid Environ. 24:387-395.

Arad Z, Avivi TR (1998). Ontogeny of resistance to desiccation in the bu sh-dwelling snail -*Theba pisana* (helicidae). J. Zool. 244:515-526.

Arad Z, Goldenberg S, Heller J (1998). Short and long term resistance to desiccation in a minute litter-dwelling land snail *Lauria cylindracea* (Pulmonata, Pupillidae). J. Zool. 246:75–81.

ARD (2008). Generating global public good to promote sustainable pro-poor growth in the Livestock Sub-Sector. GL-CRSP/World Bank Planning Initiative.

Baba KM, Adeleke MT (2006). Profitability of snail production in Osun State. Nig. J. Agric. Food Sci. 4:147-155.

Blay ET, Ofori BD, Heloo J, Ofori JB, Nartey E (2004). Agrodiversity within and without conserved forests for enhancing rural livelihoods.

In: Gyasi, E. A., Kranjac-Berisavljevic, G., Baly ET, Oduro W (eds) Managing agrodiversity the traditional way: Lessons from West Africa in sustainable use of biodiversity and related natural resources. Tokyo: United Nations University Press, pp. 203-214.

Blondel J, Aronson J (1999). Biology and wildlife of the Mediterranean region. Oxford University Press, Oxford.

Clarke IP (1998). Recruitment dynamics in a Southern calcareous grass land: Effects of climate change. PhD Thesis, University of London, London.

Cobbinah JR (1993). Snail farming in West Africa. A practical guide. CTA publication.

Cohen RDH, Sykes CD, Wheaton EE, Stevens JP (2002). Evaluation of the effects of climate change on forage and livestock production and assessment of adaptation strategies on the Canadian *Prairies*. University of Saskatchewan, Saskatoon.

Cook A (2001). Behavioural ecology, on doing the right thing, in the right place at the right time. In: Barker GM (eds), The Biology of Terrestrial Molluscs. CABI Publishing, Oxon, UK, pp. 447-487.

ECOWAS-SWAC/OCED (2008). Climate and climate change. The Atlas on Regional Integration in West Africa. Environment Series. January 2008. www.atlas-westafrica.org.

Ejidike BN Afolayan TA, Alokan JA (2004). Observations on some climatic variables and dietary influence on the performance of cultivated African giant land snail (*Archachatina marginata*): notes and records. Parkistan J. Nutr. 3:362-364.

Etchu KA, Mafeni JM, Ngenwi AA (2008). Comparative performance of three edible snail species using intensive cage housing system in Cameroon. Bull. Anim. Health Prod. Afr. 56:345-352.

FAO (1986). Farming Snails, FAO Better Farming Series, 3/33 Rome, Italy.

FAO (2008). Climate-related trans-boundary pests and diseases including relevant aquatic species. Expert meeting, FAO. February, 2008.

FAO (2009). The state of food and agriculture, livestock in a balance. Rome.

Hoffman AA, Parsons PA eds. (1991). Evolutionary genetics and environmental stress. Oxford University Press, Oxford.

Imevbore EA (1990). Management techniques in rearing African giant land snail (*Archachatina marginata*). Ph.D Thesis, University of Ibadan, Nigeria

IPCC (2007). Climate change 2007: Impacts, Adaptation and Vulnerability. Contribution of working group II to the Fourth Assessment Report of the Intergovernmental Panel on ClimateChange. Parry ML, Canziani OF, Palutikof JP, Van der Linden PJ, Hanson CE (eds) Cambridge University Press, Cambridge, UK, p. 976.

Machin J (1975). Water relationships. In: Fretter V, Peake J (eds) Pulmonates I. Academic Press, London, pp. 105–163.

Marcelo S (2000). Terrestrial gastropods and experimental climate change: A field study in a calcareous grassland. Ecol. Res. 15:73-81.

Maynard SJ (1989). The causes of extinction. Philos. T. Roy. Soc. B, 325:241-252.

Ngenwi AA, Mafeni JM, Etchu KA, Oben FT (2010). Characteristics of snail farmers and constraints to increased production in West and Central Africa. Afr. J. Environ. Sci. Tech. pp. 274-278.

NRC (1991). Micro-liveweight. Little-known small animals with a promising economic future. National Academy Press, Washington, D. C.

Okpeze CN, Omole AJ, Ajayi FT, Adebowale EA (2007). Effects of feeding adult snails *Stylosanthes guianensis* or *Lablab purpureus* as substitute for pawpaw leaf. Afr. J. Biotechnol. 6:1959-1962.

Pedler S, Fuery CJ, Withers PC, Flanigan J, Guppy M (1996). Effectors of metabolic depression in an estivating pulmonate snail (Helix aspersa), whole animal and *in vitro* tissue studies. J. Comp. Physiol. B. 166:375–381.

Peters RL, Lovejoy TE (1992). Global warming and Biological Diversity. Yale University Press, New Haven.

Rees BB, Hand SC (1993). Biochemical correlates of estivation tolerance in the mountain snail Oreohelix (Pulmonata, Oreohelicidae). Biol. Bull. 184:230-242.

Riddle WA (1983). Physiological ecology of land snails and slugs. In: Russell-Hunter, W.D. (eds) The Mollusca 6. Academic Press, London, pp. 431-461.

Sinos G, Panayiotis K, Panayiotis P, Efstratios V (2007). Relictual physiological ecology in the threatened land snail Codringtonia helenae: A cause for decline in a changing environment? ACTA Oecol. 32:269-278.

Sternberg M, Brown VK, Masters GP, Clarke IP (1999). Plant community dynamics in a calcareous grassland under climate change manipulations. Plant Ecol. 143:29-37.

Storey KB (2002). Life in the slow lane, molecular mechanisms of aestivation. Comp. Biochem. Physiol. A. 134:733-754.

Wieser W, Wright E (1979). Effects of season and temperature on D-Lactate dehydrogenase, pyruvate kinase and arginine kinase in the foot of *Helix pomatia* L. Hoppe-Seylers Zeitschr. Physiol. Chem. 360:533-542.

12

Variability and predictability of productive and body traits of Fulani ecotype chicken

Jesuyon, Oluwatosin M. A. and Salako, Adebowale E.

Department of Animal Science, University of Ibadan, Oyo State, Nigeria.

Improvement of the domestic chicken has dominated the effort of indigenous breeders in Nigeria in recent times. Three hundred and fifty-seven (357) Fulani ecotype chickens were surveyed and assessed for their phenotypic and productive characteristics in Osun State, Nigeria. Mature weight (kg), comb size, wattle size, breast length, breast width, leg length (all in cm); egg weight (gm), clutch size were examined and classified. A higher level of variability was revealed in comb size, wattle size, egg weight, and clutch size within the hen population of each local government; and in leg length within Orolu cocks and hens as shown by their coefficient of variation (CV), respectively. Strong and significant association was observed between mature weight and wattle size, mature weight and comb size, and between wattle size and comb size in the cocks; and between egg weight and egg clutch size in hens. Breast length and breast width were best predictors of each other while mature weight was best predicted by wattle size in cocks. Breast length was predicted by mature weight and leg length, breast width was predicted by mature weight and, mature weight was predicted by the combination of leg length, breast length and breast width in hens. All parameters were more accurately predicted in cocks than in hens. Statistical modelling revealed sexual dimorphism on all equations.

Key words: Body parameters, Fulani chicken, predictability, settlements, variability.

INTRODUCTION

Domestic chickens contribute highly to the socio-economic condition of Nigerians. A survey (Aphca news, 2006) reported the population of backyard poultry in Nigeria to be 84 million, which amounts to about 60% of the total poultry population in Nigeria.

These indigenous chickens are largely unimproved and uncharacterized. Their management, feeding and housing is simple free-range. There are three chicken ecotypes in Nigeria namely the Fulani, the Eastern and the Yoruba ecotypes. The Fulani ecotype is prevalent in the middle and northern parts of the country but there are pockets of this ecotype among the Fulani descendants in the rural settlements of Osun State. Various studies on growth (Ibe, 1993; Sola-Ojo and Ayorinde, 2009); egg traits, fertility and hatchability (Peters et al., 2004; Fayeye and Oketoyin, 2006) and phenotypic variations (Mancha et al., 2006; Ajayi and Agaviezor, 2009) have been

reported for local chickens in Nigeria. But there is paucity of information on the physical characteristics of this chicken ecotype in Osun State, Nigeria. Furthermore, not much is known about their body characteristics, distribution and potential for improvement. Latshaw and Bishop (2001) submitted that for livestock, it is more common to estimate weight by measuring a part of the body trunk rather than an extremity.

Thus, Nwosu et al. (1985) observed in their biometrical study of the conformation of the native chicken in Nsukka that, there was very little variability in shank length among native chickens. Besides, not much study has been conducted into the variability that exists within the body parameters and between the various existing ecotypes. This study therefore was conducted to examine the ecological distribution, body parametric indices, variability and predictability of these biometric parameters in indigenous Fulani ecotype chicken in Osun State, Nigeria. Body traits measurement analyses are important for estimating the relative standard deviation of body parameters while regression analyses could reveal the predictability and the predictors of parameters of interest, thus characterizing each ecotype within and between sexes. The results from this study could be useful to all poultry researchers and breeders for further studies, inclusion in international data bank.

MATERIALS AND METHODS

This survey was conducted in five local Fulani settlements namely, Gaa Baba-Bayo, Gaa Abu, and Aba Aro in Ifon (Orolu Local Government Area, LGA); Gaa Koko, Ire (Ire Local Government Area, LGA) and Gaa Power-line Osogbo (Osogbo Local Government Area, LGA) of Osun State. These settlements are situated in the mid South-West region of Nigeria.

Three hundred and fifty-seven mature chickens in ratio 3 females to one male were assessed using the random sampling technique to pick famer-owners. The sampling size, n, was made large to compensate for the small population, N of this ecotype in the region. All birds were examined with the help of owner-farmers early in the mornings before they were released for scavenging. Body measurements taken were comb size, wattle size, breast length, breast width, leg length (all in cm), egg weight (gm), mature weight (kg) and number of eggs per clutch per hen (Clutch size).

Comb size was measured as length along the base, from the beak end to the end of the comb, wattle size was taken as length from the topmost part below the beak to the end of the wattle, breast length was measured as length from the tip of the breast at the sternum along the mid region to the end of the keel bone on the belly (Wikipedia.org, 2012), breast width was length was length across the breast over the tip of the sternum (pectus) from one edge of the wing to the other (Momoh and kershima, 2008), the Leg length was measured as length from the hip bone to the tarso-metatarsus joint (Semakula et al., 2011), egg weight of individual hens were taken and recorded individually for each hen, mature weight was taken as weight of birds that have roosted or layed eggs at least once while the clutch size was taken as number of eggs layed at a single stretch before the hens start to sit upon the eggs for brooding. Measurements were made with a cotton thread and the lengths measured were determined on a metric ruler. Egg weight was measured with a portable egg weighing scale while the weight of birds were determined with the outdoor 5 kg-capacity platform scale.

Data analysis was performed using ANOVA and descriptive procedure of the SPSS (Version 10.0 of 2001) software Pearson's correlation and stepwise regression analyses were done using the SAS (Version 8 for Windows of 1999) procedures ($p = 0.05$). Stepwise regression procedures was adopted for predicting each of comb size, breast length, breast width from mature weight as dependent parameter against all others as explanatory variables. All variables utilized in the model were those that met the 0.05 significance level for entry into the model.

RESULTS

Table 1 shows the mean of parameters studied and ecological distribution of the Fulani chickens surveyed among the three LGAs in Osun State, Nigeria. ANOVA (p < 0.05) shows that cocks differed from hens in all parameters measured. Among cocks, Ire sub-type had the longest breast (23.1 cm), the Orolu and Osogbo sub-types showed the widest breast (15.5 cm and 15.2%) while Osogbo cocks exhibited the longest legs (29.3 cm). Among hens, Osogbo sub-type had the longest breast (18.1 cm) and legs (13.2 cm); and a wider breast (14.2 cm). All settlements showed significant difference (p ≤ 0.05) in egg weight: Orolu and Osogbo sub-types had the heaviest eggs (39.0 and 37.9 gm), followed by Ire (35.6 gm) sub-type, respectively. Mean mature weight was 2.29 and 1.44 kg for cocks and hens, respectively while Osogbo subtype cocks and hens displayed the heaviest mature weights of 2.40 and 1.50 kg, respectively. The hens revealed egg-clutch ranges of either 2 to 6 or 7 to 12 eggs per hen. Each L.G.A. had birds in both clutch ranges. Ire hens gave the highest prevalence of 4 to 6 egg clutch range of (62.5%) while Ifon hens submitted the highest record of 7 to 12 egg clutch range (55.6%).

Table 2 shows the CV of the various biometric parameters of the Fulani ecotype chicken. This table revealed low to medium level of variations within parameters within sex ranging from 0.006 to 0.476 and 0.025 to 0.376, respectively for cocks and hens. However, medium variability (0.476) was obtained in leg length among Orolu cocks, while a low level of variability was obtained in wattle size (0.314 to 0.376), comb size (0.300 to 0.334), egg clutch size (0.326 to 0.370) and egg clutch range (0.314 to 0.376) within each LGA ecological hen population. Whereas leg length (0.303) and egg weight (0.360) were comparatively variable among Orolu hens.

Table 3 shows the matrix of Pearson's phenotypic correlation coefficients between body and production parameters in both Fulani cock and hen populations, respectively in Osun State, Nigeria. Most pairs of parameters exhibited weak associations. Nevertheless, there were strong associations between wattle size and comb size (r = 0.91 and 0.32), wattle size and mature weight (r = 0.89 and 0.31) and, between breast length and breast width (r = -0.81 and 0.38). A highly positive association (r = 0.82) was obtained between egg weight and egg clutch size. These correlation coefficients were significant at 0.05 levels.

Table 4 reveals the stepwise regression equations for prediction of breast length, breast width, comb size and mature weight in both cocks and hens. The models were highly significant (p < 0.05) R^2 ranging from 0.66 to 0.82 in the cocks, to very low R^2 from 0.10 to 0.38 for all parameters predicted in the hens. All the equations regressed for cock parameters had higher intercepts or constants (47.5 vs. 12.98; 22.27 vs. 9.52; 2.23 vs. 1.05; 1.33 vs. 0.55) than those of same traits in the hens. This further demonstrated sexual dimorphism already observed between sexes of Fulani chicken in Osun and in the results. Thus, the stepwise model predicted all parameters better for cocks (R^2 = 0.66-0.82) than hens (R^2 = 0.10-0.38).

Table 1. Mean of body and productive parameters of Fulani ecotype chickens on free-range management in three selected Local Government Areas of Osun State, Nigeria

Parameter	Sex type	Mean (SE)	Ifon (SD)	Ire (SD)	Osogbo (SD)
Comb size,	Cocks	5.17(0.23)[a]	5.84 (1.01)	5.03 (0.00)	5.72(0.18)
Cm	Hens	1.65(0.09)[b]	1.48 (0.47)[b]	2.47 (0.18)[a]	1.41 (0.42)[b]
Wattle size,	Cocks	4.87(0.33)[a]	4.82 (1.40)	4.40 (1.03)	5.72 (0.18)
Cm	Hens	1.48(0.07)[b]	1.51 (0.50)	1.33 (0.43)	1.41 (0.42)
Breast,	Cocks	19.9(0.54)[a]	19.1 (1.10)[b]	23.1(0.64)[a]	19.1 (0.26)[b]
length cm	Hens	17.4 (0.24)[b]	17.3 (1.75)[b]	16.7 (0.74)[b]	18.1 (1.55)[a]
Breast,	Cocks	15.1 (0.24)[a]	15.5 (0.56)[a]	13.7 (0.44)[b]	15.2 (0.46)[a]
width cm	Hens	14.1 (0.23)[b]	14.6 (1.32)[a]	13.1 (1.86)[b]	13.7 (1.16)[b]
Leg length,	Cocks	21.0 (2.19)[a]	19.7 (9.38)[b]	19.4 (1.46)[b]	29.3 (0.18)[a]
Cm	Hens	12.1 (0.43)[b]	11.4 (3.45)[b]	13.1 (1.11)[a]	13.2 (0.34)[a]
No. of	Cocks	4.93 (0.07)[a]	4.89 (0.33)	5.00 (0.00)	5.00 (0.00)
Digits	Hens	4.02 (0.02)[b]	4.00 (0.00)	4.13 (0.35)	4.00 (0.00)
Mature	Cocks	2.29 (0.07)[a]	2.28 (0.31)	2.27 (0.31)	2.40 (0.14)
Weight, kg	Hens	1.44 (0.03)[b]	1.43 (0.19)	1.40 (0.21)	1.50 (0.14)
Egg weight, gm	Hens	38.2 (1.81)	39.0 (14.1)[a]	35.6 (8.63)[b]	37.9 (7.49)[a]
Mean of eggs /clutch/hen	Hens	6.93 (0.34)	7.33 (2.30)	6.25 (2.32)	6.33 (2.06)
Egg clutch Range	Range %	2-6 eggs 50	2-6 44.4	4-6 62.5	2-6 55.6
	Range %	7-12 eggs 50	7-12 55.6	9-12 37.5	7-10 44.4

[a, b] superscripts associated with Mean in each column and row indicate significant differences at P < 0.05; SE means Standard Error. SD means Standard Deviation.

DISCUSSION

The mean mature weight range values obtained within sex were close (2.28 to 2.40 and 1.40 to 1.50 kg) for cocks and hens, respectively. These figures were higher than 1.47 to 1.77 and 0.85 to 1.44kg reported by Nwosu et al. (1985) for Nsukka cocks and hens and 1.38 to 1.55 kg and 0.86 to 1.45 kg range for Owerri chickens, respectively. Results are also higher than Bayelsan (1.50 and 1.23 kg; Ajayi and Agaviezor, 2009); Tanzanian (1.95 and 1.35 kg; Goromela et al., 2009) and Central Mali (1.60 and 1.02 kg; Wison et al., 1987) cocks and hens. The egg weight was higher than 34.4 gm layed by Central Mali hens (Wison et al., 1987) while the clutch range of 7 to 12 was comparable to the mean clutch size of 12 reported by Tanzanian Village hens (Goromela et al., 2009) but ht mean clutch size was lower than 8.8 egg/hen layed by Malian hens (Wilson et al., 1987). This result showed that the Fulani chicken was heavier than other ecotypes in Nigeria. Results also showed that cocks were superior in weight to the hens in all body parameters thus demonstrating sexual dimorphism. This has been reported by Olawumi et al. (2008) and Gueye et al. (1998) for chickens. The heavier mature weight of the cocks was attributed to the ability of the males to secrete more quantity than females of sex hormones responsible for muscle development (Semakula et al., 2011).

This study showed low level of variability among Fulani Chickens and this seemed to be an adaptive feature to their natural environment. This suggested reasons for the slow progress in efforts to improve local chicken but this also directs efforts towards rapid improvement through crossbreeding with chicken possessing high level of variation in desired traits to bring about remarkable variations in their gene pool that will allow for selection. In this environment foundation selection favours Orolu cocks

Table 2. The coefficient of variation (CV) of body and productive parameters of Fulani ecotype chicken in three selected Local Government Areas of Osun State, Nigeria.

Parameters	COCKS				HENS			
	Total pop.	Orolu	Ire	Osogbo	Total pop.	Orolu	Ire	Osogbo
Wattle size	0.255	0.290	0.233	0.031	0.337	0.314	0.376	0.365
Comb size	0.160	0.198	0.000	0.031	0.323	0.334	0.319	0.300
Leg length	0.389	0.476	0.076	0.006	0.237	0.303	0.085	0.025
No of digits	0.054	0.068	0.000	0.000	0.038	0.000	0.086	0.000
Breast length	0.101	0.058	0.058	0.063	0.092	0.101	0.045	0.086
Breast width	0.060	0.036	0.032	0.030	0.106	0.090	0.143	0.084
Mature weight	0.120	0.135	0.135	0.059	0.129	0.134	0.153	0.094
Egg weight					0.314	0.360	0.242	0.198
Mean of eggs/clutch/hen					0.337	0.326	0.370	0.326
Egg clutch range					0.337	0.314	0.376	0.365

Total pop. means Total Population of Chicken by sex. OROLU, IRE and OSOGBO are the LGA of Settlements chosen for the survey.

Table 3. Pearson's correlation of body and production parameters in Fulani ecotype chickens on free-range management from selected Local Government Areas of Osun State, Nigeria.

	COCKS							
	Parameters	Egg clutch	Egg weight	Wattle size	Comb size	Breast length	Breast width	Mature weight
Hens	Egg weight	0.82**						
	Wattle size	0.22	0.30		0.91**	0.06	-0.12	0.89*
	Comb size	-0.03	0.08	0.32*		0.29	-0.32	0.83**
	Breast length	0.16	0.06	0.36*	-0.06		-0.81**	0.21
	Breast width	0.23	0.09	0.07	-0.16	0.38*		-0.13
	Mature weight	0.16	0.20	0.31*	0.20	0.44**	0.40**	

Note: * means $P \leq 0.05$; ** means $P \leq 0.01$.

and hens as starting genetic material for improvement of the Fulani chicken because of their attributes in leg length, egg weight, number of eggs/clutch and the clutch range. Orolu cocks and hens were also found to be taller than chickens fro other LGAs. This result thus substantiated the observation of Nwosu et al. (1985) that there was little variability in shank length among native chickens, although this could be an adaptation for ranging the environment. Those traits that exhibited lower standard deviation values such as number of digits, breast length, breast width and mature weight were thought to be least sensitive to the environment and were expected to be highly genetically influenced traits. Therefore, they could be useful for characterization purpose in Fulani chicken. The values of the CV obtained in this study were lower than 4.11 to 16.93% published by Okon et al. (1997) for various body parameters of Lohmann brown broilers in the humid tropics.

The result on correlation signified a high level of association between body traits at mature live weight. These traits could be used to predict one another or select for correlated body traits. The highly significant association obtained between egg clutch size and egg weight may indicate that improvement in egg size could result in high and stable clutch size. Also, an improvement in breast length (r = 0.44) and breast width (r = 0.40) could have positive influence on the mature weight of the Fulani hen. The cocks could be developed as a meat type equivalent of the exotic broiler. However, the traits correlated in this study were different from those reported by Essien and Adeyemi (1999) where different traits were correlated with body weight. The coefficients of mature weight with other traits in this work were lower (0.16 to 0.44) than that (0.25 to 0.59 at 12 weeks and 0.48 to 0.62 at 9 weeks) reported by Okon et al. (1997) where they correlated body weight with body girth, body

Table 4. Stepwise regression equations for prediction of breast length, breast width, comb size and mature weight in Fulani ecotype chickens on free-range management from three selected Local Government Areas of Osun State, Nigeria.

Dependent parameter	Sex type	Model selection step	Explanatory body measurement	Intercept (a)	Independent b – values	Standard error of model	R^2	% model significance (p)
Breast length, cm	Cocks	1	BRW	47.531	-1.829	6.126	0.659	0.0004
	Hens	1	MWT	11.865	3.826	2.935	0.195	0.0027
		2	LL	12.975	-0.181	-	0.195	-
			MWT		4.569	2.964	0.096	0.0009
Breast width, cm	Cocks	1	BRL	22.269	-0.360	1.571	0.659	0.0005
	Hens	1	MWT	9.516	3.204	2.794	0.158	0.0076
Comb size, cm	Cocks	1	WS	2.225	0.605	0.492	0.820	0.0001
		2	WS	0.595	0.267	-	0.820	-
			BRL		0.101	1.025	0.060	0.0001
	Hens	1	WS	1.045	0.407	0.476	0.101	0.0358
Mature weight, kg	Cocks	1	WS	1.328	0.198	0.178	0.786	0.0001
	Hens	1	BRL	0.552	0.051	0.293	0.195	0.0027
		2	LL	0.141	0.023	-	0.195	-
			BRL		0.059	0.422	0.188	0.0004
		3	LL	-0.133	0.023	-	0.195	-
			BRL		0.047	-	0.124	-
			BRW		0.034	0.357	0.064	0.0001

Note: BRW = Breast Width (cm); BRL = Breast Length (cm); MWT = Mature Weight (kg); WS = Wattle Size (cm); LL = Leg Length (cm); P = model significance; R^2 = Co-efficient of multiple determination.

length, keel length, shank length and shank width. The focus of this study also differs from that of Ojedapo et al. (2008) where they studied the phenotypic correlation between internal and external egg qualities of a commercial layer strain in which they reported low indices (0.01 to 0.48) within and between egg internal and external quality traits; but Oleforuh et al. (2008) obtained higher coefficients (0.02 to 0.83) within egg internal and external quality traits at p < 0.01. The result from this study was comparable to the correlation indices (0.15 to 0.77) obtained between body weight and different egg traits in a commercial layer reported by Ayorinde et al. (1988), although they were generally lower than the correlated coefficients reported by Ajayi and Agaviezor (2009).

The stepwise regression model revealed that cock parameters were predicted in Step 1 of the procedure with higher R^2 of 0.66, 0.66, 0.82 and 0.79 for breast length, breast width, comb size and mature weight, respectively. All predictive equations for hen parameters had low R^2 (0.29, 0.16, 0.38 and 0.10), while optimum equations were obtained at the Step 2 of the stepwise procedure for hen parameters, with very low constants (12.98, 9.52, 0.60 and 0.14), respectively.

These equations also indicated that the best predictors of mature weight were wattle size and leg length and breast length in cocks and hens. Breast width was best predicted by breast length in cocks and mature weight in hens. Breast length was optimally predicted by breast width in males, and jointly by leg length and mature weight in females. It was believed that biological components of the measured parameters have contributed to their predictability.

Similarly, the high model significance ($p < 0.05$) of the all equations indicated their adequacy for predictive purposes. The R^2 range obtained from the cock equations (0.66 to 0.82) were higher and closer compared with the wide range (0.16 to 0.95) obtained by Oni et al. (2001a) from parabolic – exponential, and gamma type functions of Wood and McNally (2001), while the R^2 from the hen population were lower than that of cocks but close in range (0.10 to 0.38, $p < 0.05$). This signified that the stepwise model fits the cock better than hen data (Oni et al., 2001b). The R^2 of 0.98 that was obtained by Essien and Adeyemi (1999) from the predictive equations of body weight on age in Lohmann brown and Annak broilers was higher than that obtained ($R^2 = 0.79$) from the regression of mature weight on wattle size in cocks in this study. The R^2 of 0.347 to 0.575 obtained from stepwise regressions by Okon et al. (1997) were intermediate between that of the cocks and the hens in this study. The R^2 from the hen population in this study were lower but closer in range than that (0.11 to 0.86) reported by Abdulrazaq et al. (2010).

Conclusion

The low level of variability observed among Fulani chicken population in Osun implied that a high level of adaptability to the natural environment existed within this ecotype. This study also revealed a lower level of variability within traits compared to exotic broiler strains probably due to high level of inbreeding within the local free-range population. The R^2 also indicated the adequacy of the stepwise model for predicting breast length, breast width, comb size and mature weight better in cocks than in hens. The higher R^2 and constant in male equations suggested a basic difference in the biological mechanism for growth between cocks and hens.

REFERENCES

Abdulrazaq OR, Joseph UI, Ibrahim DK (2010). Regression models for estimating breast, thigh and fat weight and yield of broilers from non invasive body measurements. Agriculture and Biology Journal of North America. ScienceHuβ, Retrieved on July 20, 2012 from http://www.scihub.org/ABJNA. ISSN Print: 2151-7517, ISSN Online: 2151-7525.

Ajayi FO, Agaviezor BO (2009). Phenotypic characteristics of indigenous chicken in Selected Local Government Areas in Bayelsa state, Nigeria. In: Proceeding of the third Nigerian International Poultry Summit. 22-26 February 2009. Abeokuta, Nigeria. Ola, SI; Fafiolu, AO and Fatufe, AA (eds), pp. 75-78.

Aphca News (2006). Bird flu advances in Nigeria. Retrieved in March2008 from http://www.aphca .org/news/news records/news2006/22Feb06 AL.html.

Ayorinde KL, Toye AA, Aruleba, TP (1988). Association between body weight and some egg traits in a strain of commercial layer. Nig. J. Anim. Prod. 15:119-125.

Essien AI, Adeyemi JA (1999). Comparative growth characteristics of two broiler strains raised in the wet humid tropics. Trop. J. Anim. Sci. 1(2):1-8.

Fayeye TR, Oketoyin AB (2006). Characterization of the Fulani-ecotype chicken for Thermoregulatory feather gene. Livestock Research for rural development. 18. Article #3 Retrieved in March 2010 from http://www.cipav.org.co.lrrd 18/3/faye18045.

Goromela EH, 'Kwakkel FP, 'Verstigen MWA, Katule, AM and 'Sendalo, DSC (2009). Feeding System and Management of Village poultry in Tanzania. In: Proceedings of the 3rd Nigeria International Poultry Summit. 22-26 February 2009. Abeokuta, Nigeria, pp. 193.198.

Gueye EF, Ndiaye A, Branckaert DR (1998). Prediction of Body weight on the basis of body measurements in mature indigenous chickens in Senegal. Livestock Research for Rural Development . 10(3). Retrieved on July 20, 2012 from http://www.lrrd.org/lrrd 10/3/sene103.htm.

Ibe SN (1993). Growth performance of normal, frizzle and naked neck chickens in a Tropical Environment. Nig. J. Anim. Prod. 20:25-31.

Latshaw JD, Bishop BL (2001). Estimating Body Weight and Body Composition of Chickens by Using Noninvasive Measurements. Poult. Sci. J. 80(7):868-873. Retrieved in July 2012 from http://ps.fass.org/content/80/7/868.full.pdf.

Mancha YP, Mbap ST, Abdul SD (2006). Phenotypic characterization of local Chickens in the Northern Region of Jos plateau. Trop. J. Anim. Sci. 9(12):47-55.

Nwosu CC, Gowen F, Obioha FC, Akpan IA, Onuora GI (1985). A Biometrical study of the Conformation of the native chicken. Nig. J. Anim. Prod. 12(2):141-146.

Ojedapo LO, Adedeji TA, Olayemi TB (2008). Phenotypic correlation between internal and external egg quality traits of a commercial layer strain. In: Proceedings of the 33rd Annual Conference of Nigerian Society for Animal Production, pp. 31-33.

Okon B, Ogar IB, Mgbere OO (1997). Interrelationship of live body measurements of broiler chickens in a humid tropical environment. Nig. J. Anim. Prod. 24(1):7-12.

Olawumi OO, Salako AE, Afuwape AA (2008). Morphometric differentiation and assessment of function of the Fulani and young ecotype indigenous chickens of Nigeria. Intl. J. Morphol. 26(4):975-980.

Oleforuh VU, Okoleh AI, Adeolu AI, Nwosu CC (2008). Egg quality traits of the Nigerian Light local chicken ecotypes. In: P croceedings of the 33rd Annual Conference of Nigerian Society for Animal Production, pp. 1-2.

Oni OO, Dim NI, Abubakah BY, Asiribo OE (2001a). Egg production curve of Rhode Island Red Chickens. Nig. J. Anim. Prod. 28(1):78-83.

Oni OO, Abubakah BY, Dim NI, Asiribo OE, Adeyinka IA (2001b). Predictive ability of egg production models. Nig. J. Anim. Prod. 28(1):84-88.

Peters SO, Omidiji EA, Ikeobi CON, Ozoje MO, Adebambo OA (2004). Effects of Naked neck and frizzled genes on egg traits, fertility and hatchability in local chicken. Proceeding. Annual conference of Animal Science Association of Nigeria (ASAN), September 13-16, 2004. Ebonyi State University, Abakaliki, pp. 262-264.

Semakula J. Lusembo P, Kugonza DR, Mutetikka D, Ssennyonjo J, Mwesigwa M (2011). Estimation of live body weight using zoometrical measurements for improved marketing of indigenous chicken in the Lake Victoria basin of Uganda. Livestock Research for Rural Development. 23(8). Retrieved on July 21, 2012 from http://www.lrrd.org/lrrd23/8/sema23170htm.

Sola-Ojo FE, Ayorinde KL (2009). Characterization of growth potential of the Fulani ecotype chicken. In: Proceedings of the Third Nigerian International Poultry Summit. 22-26 February 2009, Abeokuta,

Nigeria, pp. 79-83.

SAS/STAT (1999). Statistical Analytical Systems Computer software. Version 8 for windows. SAS Institute Incorporated, N C, USA.

Statistical Package for Social Sciences (SPSS). (2001). Version 10.0. SPSS Incorporated. Illinois. USA.

Wikimedia commons (2012). Retrieved in July 2012 from http://commons.wikimedia.org/wiki/file:squelette_oiseau.svg

Wilson RT, Traore A, Kuit HG, Slingerland M (1987). Livestock production in Central Mali: reproductive, growth and mortality of domestic fowl under traditional management. Tropical Animal health and production. 19(4):229-236. In: Animal breeding Abstract (1988). 56(1).

13

Production performance of local and exotic breeds of chicken at rural household level in Nole Kabba Woreda, Western Wollega, Ethiopia

Matiwos Habte[1], Negassi Ameha[2] and Solomon Demeke[3]

[1]Departement of Animal and Range Sciences, Dilla University, P.O. Box 419, Dilla Ethiopia.
[2]School of Animal and Range Sciences, Haramaya University, P. O. Box 138, Dire Dawa, Ethiopia.
[3]Departement of Animal and Range Sciences, Jimma University College of Agriculture, P.O. Box 307, Jimma Ethiopia.

Production performance of rural chickens was studied in three agro-ecologies of Nole Kabba Woreda of west Wollega zone, about 32, 29 and 31 from mid-highland, highland and lowland agro-ecological zones, respectively and a total of 92 households and 18 key informants were used for the survey work. Finally, all the data collected were subjected to Statistical Package for Social Science (SPSS) version 17.0 and SAS (2002). The results obtained revealed that, statistically significant difference between the agro-ecologies in sexual maturity of indigenous chickens as measured by age at first mate (P < 0.05). There was significant difference between local and exotic chickens in rate of egg production (P < 0.05). Mean age at first egg of about 7.02 ± 0.220 and 5.66 ± 0.116 months was calculated for indigenous and exotic pullets, respectively. Overall percent hatchability of 82 and 44.36% was reported from eggs of indigenous and exotic chicken, respectively. In summary the results of this study tends to indicate that improvement in exotic chicken eggs hatchability under scavenging condition seems to be the future direction of research in the Nole Kabba Woreda.

Key words: Chicken, indigenous, exotic, household, agro-ecology.

INTRODUCTION

The Ethiopian indigenous chickens are none descriptive breeds closely related to the Jungle fowl and vary in color, comb type, body conformation, weight and may or may not possess shank feather (Alemu and Taddele, 1997). Broodiness (maternal instinct) is pronounced, it is by natural brooding that baby chicks are raised all over rural Ethiopia. The broody hen rearing and protecting few chicks ceases laying during the entire incubation and brooding periods of up to 81 days. Yet the success of the brooding process depends on the maternal instinct of the broody hen and the prevalence of predators such as

birds of prey, pets and some wild animals, all of which are listed as the major causes of premature death of chicks in Ethiopia (Solomon, 2007a). They are characterized by slow growth, late maturity and low production performance. According to FAO (2004), indigenous chickens lay about 36 eggs in three clutches of 16 days each with about 12 to 13 eggs per clutch. Egg laying period and number of eggs laid per period are to some extent higher in urban than in rural areas (CACC, 2003).

The low productivity of the indigenous stock could also

partially be attributed to the low management standard of the traditional production system. It has been seen that the provision of vaccination, improved feeding, clean water and night time enclosure improves the production performance of the indigenous chickens, but not to an economically acceptable level (Solomon, 2007b). In Ethiopia, the idea of distributing exotic chickens particularly Rhode Island Red (RIR) was to improve the productivity of local birds by mating them with improved cocks. According to Permin (2008), this scheme usually failed to work due to the fact that the introduced breeds could not adapt to the hot climate, low feeding, and extensive management.

Furthermore, the improved cocks were not as lively and active under village conditions as the local cocks and therefore lost in the mating competition for the hens. When reproduction succeeded, the first generation of these cocks often showed a slight increase in production, but as no strict breeding scheme was maintained, the effect was gone after a few generations. The other important potential disadvantage was loss of broodiness, reduced scavenging capability and survival. Solomon (2003) showed that there was no difference between White Leghorn and local chickens raised under scavenging condition in mean daily body weight gain at 2 months.

A comparative study of the egg production performance of six different exotic breeds, namely: Brown Leghorn, White Leghorn, RIR, New Hampshire, Light Sussex, and Barred Rock was carried out at Debre Zeit Agricultural Research Centre. Egg production, hatchability and mortality data were collected and evaluated over several years. The White Leghorn was rated the best in terms of egg production, adaptability, disease resistance and efficiency (DZARC, 1984).

Solomon (2007a) noted that sexual maturity in White Leghorn under intensive and extensive management ranged from 149 to 169 days, while in RIR and Fayoumi crosses under intensive management ranged from 147 to 151 days (Rahman et al., 2004). Abraham and Yayneshet (2010) reported that hatchability more than 70% of the indigenous and White Leghorn eggs set were hatched.

The relatively higher proportion of eggs hatched by the indigenous birds may be attributed to a number of factors such as lighter egg weight, small clutch size, and the presence of higher mortality of indigenous chicks that forced the farmers to restock the lost birds (Tadelle et al., 2003). Only 39% of the eggs produced by RIR hens were hatched, probably due to a negative correlation between heavier egg weight and its hatchability (Yassin et al., 2008). Hatchability of eggs is a function of both maternal and paternal components, and the former has an overriding effect on genetic variation in hatchability of a fertile egg, which is attributed to the quality (external and internal) of the laid egg. Eggs stored for a longer period of time and collected from older age flocks are known to have lower hatchability (Yassin et al., 2008). Poultry development initiatives have been made in the Nole Kabba Woreda of west Wollega zone, Ethiopia, focusing on Hop-cock and RIR breeds. Still, along distribution of different breeds of poultry to rural household farmers in the Woreda, no attempts have been made to assess their production performances. The present work aimed to generate information on comparative production performance of indigenous and exotic chicken under traditional management system.

MATERIALS AND METHODS

Description of the study area

This study was conducted in Nole Kabba Woreda of western Wollega Zone of Oromia Regional State, located at 491 km west of Addis Ababa. The altitude of Nole Kabba ranges between 1400 and 2576 m.a.s.l. and the Woreda is predominantly classified as mid-highland (Woinadega). The mean annual temperature ranges from 13.5 to 27.5°C. The annual rainfall of the study area ranges between 1600 and 2000 mm. Nole Kabba has high potential for livestock production and the total chicken population of the Woreda is estimated at 42,075 heads.

Selection of study site and households

Nole Kabba Woreda was stratified into 3 agro-ecological zones based on altitude. One kebele from each agro-ecological zone were purposely selected based on poultry, and human population and total area coverage. A total of 92 households and 18 key informants were used to assess the production and productivity of local and exotic breeds of chicken. During household selections the household in the agro-ecology were randomly selected.

Data collection

Structured questionnaire was used to collect data from primary source which mainly comprised of households, development agents and key informants followed by review of the available secondary data source. A visit to physical facility of live bird and egg markets and open discussion with poultry farmers was also made. Finally, data on poultry production performance (egg production, number of clutches, age at first egg and hatchability) including the performance of the distributed exotic chickens were collected using the questionnaire prepared. Appropriate timing for data collection was fixed after negotiation with respondent, placing special emphasis on women, while interviewing the households.

Statistical analysis

All data collected were analyzed by using Statistical Package for Social Science (SPSS, 2009) version 17.0 for windows. Mean difference was assessed by Duncan's multiple range test, where F-values were significant (Duncan, 1955). Chi-square procedure was carried out to examine significance difference of some parameters. Analysis of variance (ANOVA) was carried out to examine variance of the data collected. For qualitative factors, descriptive statistics was used. Standard error of mean (SE) was used while describing mean.

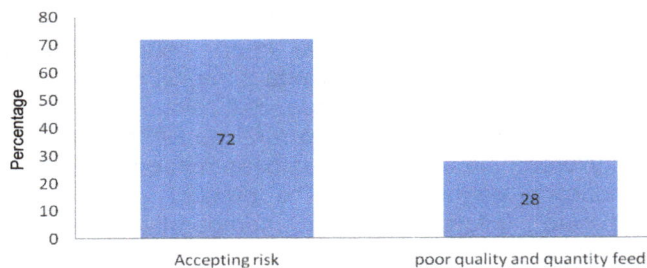

Figure 1. Reasons of poor exotic breed preference's (% HH).

RESULTS AND DISCUSSION

Production and reproduction performance

Egg production performances

The respondents were requested to rank the comparative production performance of the indigenous and exotic chickens kept in the study area. It has been found that the number of exotic chickens and their crosses are low in all the agro-ecologies studied as compared to the number of indigenous chickens attributed to low adaptability to the local conditions of the study areas. About 71.7% of all the respondents indicated that local chickens performed well than exotic chickens and their crosses in terms of survivability, disease résistance and alertness against predators. About 72 and 28% of all the respondents reported that risk of disease and predation and lack of supplementary feeding are the major limitations to the productivity of exotic chickens and their crosses under rural household conditions respectively (Figure 1). On the other side, the result of the discussion made with key informants indicated that all the respondents agree that the egg production performance of exotic chickens and their crosses is superior to the egg production performance of indigenous chicken under improved management system. This result is in agreement with that of Alemu and Tadelle (1997) who reported that indigenous flocks are considered to be very poor in egg production performance attributed to low genetic potential, poor management and long natural reproductive cycle.

All the respondents reported the existence of mating preference by both exotic and indigenous cocks which might have negatively affected fertility and hatchability of eggs collected from exotic chickens and their crosses (Table 1). The results of the discussion made with key informant showed that household members of the study area purchase breeding cocks of pure exotic or crossbred on the basis of size and body conformation. According to the respondents' responses, the farmers remove their local cocks before introducing pure exotic cocks. The respondents confirmed that the first crosses are good layers but need to have the second cock for back crossing to maintain broodiness. All the respondents said

to prefer pure and cross of RIR to local chickens in terms of egg production. However, pure RIR and their crosses are said to be susceptible to disease and predators. The pure RIR breeds of chicken are reported to be characterized by seasonal fluctuation in egg production and poor in alertness against predation and scavenging ability compared to indigenous chickens. There was statistically significant difference between the agro-ecologies in sexual maturity of indigenous chickens as measured by age at first mate ($P < 0.05$). Mean age at first mate of about 7.0 and 6.0 months was calculated for indigenous cock of mid-highland and lowland, respectively the values of which are significantly ($P < 0.05$) longer than that of highland (5.69 months). There was no statistically significant difference between the agro-ecologies in sexual maturity of indigenous chickens as measured by age at first egg ($P > 0.05$). However, mean age at first egg of about 6.73 and 6.08 months was calculated for indigenous pullets of mid-highland and highland respectively, the values of which are significantly ($P < 0.05$) longer than that of crossbreds (5.66 months). The results of this study indicated that the mean age at first egg of indigenous pullets of mid-altitude (6.73 months) is longer than that of low and highlands indicating that pullets of mid-altitude are characterized by late sexual maturity compared to the pullets of low and high lands. This variation might be associated with the variation in feed availability. The mean age at first egg recorded from this study is shorter than those recorded by Udo et al. (2001) and Tadelle et al. (2003) who reported average age at first lay of 8 and 6.8 months, respectively.

As indicated in Table 2, there was significant difference between local and exotic chickens in rate of egg production ($P < 0.05$). It is revealed that the numbers of eggs produced/clutch/hen was higher for the crossbreds (31.7) as compared to that of indigenous chickens (11.23).

Breaking broodiness

The traditional methods of breaking broodiness are shown in Table 3. All the respondents reported to exercise traditional methods of breaking broodiness aimed at increasing egg productivity. About 30.4% of the respondents were reported to exercise disturbing of the broody hen in the laying nest including replacing of eggs with some other foreign materials. About 29.3% of the respondents tie both legs of broody hen and suspend it from branches of trees in upside down position for 3 to 4 days. About 16, 15 and 9% of the respondents exercise piercing shank feather into nostril of the broody hen, hanging of broody hen upside down position and taking broody hen to neighborhoods for 3 to 4 days, respectively. According to the results of the discussion made with key informants, disturbing the laying nest (Place some material on egg laying place), and tying the legs of broody hen and hanging of the hen upside down

Table 1. Mating preference toward exotic breed of chickens (% of HH).

Preferred sex	Mid-highland	Highland	Lowland	Overall
Exotic cocks towards local hen				
Good	46.9	48.3	54.8	50.0
Poor	21.9	20.7	32.3	25.0
No preference	31.3	31.0	12.9	25.0
Local cocks towards exotic hens				
Good	43.8	41.4	29.0	38.0
Poor	21.9	24.1	38.7	28.3
No preference	34.4	34.5	32.3	33.7

Table 2. Age of sexual maturity (months) of crosses and local chickens and egg produced/clutch/hen of local, crosses and exotic chickens.

Parameter	Mid-highland	Highland	Lowland	Overall
Age of sexual maturity				
Female(local)	7.44 ± 0.391	6.76 ± 0.396	6.82 ± 0.353	7.02 ± 0.220
Male(Local)	7 ± 0.37[a]	5.69 ± 0.290[b]	6 ± 0.278[b]	6.25 ± 0.191
Crosses (Female)	5.77 ± 0.178	5.79 ± 0.250	5.38 ± 0.145	5.66 ± 0.116
overall	6.73 ± 0.21	6.08 ± 0.189	6.11 ± 0.175	6.32 ± 0.111
Egg produced/clutch/hen				
Locals	11 ± 0.51	11.17 ± 0.653	11.52 ± 0.601	11.23 ± 0.336
Exotics	25.66 ± 0.718	26.14 ± 1.505	27.2 ± 1.935	26.14 ± 0.710
Crosses	32.44 ± 1.142[ab]	28.41 ± 1.783[b]	35.23 ± 1.942[a]	31.66 ± 0.923
Overall	23.03 ± 1.034	20.88 ± 1.191	20.73 ± 1.5	21.75 ± 0.697

[ab]Means in the same row for each parameter with different letter superscripts are significantly different (P < 0.05).

Table 3. Practices of breaking broodiness in the study area (% of HH).

Practice to break broodiness	Mid-highland	Highland	Lowland	Overall
Tying	25.0(8[NS])	34.5(10[NS])	29.0(9[NS])	29.3(27*)
Piercing feather in the nose	21.9(7)	13.8(4)	12.9(4)	16.3(15)
Place some material on egg laying place	31.3(10)	27.6(8)	32.3(10)	30.4(28)
Hanging upside down	15.6(5)	13.8(4)	16.1(5)	15.2(14)
Taking into neighborhoods	6.3(2)	10.3(3)	9.7(3)	8.7(8)
X^2_Value	5.8125	6.3448	6.2581	16.587

*P < 0.05 level; NS-Not-significant across the column; Value in the Parenthesis are the numbers of respondent responded in each parameters.

position could break broodiness within 3 to 4 days depending on the degree of strength of broodiness which vary from hen to hen. Similarly, Dereje (2001), Tadelle (2003) and Mammo (2006) reported that piercing the nostril with feather, moving the bird to a nearby house for a couple of days and hanging upside down are effective in breaking broodiness within 3 to 4 days. All the respondents confirmed that the hens resume laying soon after breaking broodiness resulting in increase in total annual egg production. This result is in agreement with that of Rushton (1996) as cited by Kitalyi (1998) who reported higher egg productivity (143 eggs/hen/year) by the Ethiopia indigenous chickens with the proper management of broody hen. These results also agree with that of Tadelle (1996), Dereje (2001), Tadelle et al. (2003) and Resource-Center (2005) who reported that

Table 4. Egg selection, incubation and season of hatching in the study area (% of HH).

Parameter	Mid-highland	Highland	Lowland	Overall
Season of hatching				
Dry season only	82.8	72	86.2	80.7
Both season	17.2	28	13.8	19.3
Selecting eggs for incubation				
Selecting	55.2	40	48.3	48.2
No selecting	44.8	60	51.7	51.8
Size of eggs selected for incubation				
Large size	68.8	80	64.3	70
Medium size	31.3	20	35.7	30
Source of incubating eggs				
Lay at home	62.1	68	65.5	65.1
Both lay at home and purchase	37.9	32	34.5	34.9

Table 5. Storage place of table and hatching eggs (% of HH).

Storage places	Mid-highland	Highland	Lowland	Overall
Inside laying nest	37.5	27.6	19.4	28.3
Clay pot	21.9	24.1	35.5	27.2
Inside teff grain	25	31	29	28.2
Any place	15.6	17.2	16.1	16.3

households traditionally attempt to break broodiness to resume egg laying with final goal of increasing egg productivity.

Hatchability

Hatchability and rate of chick survival are one of the major determinant factors of productivity in poultry. The results of the hatchability and related factors obtained in this study are shown in Table 4. All the respondents said that they commonly incubate eggs during dry seasons and use " hammattu" (clay pot with straw bedding) as an incubation box. This result is in agreement with that of Solomon (2007a), who reported that it is by natural incubation and brooding that chicks are hatched and raised all over the rural Ethiopia. A broody hen hatching, rearing and protecting few number of chicks (6 to 8) ceases egg laying during the entire incubation and brooding periods of 81 days. Yet the successes of the hatching and brooding process depends on the maternal instinct of the broody hen and prevalence of predators in the area, such as birds of prey, pets and some wild animals, all of which are listed as the major causes of premature death of chicks in Ethiopia. This result is also in line with the report of Kyvsgaard et al. (2002), who

found that most women preferred hatching in dry seasons but disagreed with that of Maphosa et al. (2004) who noted that, there was no seasonal effect on eggs hatchability.

According to the results of the discussions made with key informants, the number of eggs set per hen depends on availability of eggs, size of eggs and size of broody hen and the maternal instinct of the broody hen. The overall mean number of eggs incubated in the study area was reported to be 11.32 eggs with minimum of 6 and maximum 20 eggs per hen, the value of which agrees with Sonaiya and Swan (2004), Udo et al. (2001) and Tadelle et al. (2003) in Ethiopia who indicated that the average number of eggs set per hen is about maximum 16, 14 and 13, respectively.

About 55.4% of the respondents reported to store both hatching and market eggs either in clay pot or inside teff grain for up to 10 days (Table 5) without considering storage position. All respondents believe that temperature of clay pot and teff grain is not detrimental in terms of hatchability and nutritional characteristics. In agreement with this results Sonaiya and Swan (2004) reported safe egg storage in clay pot, while Dereje (2001) and Tadelle et al. (2003) indicated the practice of egg storage for about 2 weeks in the grain-store (especially in teff), without considering egg storage positions.

Table 6. Methods of identifying normal eggs from spoiled (% of HH).

Method	Mid-highland	Highland	Lowland	Overall
By shaking	46.9	55.2	41.9	47.8
Floating techniques	31.3	17.2	25.8	25.0
Visual examination	21.9	27.6	32.3	27.2

Table 7. Hatchability of eggs in the study area (% of HH).

Parameter	Mid-highland	Highland	Lowland	Overall mean
Indigenous				
No. of eggs/incubation	11.28 ± 0.409	11.2 ± 0.574	11.45 ± 0.432	11.32 ± 0.267
No. of chicks hatched/incubator	9.31 ± 0.330	8.88 ± 0.578	9.81± 0.339	9.36 ± 0.239
Hatchability %age (%)	82.57	79.28	85.65	82.74[a]
Exotics				
No. of eggs/incubation	9.54 ± 0.595	9.47 ± 0.515	9.6 ± 0.400	9.52 ± 0.243
No. of chicks hatched/incubator	4 ± 0.519[ab]	4.88 ± 0.392[a]	3.25 ± 0.403[b]	4.22 ± 0.224
Hatchability percentage (%)	41.935[ab]	51.55[a]	33.85[b]	44.36[b]
Age of eggs used for incubation (Day)	10.17 ± 0.643	10.56 ± 0.653	11.07 ± 0.438	10.34 ± 0.365

[ab]Means in the same row for each parameter with different letter superscripts are significantly different (P < 0.05).

The respondents reported to select hatching eggs on the basis of size and shell structure. Large and medium sized egg with smooth shell are said to be the preferred ones for incubation. This practice of selection is in agreement with the report of Dereje (2001) who stated that large and medium sized eggs are selected for incubation in Ethiopia. There was no reported practice of cleaning and treating hatching eggs. In contrast to the results of this study, Mammo (2006) reported that externally dirty and contaminated eggs are cleaned using dry materials (cloth). According to Sonaiya (2004), rubbing slightly the dirty eggs with a rough cloth is better than wet cleaning.

As shown in Table 4, about 65.1% of the respondents use home laid eggs for incubation, while about 34.9% of the respondents reported to have used eggs purchased from neighborhoods for incubation. About 47.8% of the respondents are capable of identifying spoiled eggs by shaking, visual appraisal and floating in water (Table 6). It is indicated that broody-hen sitting on hatching eggs is placed in hidden and protected areas characterized by minimum disturbances. About 94, 91 and 86% of the respective respondents of lowland, mid-highland and highland, said to have regularly practice incubation and hatching.

Hatching on regular basis indicating that there was no statistically significant difference (P < 0.05) between the kebeles (agro-ecologies) in the frequency of incubation and hatching, even though respondents categorized as poor households tended to frequently practiced incubation and hatching. Overall mean percent hatchability of 82%

was reported from eggs of indigenous chicken and there was no statistically significant difference (P > 0.05) between the study sites agro-ecologies in hatchability of eggs collected from indigenous chickens. The highest percent hatchability (86%) was reported from eggs of indigenous chickens of lowland whereas the lowest percent hatchability of 79% was reported from eggs of indigenous chickens of highland (Table 7). The results of hatchability reported from eggs of indigenous chickens in this study is comparable to those reported from different parts of Ethiopia, with the exception of that of Jimma (Brännänng and Pearson, 1990; Tadelle, 1996).

Meseret (2010) reported mean percent total hatchability of 22% from eggs of indigenous chickens of Gomma Woreda of Jimma zone, indicating that hatchability is one of the detrimental factors limiting poultry production in Gomma Wereda.

Overall mean hatchability of 44% was reported from eggs of exotic chickens as reported by all the respondents of the three sites. There was significant difference between the study sites (agro-ecologies) in hatchability of eggs collected from exotic chickens (P < 0.05). The highest hatchability of 52% was reported from eggs of exotic chickens of highland, while the lowest hatchability of 33% was reported from eggs of exotic chickens of lowland.

The results of this study clearly showed that hatchability of exotic chickens was significantly lower than that of the indigenous chickens (P < 0.05). The results of hatchability obtained in this study seem to agree with that of Sonaiya and Swan (2004) who reported that hatchability using broody hen is around 80% to be normal, but a range of 75 to 80% is considered to be satisfactory. In agreement to the result of this study, Abraham and Yayneshet (2010) revealed that 76 and 39% of hatchability of egg collected from indigenous and RIR, respectively in the semi-arid Tigrayi region of Northern Ethiopia.

Conclusion

The result of the current study revealed that, mean age of sexual maturity for local stock is reported to be longer (7 months) than that of crossbred (6 months) and exotic chickens (5 months). Numbers of eggs per clutch per hen per year for indigenous, exotics and crossbreds were; 11, 26 and 31 eggs, respectively. All the respondents practice natural incubation mainly in dry seasons and reported to attain hatchability of about 83 and 44% through the incubation of indigenous and exotic eggs, respectively.

All the respondents said to prefer pure and cross of RIR to local chickens in terms of egg production. The results of this study indicated that the mean age at first egg of indigenous pullets of mid-altitude (6.73 months) is longer than that of low and highlands indicating that pullets of mid-altitude are characterized by late sexual maturity compared to the pullets of low and high lands. All the respondents reported to exercise traditional methods of breaking broodiness aimed at increasing egg productivity. There was significant difference between the study sites (agro-ecologies) in hatchability of eggs collected from exotic chickens (P < 0.05). The result of this study tends to indicate that production performance local chicken is low, not only because of poor egg production performance but also due to long hatching and brooding periods. Therefore, the use of hay box brooder was found to be effective in reduction of mortality and releasing the broody hen to go back to laying. Popularization of the technology within the farming population including the provision of constructional and operational manual in local language seems to be desirable. The critical analysis of poor hatchability of eggs collected from RIR is the future line of work.

REFERENCES

Abraham L, Yayneshet T (2010). Performance of exotic and indigenous poultry breeds managed by smallholder farmers in northern Ethiopia. Livestock Research for Rural Development. 22, Article #133. Retrieved February 6, 2011, from http://www.lrrd.org/lrrd22/7/leml22133.htm

Alemu Y, Tadelle D (1997). The Status of Poultry Research and Development in Ethiopia, In: Fifth National Conference of Ethiopian Society of Animal Production (ESAP), 15-17 May 1997, Addis Ababa Ethiopia. pp. 40-60.

Brännänng E, Person S (1990). Ethiopian Animal Husbandry Uppsala, University of Sweden, 127 p central high land of Ethiopia M.S.C. Thesis Swedish University of agriculture.

CACC (Central Agricultural Census Commission), 2003. Statistical report on farm management practices, livestock and farm managments Central Statistical Authority report of 2004-2005, II, Addis Ababa, Ethiopia.

Debre Zeit Agricultural Research Center (DZARC) (1984). Debre Zeit Agricultural Research Center, Annual Report 1984. Debre Ziet, Ethiopia.

Dereje D (2001).The effect of some common methods of storage and duration on egg quality and hatchability in East Wolegga, Ethiopia. M. Sc. Thesis Submitted to School of Graduate Studies Alemaya University. P. 83.

FAO (Food and Agriculture Organization) (2004). Small- Scale Poultry Production: Animal Production and Health technical shed, No. 1, FAO. Rome Italy.

Kitalyi A (1998). Village chicken production systems in rural Africa. Households food and gender issues. Food and Agriculture Organization of the United Nations: Rome Italy. P. 81.

Kyvsgaard, Niels C, Luz Adilial U, Peter N (2002). Analysis of traditional grain and scavenge based poultry systems. Poultry as a tool of in poverty eradication and promotion of gender equality: proceeding workshop Nicaragua. (mailto:nck@kvl.dk).

Mammo M (2006). Survey on village chicken production under traditional management systems in Jamma woreda, south Wollo, Ethiopia. M.Sc. Thesis Presented to School of Graduate Studies of Alemaya University, Ethiopia.

Maphosa TJ, Kusina NT, Makuza S, Sibanda S (2004). A monitoring study comparing production of village chickens between Communal (Nharira) and small-scale commercial (Lancashire) farming areas in Zimbabwe. University of Zimbabwe. Livestock Research for Rural Developments16/7/2004. (mkhonto@avu.org).

Meseret M (2010). Characterization of village chicken production and marketing system in gomma wereda, jimma zone, Ethiopia. M. Sc. Thesis Presented to School of Graduate Studies of Jimma University, Ethiopia.

Permin A (2008). Good practices in small scale poultry production: a manual for trainers and producers in east Africa. A consultancy report to FAO, Addis Ababa, Ethiopia.

Rahman MM, Baqui MA, Howlider MAR (2004). Egg production performance of RIR x Fayoumi and Fayoumi x RIR crossbreed chicken under intensive management in Bangladesh. Livestock Research for Rural Development 16(11):92. Retrieved May 23, 2008, from http://www.lrrd.org/lrrd16/11/rahm16092.htm.

Resource-Center (2005). Improved management of indigenous chicken. Kenya Agricultural Institute. Kenya, (resource center@Kari.org).

Rushton J (1996). Emergency assistance to Newcastle Disease control in Zimbabwe, consultant report, projects TCP/ZIMB/4553. Rome, FAO.

Solomon D (2003b). Growth Performance and Survival of Local and White Leg Horne chicken under scavenging and intensive System of management in Ethiopia. Jimma College of Agriculture. Jimma Ethiopia.

Solomon D (2007a). Suitability of hay box brooding technology to the rural household poultry production system. International Journal for Research into Sustainable Developing World Agriculture. CIPAV, Cali, Colombia.

Solomon D (2007b). Poultry sector country review. HPAI prevention and control strategies in Eastern Africa, The structure, marketing and importance of the commercial and village poultry industry: An analysis of the poultry sector in Ethiopia. FAO animal production and health division.

Sonaiya EB (2004). Direct assessment of nutrient resources in free-range and scavenging systems. World's Poult. Sci. J. 60(4):523-535.

Sonaiya EB, Swan ESJ (2004). Small scale poultry production technical guide. Animal Production and Health, FAO of United Nations. Rome Italy, 2004. P. 114.

SPSS (Statistical Package for Social Science, 2009). 17.0 versions. SPSS user's guide.

Tadelle D (1996). Studies on village poultry production systems in the central highlands of Ethiopia. M.Sc. Thesis, Swedish University of Agricultural Sciences. Uppsala, Sweden.

Tadelle D (2003). Phenotypic and genetic characterization of local ecotypes in Ethiopia. PhD Thesis, Humboldt University of Berlin, Germany.

Tadelle D, Million T, Alemu Y, Peters KJ (2003). Village chicken production system in Ethiopia: Paper 1. Flock characteristics and Performances. Livest. Res. Rural Dev. 15(1):4-8. http://www.cipav.org.co/irrd/irrd15/1/tadaa 151.htm.

Udo HMJ, Asgedom AH, Viets TC (2001). Modeling the impact of intervention in village poultry productions. Livestock Community and Environment. Proceeding of the 10th Conference of the Association of Institution for Tropical Veterinary Medicine Copenhagen, Denmark. Mekele University College, Ethiopia. henk.udo@dpsvh.wau.nl.

Yassin H, Velthuis AGJ, Boerjan M, van Riel J, Huirne RBM (2008). Production, modeling, and education: Field study on broiler eggs hatchability. Poult. Sci. 87:2408-2417 http://ps.fass.org/cgi/reprint/87/11/2408.

14

Influence of physiological stage and parity on energy, nitrogen and mineral metabolism parameters in the Ouled Djellal sheep in the Algerian Southeast arid area

[1]   [2]   [2]   [3]
DEGHNOUCHE K. , TLIDJANE M. , MEZIANE T. and TOUABTI A.
[1]Department of Agronomy, Mohamed Kheider University, Biskra 07000, Algeria.
[2]Department of Veterinary, Hadj Lakhder University, Batna 05000, Algeria.
[3]Biochemistry Laboratory, CHU Setif 19000, Algeria.

Investigations were conducted to determine the influence of physiological stage and parity on some indicators of energy, nitrogen and mineral metabolisms in sheep Ouled Djellal living in the Algerian Southeast arid area. The study was carried out on 100 clinically healthy multiparous and primiparous ewes, aged 2 to 7 years. The animals were divided into three ewe groups: Pregnant (P), lactating (L), and empty (E). Biochemical analysis of blood samples concerned the determination of the values of 11 metabolites [glucose, cholesterol, triglycerides, urea, total protein, albumin, calcium (Ca), phosphates (PO_4), sodium (Na), potassium (K), and magnesium (Mg)]. The results showed that pregnant ewes and multiparous ones have the lowest blood glucose levels and the highest proteinemia. However, the lowest total protein and albumin values were found in empty sheep. Cholesterol and triglyceride levels were the highest in lactating and primiparous ewes. In this study, the most important calcium levels were recorded in lactating ewes which showed the lowest magnesium levels. The pregnant ewes had the highest sodium levels and low phosphatemia. Potassium levels were comparable in all animals. Statistical analysis showed that the physiological stage has a significant influence ($p < 0.05$) on serum glucose, triglycerides, phosphorus and potassium levels and a highly significant effect ($p < 0.001$) on urea, and magnesium levels. No parity effect has been observed in our study.

Key words: Ouled Djellal ewes, physiological stage, parity, energy, nitrogen, mineral metabolisms.

INTRODUCTION

Gestation and lactation are the two most critical periods in sheep feeding. The energy requirements of pregnant ewes increase significantly towards the end of gestation, during which 70 to 80% of foetal growth occurs (Sormunen-Cristian and Jauhiainen, 2001). Similarly, during the first 2 weeks of lactation, the export of nitrogen and energy in milk is very high and the animals can not ensure it without the mobilization of their reserves (Tissier and Thériez, 1978). Ewes should be in good health during and after pregnancy so as to produce viable lambs.

The identification of metabolism change of such sheep in various reproduction phases, the determination of metabolic blood profiles, including serum mineral and biochemical parameters is necessary to study ruminant metabolism disorders and can provide useful information on the animal nutritional status (Balikci et al., 2007). Correspondingly, mineral content and biochemical indicators in the blood of sheep are widely used (Sykes and Field, 1974; Hajdarevié et al., 1989).

A significant number of authors describe the mineral and biochemical indicators in the blood of the sheep (Pastrana et al., 1991a, b; Shinde et al., 1995; Klinkon and Zadnik, 1997) reported by Antunovié et al. (2004). But, only a very small number of authors report about the influence of the reproductive status (Baumgartner and Penthaner, 1994; Ramos et al., 1994; EL-Sherif and Assad, 2001; Antunovié et al., 2002, 2004) and the parity as the important prerequisites for the biochemical indicators' interpretation in sheep. In order to establish the metabolic profile of the sheep, it is necessary to know the influence of reproductive status as well as the parity. In Ouled Djellal breed (main breed of sheep in Southeastern Algeria), information about the determination of normal biochemical and mineral values remain insufficient; therefore, the aim of this research is to show the changes of the serum mineral and biochemical indicators in the blood of ewes depending on reproductive status and the parity.

MATERIALS AND METHODS

Choice of farms and animals

Investigations were carried out on farms located in the agricultural region of El Doucen, in the Algerian Southeast arid area. This region is characterized by a dry climate, low rainfall, an average summer temperature of 42°C and dry pastures.

The choice of these farms was made on the basis of availability of information on studied animals, and the presence of a large number of sheep.

The study involved 100 multiparous and primiparous ewes of Ouled Djelal breed; animals were divided into three ewe groups: P (pregnant n = 34), L (lactating n = 33) and E (empty n = 33).

Sampling and analysis of blood samples

Blood samples were taken from the jugular vein into dry and heparinized tubes; the operation took place in the morning (7:00 am) before food intake. Analysis concerned biological constants (glucose, cholesterol, triglycerides, urea, total protein, and albumin) as well as, [calcium (Ca), phosphates (PO_4), sodium (Na), potassium (K), and magnesium (Mg)]. These analysis were performed by spectrophotometer (BOEHINGER5010).

Statistical analysis

Data statistical analysis was performed using Epi Info software (version 6.04, 2003)

RESULTS

The lowest blood glucose values are in pregnant ewes, and in multiparous ones, the study showed statistical significant differences (P < 0.05) between ewes:

(pregnant versus empty), and (lactating versus empty), however, these values remain within the range of international standards cited by most authors (Nelson and Guss, 1992; Radostits et al., 2000; Brugere-Picoux, 2002; Dubreuil et al., 2005) Table 1.

Cholesterol levels for the various batches are relatively lower than the standards cited by Brugere-Picoux (2002), and Ndoutamia and Ganda (2005).

However, the highest cholesterol levels were observed in lactating and primiparous sheep, those levels still within the limits of values reported by these authors.

The statistical study revealed no significant differences between different batches. Our animals have relatively lower triglyceride values than those described by Ndoutamia and Ganda (2005); however, they remain within the range of standard values referenced by Mollereau et al. (1995). The statistical study showed a significant difference (p < 0.05) between ewes: (lactating versus empty) for triglycerides.

Uremia, proteinemia and albuminia rated among the various groups are within the standards described by the authors.

The comparison of urea mean levels showed highly significant differences (p < 0.0001) between ewes: (pregnant versus empty); (lactating versus empty) and also between the three batches: (pregnant versus lactating versus empty).

But the difference is not significant for total protein and albumin rates. Furthermore, the highest proteinemia levels were observed in pregnant and multiparous ewes, and the lowest among empty sheep.

Calcium and sodium levels obtained in our study were below the physiological standards cited by most authors (Jelinek et al,, 1996; Brugere-Picoux, 2002; Dubreuil et al., 2005), however, they remain within the range of values described by Baumgartner and Pernthaner (1994) for the calcium. The statistical study showed significant differences (p < 0.05) between ewes (pregnant versus lactating) Table 2.

The phosphatemia found in pregnant and in multiparous ewes were below the standards reported by most authors (Brugere-Picoux, 2002; Dubreuil et al., 2005).

Serum magnesium concentrations correspond to the standards reported in the literature, and the values recorded for potassium are within the range of international standards described by Brugere-Picoux (2002), but they are below the values reported by Jelinek et al. (1996), and Dubreuil et al. (2005).

In addition, the comparison of mean levels of serum phosphorus and potassium revealed significant differences (p < 0.05) between ewes (pregnant versus lactating), also highly significant differences (p < 0.01) were found for magnesium levels between ewes (lactating versus empty). In this study, no significant difference was found between the multiparous and primiparous ewes.

Table 1. Biochemical indicators of ewes depending on the reproductive status and parity.

Biochemical indicators (g/L)	(1) + (2)	Pregnant (n = 34)	Lactating (n = 33)	Empty (n = 33)	Multiparous (n = 51)	Primiparous (n = 49)
Glucose	0.42 - 0.76 (1)	0.39[b]* ± 0.19	0.41[c]* ± 0.18*	0,47* ± 0.10	0,39[ns]* ± 0.17	0.43* ± 0.20
Cholesterol	0.52 - 0.76 (1)	0.49[ns]* ± 0.19	0.51* ± 0.21	0,48[ns]* ± 0.06	0,48[ns]* ± 0.19	0.56* ± 0.23
Triglycerides	0.50 ± 0.19 (2)	0.34* ± 0.29	0.35[c]** ± 0.20	0,27* ± 0.12	0,34[ns]* ± 0.23	0.35* ± 0.30
Urea	0.20 - 0.30 (1)	0.29[b,e]*** ± 0.09	0.32[c]**** ± 0.14	0,19* ± 0.05	0,31[ns]* ± 0.11	0.29* ± 0.13
Total protein	60 - 79 (1)	67.17* ± 74.02	64,19* ± 15.40	58,80* ± 5.21	67,48[ns]* ± 60.47	59.78* ± 24.03
Albumin	24 - 30 (1)	25.65* ± 12.72	24,54* ± 4.47	23.13* ± 3.50	25,33[ns]* ± 10.64	24.56* ± 6.90

(1)Brugere -Picoux (2002); (2) Ndoutamia and Ganda (2005). *P < 0.05; ** P < 0.01; *** P < 0.001. [a], Différences (pregnant versus lactating); [b], différences (pregnant versus empty); [c], différences (lactating versus empty); [d], différences (multiparous versus primiparous); [e], différences (pregnant versus lactating versus empty).

Table 2. Mineral indicators of ewes depending on the reproductive status and parity.

Mineral indicators	(1)+(2)	Pregnant (n = 34)	Lactating (n = 33)	Empty (n = 33)	Multiparous (n = 51)	Primiparous (n = 48)
Ca (mg/L)	80 - 100 (1)	83.29[a]** ± 14.32	91.2* ± 11.58	69.60* ± 4.27	87.15* ± 14.43	86.94* ± 10.95
PO$_4$ (mg/L)	50 - 73 (2)	45.87[a]** ± 19.13	54.41* ± 19.59	55.86[b]* ± 11.71*	48.59* ± 17.61	54.52* ± 25.79
Na (mEq/L)	145 (2)	135.2[a]** ± 7.32	130.6* ± 10.75	132.73* ± 10.14	132.54* ± 9.77	134.76* ± 7.66
K (mEq/L)	4.5 (4-5) (2)	4.17[a]** ± 0.65	4.46* ± 0.47	4.26* ± 0.74	4..34* ± 0.57	4.19* ± 0.66
Mg (mg/L)	17 - 29 (1)	22.77[a]*** ± 4.16	18.95* ± 5.85	22.55* ± 1.95	20.42* ± 5.56	22.72* ± 4.29

(1) Baumgartner and Pernthaner (1994); (2) Brugere-Picoux (2002).

DISCUSSION

Glucose level reported in sheep is between 35 and 45 mg/dl (Nelson and Guss, 1992) and could be influenced by the physiological stage (Firat and Ozpinar, 1996) and diseases (Symonds et al., 1986; Ford et al., 1990). Our results highlight a significant influence (p < 0.01) of physiological stage on blood sugar which is consistent with the observations of Hamadeh et al. (1996) who concluded that it has lower values in pregnant ewes compared with lactating or empty ones, however, Firat and Ozpinar (1996) did not mention any significant difference in blood glucose during pregnancy or during lactation; this observation is supported also by Radostits et al. (2000), who reported lower values than those reported by Shetaewi and Daghash (1994).

In case of pregnancy toxemia, blood glucose levels are lower than 20 mg/dl (Nelson and Guss, 1992). Antunović et al. (2004) noted high blood glucose in empty females compared to pregnant ones. This was also reported in cows (Otto et al., 2000), and in Sahal's goats studied by Sandabe et al. (2004).

The decrease in blood sugar during pregnancy can be explained by the increase of maternal glucose permeability and use by the foetus (Tontis and Zwahlen, 1987; Sahlu et al., 1995).

Our results are inconsistent with those of Balikci et al. (2007) who reported a gradual increase (p < 0.05) of

cholesterol levels during pregnancy compared with values obtained at 45th days postpartum. Hamadeh et al. (1996) and Al-Dewachi (1999) pointed a high cholesterol levels in pregnant ewes compared to empty ones. This observation is supported by other studies that have reported high cholesterol levels [high density lipoprotein (HDL)-cholesterol and very-low-density lipoprotein (VLDL)-cholesterol] in the end of gestation (Krajnicakova et al., 1993; Hamadeh et al., 1996; Nazifi et al., 2002).

Also, no significant difference in serum cholesterol has been reported between pregnant ewes and empty ones (Ozpinar and Firat, 2003; Tanaka et al., 2007).

Hamadeh et al. (1996) noted that ewes giving birth to two lambs presented higher cholesterol levels than the ones with a single lamb, the same result is reported by Balkici et al. (2007) at 100 and 150th days of gestation.

Antunovie et al. (2002) spoke about insignificant increase in plasma cholesterol in non-pregnant females compared with lactating ones; in a later study, they reported a higher cholesterol levels in lactating females than empty ones (Antunovie et al., 2004) which is consistent with our results.

The increase in triglyceride levels among lactating females may be due to insulin, which plays a direct role in adipose tissue metabolism during pregnancy and its responsiveness is significantly reduced in ewes during late pregnancy (Jainudee and Hafez, 1994; Schlumbohm et al., 1997). The diminished responsiveness of the target tissue to insulin during late pregnancy predisposes the ewes to increase of cholesterol, triglyceride and lipoproteins concentrations (Schlumbohm et al., 1997).

Furthermore, Krajnicakova et al. (1993), Hamadeh et al. (1996), and Nazifi et al. (2002) have reported high levels of triglycerides during late gestation; a similar result is underscored by Balikci et al. (2007) who noted a significant increase (p < 0.05) of triglyceride levels during pregnancy compared to 45th days post partum, On 100 and 150th days of gestation, these rates were higher among ewes who had two lambs, on the other hand, Tanaka et al. (2007) found no significant difference of serum triglycerides during lactation or dry period.

West (1995) recorded higher uraemia in pregnant ewes than in lactating or empty ones; other authors have found no effect of pregnancy on uremia (Scott and Robinson, 1976; Brozostowski et al, 1996; Meziane 2001).

Furthermore, Antunovié et al. (2002) reported high serum concentrations of urea during the last trimester of gestation and during lactation. In a subsequent study, they described significant differences of uremia between empty and pregnant females (Antunovié et al., 2004).

Our results are supported by Antunovié et al. (2002) who reported high serum concentrations of total protein during the last trimester of gestation and during lactation the same result is also emphasized by El-Sherif and Assad (2001) and Meziane (2001), who described a significant increase of proteinemia in pregnant ewes, unlike Brozostowski et al. (1996) who showed a decrease in proteinemia during late gestation.

In our study, neither the physiological stage nor parity had effect on albumin. This result is in contradiction with observations made by Shetaewi and Daghash (1994) who showed a decrease in serum albumin during lactation compared to gestation.

Our results are consistent with those described by Elias and Shainkin-Kestenbaum (1990), who reported hypocalcaemia in ewes during late gestation; they attributed this to the increasing calcium needs of the foetus.

Also, according to Liesegang et al. (2006), the decrease of calcemia in females is probably explained by the loss of calcium during various reproductive stages.

However, we noted that the highest average levels is observed in the lactating ewes; this is in contradiction with the observations of Sykes and Field (1974), Alonso et al. (1997), and Antunovie et al. (2002), who noted higher serum calcium in ewes in late gestation compared with ewes in lactation.

Contrary to our results, Antunovié et al. (2002) described an increase in the concentration of sodium in pregnant and lactating ewes and a high serum levels of phosphate and potassium at the end of gestation; the latter could be attributed to metabolic disorders that may occur in this period and this may, in his turn lead to various pathological deviations of metabolites in the blood (Hajdarevié et al., 1989).

Barlet et al. (1971) noted that unlike what happens in cows and goats, ewes do not present a significant hypocalcemia nor hypophosphatemia after giving birth. Our results are in disagreement with Sansom et al. (1982) who reported that during late gestation in sheep a high concentration of serum magnesium levels, decrease towards lambing and 3 weeks postpartum.

This study showed that the physiological stage has a significant influence on serum levels of glucose, triglycerides, urea, and macronutrients: calcium, phosphates, sodium, potassium and magnesium; however, no effect of parity has been found in our research.

REFERENCES

Al-Dewachi OS (1999). Some biochemical constituents in the blood serum of pregnant Awassi ewes. Iraqi. J. Vet. Sci. 12:275–279.

Alonso AJ, De Teresa R, Garcia M, Gonzalez JR, Vallejo M (1997). The effects of age and reproductive status on serum and blood parameters in Merino breed sheep. J. Vet. Med. Assoc. 44:223–231.

Antunovié Z, Sencic D, Speranda M, Liker B (2002). Influence of the season and reproductive status of ewes on blood parameters. Small Rumin. Res. 45:39-44.

Antunovié Z, Peranda M, Steiner Z (2004). The influence of age and the reproductive status to the blood indicators of the ewes. Arch. Tierz., Dummerstorf 47(3):265-273.

Balikci E, Yildiz A, Gurdocgan F (2007). Blood metabolite concentrations during pregnancy and postpartum in Akkaraman ewes. Small Rumin. Res. 67:247–251

Barlet JP, Michel MC, Larvor P, Thériez M (1971). Calcémie, phosphatémie, magnésémie et glycémie comparées de la mère et du nouveau-né chez les ruminants domestiques (vache, chèvre, brebis). Ann. Biol. Anim. Biochem. Biophys. 11(3):415-426

Baumgartner W, and Pernthaner A (1994). Influence of age, season, and pregnancy upon blood parameters in Austrian Karakul sheep. Small Rumin. Res. 13:147-151.

Brozostowski H, Milewski S,Wasilewska A, Tanski Z (1996). The influence of the reproductive cycle on levels of some metabolism indices in ewes. Arch. Vet. Polonic. 35:53–62.

Brugere-Picoux J (2002). Maladies métaboliques des ruminants cours 2004. ENV. Alfort, France.

Dubreuil P, Arsenault J, Bélanger D (2005). Biochemical reference ranges for groups of ewes of different ages. Vet Rec. 2005 May 14;156(20):636-638.

El-Sherif MMA, Assad F (2001). Changes in some blood constituents of Barki ewes during pregnancy and lactation under semiarid conditions. Small Rumin. Res. 40:269–277.

Elias E, Shainkin-Kestenbaum R (1990).Hypocalcaemia and serum levels of inorganic P, Mg, parathyroid and calcitonin hormones in the last month of pregnancy in Awassi fat-tail ewes. | Reproduction, Nutrition, Development; 30(6):693-699.

Firat A, Ozpinar A (1996). The study of changes in some blood parameters (glucose, urea, bilirubin, AST) during and after pregnancy in association with nutritional conditions and litter size in ewes. Tr. J. Vet. Anim. Sci. 20:387–393.

Ford EJ, Evans J, Robinson I (1990). Cortisol in pregnancy toxemia of sheep. Br. Vet. J. 146: 539-542.

Hajdarevié F, Lokvanci H, Muteveli T, Nezirovi N (1989). A clinical laboratory assessment of several biochemical and mineral parameters of late pregnant ewes. XIV Savjetovanje-Nove i sav. Metode urazmnožavanju ovacai koza. Ohrid, Macedonia, pp. 71–78.

Hamadeh ME, Bostedt H, Failing K (1996). Concentration of metabolic parameters in the blood of heavily pregnant and nonpregnant ewes. Berliner Munchener Trierarztlichewo chenschrift 109(81–86):593–605.

Jainudee MR, Hafez ESE (1994). Gestation, prenatal physiology and parturition. In: Hafez, E.S.E. (Ed.), Reproduction in Farm Animals.

Lea and Febiger, Philadelphia, pp. 247–283.

JELINEK P, GAJDUSEK S, ILLEK J (1996).Relationship between selected indicators of milk and blood in sheep. Small Ruminant Research 20:53-57.

KLINKON M, ZADNIK T (1997).An outline of the metabolic profile test (MPT) in small ruminants. Stočarstvo 51:449-454.

Krajnicakova M, Bekeova E, Heindrichovsky V, Maracek I (1993). Concentrations of total lipis, cholesterol and progesterone during oestrus synchronization and pregnancy in sheep. Vet. Med. 38:349-357.

Liesegang A, Risteli J, Wanner M (2006). The effects of first gestation and lactation on bone metabolism in dairy goats and milk sheep. Small Rumin. Res Bone 38:794–802.

Meziane T (2001). Contribution à l'étude de l'effet de la salinité de l'eau de boisson et d'un régime à base de paille chez les brebis de race Ouled Djellal dans les hauts plateaux sétifiens. Thèse Doctorat (Constantine), P. 162.

Mollereau H, Porcher C, Nicolas E, Brion A (1995). Vade-Mecum du vétérinaire formulaire. Vétérinaire et pharmacologie, de P. 1672.

Nazifi S, Saeb M, Ghavami SM (2002). Serum lipid profile in Iranian fat-tailed sheep in late pregnancy, at parturition and during the post-parturition period. J. Vet. Med. Ser. A. 49:9–12.

Ndoutamia G, Ganda K (2005). Détermination des paramètres hématologiques et biochimiques des petits ruminants du Tchad Revue Méd. Vét. 156(4):202-206

Nelson DR, Guss SB (1992). Metabolic and Nutritional Diseases Nutrition. Illinois and Pennsylvania State Universities, pp. 1–5.

Otto F, Vilela F, Harun M, Taylor G, Baggasse P, Bogin E (2000). Biochemical blood profile of Angoni cattle in Mozambique. Israel Veterinary Medical Association, .55 (3).

Ozpinar A, Firat A (2003). Metabolic profile of pre-pregnancy, pregnancy and early lactation in multiple lambing Sakýz ewes. 2 Changes in plasma progesterone, estradiol-17B and cholesterol levels. Ann. Nutr. Metab. 47:139–143.

PASTRANA R, McDOWELL LR, CONRAD JH, WILKINSON NS (1991a).Macromineral status of sheep in the Paramo region of Colombia. Small Rumin. Res. 5:9-21.

PASTRANA R, McDOWELL LR, CONRAD JH, WILKINSON NS (1991b).Mineral status of sheep in the Paramo region of Colombia. II. Trace minerals. Small Rumin. Res. 5:23-34.

Radostits OM, Gay CC, Blood DC, Hinchcliff KW (2000).Veterinary Medicine, 9th ed. Harcourt Publishers Ltd., London, pp. 1417–1420.

Sahlu T, Hart SP, Fernandez JM (1995). Nitrogen metabolism and blood metabolites in three goat breeds fed increasing amounts of protein. Small Rumin. Res. 10:281-292.

Sandabe U-K, Mustapha AR, Sambo EY (2004). Effect of pregnancy on some biochemical parameters in Sahel goats in semi-arid zones. Vet. Res. Commun, May, 28(4):279-85.

Sansom BF, Bunch KJ, Dew SM (1982). Change in plasma calcium, magnesium, phosphorous and hydroxyproline concentrations in ewes from twelve weeks before until three weeks after lambing. Br. Vet. J. 138:393–401

Schlumbohm C, Sporleder HP, Gurtler H, Harmeyer J (1997). The influence of insulin on metabolism of glucose, free fatty acids and glycerolin normo- and hypocalcaemic ewes during different reproductive states. Deutsch. Tier¨arztl. Wochenschr 104:359–365.

Scott D, Robinson JJ (1976). Changes in the concentrations of urea, glucose and some mineral elements in the plasma of the ewe during induced parturition. Res. Vet. Sci. 20(3):346-347.

Shetaewi M, Daghash HA (1994). Effects of pregnancy and lactation on some biochemical components in the blood of Egyptian coarse-wool ewes. Inst. Vet. Med. J. 30:64–73.

Sormunen-Cristian R, Jauhiainen L (2001). Comparison of hay and silage for pregnant and lactating Finnish Landrace ewes. Small Rumin. Res. 39:47-57.

Symonds ME, Bryant MJ, Lomax MA (1986). The effect of shearing on the energy metabolism of the pregnant ewe. Br. J. Nutr., 56: 635-643.

Sykes AR, Field AC (1974). Seasonal changes in plasma concentrations of proteins, urea, glucose, calcium and phosphorus in sheep grazing a hill pasture and their relationship to changes in body composition. J. Agric. Sci. Camb. 83:161–116

Tanaka Y, Mori H. Tazaki A, Imai S, Shiina J, Kusaba A, Ozawa T, Yoshida T, Kimura N, Hayashi T, Kenyon PR, Blair H, Arai T (2007). Plasma metabolite concentrations and hepatic enzyme activities in pregnant Romney ewes with restricted feeding. Res. Vet. Sci. 85:17–21.

Tissier M, Theriez M (1978). Ovins, In Alimentation des ruminants, Ed. INRA Publications, Versailles, France. pp. 403-448.

Tontis A, Zwahlen R (1987). Pregnancy toxemia of small ruminants with special reference to pathomorphology. Tierarztl Prax. 15(1):25-29.

West HJ (1995). Maternal undernutrition during late pregnancy in sheep. Its relationship to maternal condition, gestation length, hepatic physiology and glucose metabolism. Br. J. Nutr. 75:593–605.121.

15

Effects of sodium selenite and chromium sulphate as metabolic modifiers on stress alleviation, performance and liver mineral contents of feedlot Bonsmara cross steers

Luseba D.

Department of Animal Sciences, Tshwane University of Technology, Private Bag X680, Pretoria 0001, Republic of South Africa.

The objective of the study was to investigate the effects of supplemental selenium and chromium on blood cortisol and glucose levels on stress alleviation and the carry-over effects on performance and liver mineral contents of local beef cattle under feedlot conditions. Seventy-two Bonsmara cross weaned calves weighing on average 185 ± 20.729 kg were allocated to 12 pens of six animals each, three replicates per treatment and fed for 120 days either a standard diet (CON) or supplemented with 0.3 mg.kg^{-1} DM sodium selenite (SEL) 0.3 mg.kg^{-1} DM chromium sulphate (CHR) and a combination of Se and Cr (SEL/CHR). Blood cortisol and glucose levels as stress parameters, feed intake and growth performance parameters were assessed. There was no statistical difference in blood cortisol levels on d0. On d14, cortisol concentrations were lower than on d0 (P<0.05) except for treatment SEL/CHR. On d42, the values were higher than on d0 and d14 except for SEL/CHR that had very low cortisol values (P<0.05). Blood glucose concentrations followed similar trend. There was no carry-over effect of stress alleviation on growth performance though SEL/CHR tended to have better ADG (P=0,148) and predicted FCR (P=0,197). Liver tissue mineral levels were within normal ranges. However, SEL increased significantly (P<0.05) liver Ca, Mg, Co and Mn while CHR decreased Ca and Mg concentrations. Selenium was positively correlated with Cu (r=0.30, P= 0.01) and phosphorus (r=0.44, P=0.0001) while Cr was negatively correlated with Ca (r=-0.50, P=0.01) and Mg (r=-0.30, P=0.01). Liver tissue minerals did not affect performance parameters. SEL/CHR might act better on stress alleviation and growth improvement. The effects on carcass characteristics and meat quality and the use of different forms of supplemental Se and Cr warrant further research.

Key words: Selenium, chromium, stress, cortisol, glucose, feedlot, Bonsmara.

INTRODUCTION

Beef production in South Africa faces many constraints compared with other bigger producers such as the USA, Canada and Brazil. Grain to beef ratio in South Africa is estimated at 13:1 as compared to Australia with 22:1 and USA 24:1 (Ford, 1998). This situation opens the door to all types of manipulations aiming at compensating the low use of grain in feedlot animal diets. Many metabolic modifiers are still used but consumers are becoming aware of health risks related to some of these products; many of them have been banned in some regions like the European Union (EU). It is suggested that minerals and, more particularly, trace minerals constitute a natural and safe way of supplementing animal feed in order to improve animal performance.

Feedlot cattle are subjected to many stressors that is, transport, high energy diet etc. Transportation stress alters rumen function, serum biochemical constituents and serum cortisol concentration more than those fasting alone (Cole et al., 1988). In feedlot practice, stress is assessed by the degree of shrink or weight loss which is due primarily to losses of body and digestive tract water (Hutcheson, 1992). This loss is accompanied by depletion of body minerals and vitamins. Cortisol is a useful indicator of short-term stresses such as transport and handling. It is a time-dependent measure that takes 10 to 20 min to reach peak values (Grandin, 1997). That is why other indicators such as blood glucose level and tissue mineral status need to be investigated. During stress, glucose metabolism increases simultaneously with increased secretion of cortisol, as well as an elevation of blood glucose and increased urine chromium secretion (Burton, 1995). It is suggested that during stress, cortisol acts antagonistically to insulin, preventing entry of glucose into muscle and adipose tissue and sparing it for tissues of high demand (e.g. liver and brain) (Burton, 1995).

Studies on the use of both Cr and Se in animal nutrition are scarce (Dominguez –Vara et al., 2009). As far as it can be ascertained, the use of the inorganic sulphate form has not been reported in animal feeding. Chromium sulphate ($Cr_2(SO_4)_3$) is obtained by preparation of anhydrate salt by dehydration of hydrated forms. It is used in insolubilisation of gelatine and also, like the previous form, in catalyst preparation; as mordant in textile industry and in tanning (Merck and Co, 1996).

Chromium is a nutritional substance and not a therapeutic product. As such, it will act during the period of deficiency such as during stress. Because stress in a feedlot is constant, that is, more physical during the adaptation period but dietary later, it is assumed Cr would be beneficial to feedlot animals for the whole feeding period. Chromium supplementation is thought to prevent other mineral losses during stressful conditions (Moonsie-Shageer and Mowat, 1993).

The dietary supplements of Se and Cr have been described separately in animal production. This study was therefore aimed at assessing the effect of a combination of selenium and chromium on stress alleviation during the adaptation period in the feedlot; to determine the magnitude of the response to supplementation on animal performance and to determine the status and interaction of major minerals and trace elements in animal tissues.

MATERIALS AND METHODS

Experimental animals

The animals were taken care of according to the Ethics Committee of the Medical University of Southern-Africa (Medunsa, South

Africa). Seventy-two Bonsmara X Brahman X Nguni cross male weaned calves aged seven to nine months old and weighing between 150 and 180 kg were sorted into twelve groups of six animals and thereafter different groups were allocated at random to four treatments in three replicates. They were processed after two days of adaptation into a feedlot setting to the standard diet (control) and hay; this included castration, dehorning, tagging and sorting randomly into the four treatment groups. They were treated for tick control with Ectoline [ND] (Bayer, (Pty) Ltd) and dewormed with Valbazen[ND] (Pfizer, (Pty) Ltd). The dietary composition is presented in Table 1.

The diet did not include any other metabolic modifiers (e.g. monensin, antibiotics) and was kept constant for the entire period of 120 days. The animals were weighed full stomachs on d1 and fortnightly subsequently. On d120, they were weighed and transported to the Johannesburg City Deep Abattoir (approximately 90 km from Medunsa) for slaughtering the next day. The liver samples were collected and kept in 10% formalin solution for further processing and mineral analyses. The liver is the most labile body tissue for most of the minerals (Boyazoglu, 1997). Carcass characteristics and meat colour data have been published elsewhere.

The diet contained in average, that is Dry matter %: 90.41; Ash %: 6.40; Crude protein %: 14.56; Crude fat %: 2.99; crude fibre %: 13.25; Ca %: 0.55; P%: 0.47; Mg%: 0.21; Co (mg/kg): 2.7; Cu (mg/kg): 12.9; Fe (mg/kg): 562.5; Mn (mg/kg): 112.9; Se (mg/kg): 1.48; Zn (mg/kg): 52; Cr (mg/kg): 3.4.

Laboratory procedures

Blood samples were collected into evacuated glass tubes containing sodium fluoride-potassium oxalate for glucose, and plain silicone coated tubes for cortisol determinations by jugular vena puncture on d0, d14 and d42. Plasma glucose determination was made using the SYNCHRON SYNCHRO System SYNCHRO MULTI Calibrator (Beckman Instruments, 1993). Quantitative determination of cortisol levels in serum was made using the Clinical Assays GammaCoat Cortisol Radioimmunoassay Kit (Incstar Corporation - Stillwater, Minnesota, USA).

Feed samples were analysed by proximate procedures for nitrogen (N) content, as an indication of crude protein (CP) of the feed samples using the FP 428 Nitrogen Determinator (LECO Corp.); dry matter (DM), ash and crude fibre (total) as per laboratory methods by ALASA (1998); crude fat as per modified Soxhlet Method with hexane as an extracting solvent.

Mineral determinations in feed and liver were done using specialised laboratory procedures. The Ca, Mg, Cu, Fe, Mn, Co, Cr, Zn content of feed and liver samples were determined on a Flame Atomic Absorption Spectrophotometer (FAAS) (Perkin Elmer, Model 5100 PC). Phosphorus was determined by spectrometry (Sequoia - Turner Corp.). The anhydride generator for atomic absorption was used for Se determination (FIAAS 100, Perkin Elmer).

Statistical analysis

Data from two animals, which died, were removed from the analysis of growth but they were included in the stress assessment because they were still alive during this initial period. One animal died when attempt was done to repair a fistula and the second died towards the end of the trial due to acidosis (bloat). The data was analysed by Analysis of Variance (ANOVA) using the General Linear Models (GLM) procedure of SAS version 8.3 (SAS Institute Inc., 1999). It should be noted that unless stated otherwise, tabulated data are least square means. Discrepancies may arise if it is attempted to calculate one value from another through simple arithmetic.

Table 1. Diet composition in a ton of feed for feedlot cattle.

Ingredients	Inclusion (kg)	Inclusion (%)
Yellow maize meal	450	45
Wheaten bran	100	10
Yeast	50	5
Malt dust	50	5
Eragrostis meal	260	26
Molasses	50	5
Urea	10	1
Limestone powder	10	1
Monocalcium phosphate	5	5
Salt	10	1
Premix*	5	5

*Contains appropriate amounts of trace elements including Se and Cr according to treatment.

Table 2. Blood cortisol (nmol/L) and glucose (mmol/L) concentrations (±sem) of feedlot cattle fed a supplement of se and cr (±SEM).

Treatment	do		d14		d42	
	Cortisol	Glucose	Cortisol	Glucose	Cortisol	Glucose
CON	63.39± 5.8	4.88±0.2	54.05±6.6[b]	4.82±0.1[b]	88.72±7.3[b]	5.31±0.1[b]
SEL	82.28±6.0	4.91±0.2	70.39±6.8[b]	5.44±0.1[a]	82.13±7.5[b]	5.00±0.1[a]
CHR	70.86±6.0	5.13±0.2	60.83±6.8[b]	4.94±0.1[b]	82.94±7.5[b]	5.19±0.1[b]
SEL/CHR	71.94±5.8	5.04±0.2	79.17±6.6[a]	4.87±0.1[b]	61.28±7.3[a]	4.98±0.1[a]

Values with superscripts within a column differ significantly from others (P≤0.05).

Student's t test was used for comparisons of two means. The 5% probability (P≤0.05) was used as the significance level.

RESULTS AND DISCUSSION

Stress alleviation

The data related to cortisol and glucose concentrations are presented in Table 2. For data analysis, the model used includes cortisol and glucose levels as dependent variables on d0, d14, and d42. Treatments are used as independent variables. For overtime changes, the differences between two consecutive measurements e.g. cortisol d14-d0, d42-14 are considered as dependent variables.

The least-square means of the blood cortisol measurements (nmol/L of blood) across the different treatments in this trial were 73.31, 65.79 and 78.60 respectively on d1, d14 and d42. These values are higher than those reported by Moonsie-Shageer and Mowat (1993). Values higher than 70 nmol/L have been suggested by Grandin (1997) as sign of stressful handling and at a level of 90 nmol/L, it was assumed that there was an extreme stress upon that animal. But in the study by Moonsie-Shageer and Mowat (1993) the cortisol values were below that mark line of 70 nmol/L for three out four treatments. It was assumed, nevertheless, that the higher the value of cortisol for a specific treatment, the higher was its stress level. The same approach was taken in the present study.

It seems that the animals were more stressed on d0 and d42 because of the higher cortisol values. Obviously on d0, all the stressors as indicated earlier are related to transportation, handling and other management procedures (Cole et al., 1988) while on d42 it could be due to dietary stress (Anderson et al., 1997). But values may also differ according to breeds. Similarly to animals used in this study, Brahman-cross cattle have been reported to have a high cortisol level during handling. It is therefore possible that values reported here are relatively higher than those of Moonsie-Shageer and Mowat (1993) who used Charolais-crosses in their research.

Paired-T test was used to compare the different treatments. There was no statistical difference between treatments on d0. On d14, all the treatments resulted in lowered cortisol levels except for treatment SEL/CHR for which a peak in the cortisol concentration (P≤0.05) was noted. On d42, all the values were high again except for treatment SEL/CHR that had very low (P≤0.05) concentration.

The combination of Se and Cr may have been effective in lowering blood cortisol. It is also probable that the animals became used to handling (Cole et al., 1988). But after this settlement in the new environment for two weeks (d14), the increased cortisol level on d42 might have been a sign of the dietary stress (Anderson et al., 1997). With low levels recorded for treatment SEL/CHR, it can be speculated that there was a direct positive effect of this treatment on stress alleviation when the dietary stress is prominent. This can also be confirmed by lowered glucose concentration (P≤0.05). Moreover, this treatment had stable cortisol concentrations during the 42-day assessment period. It appears that the effect of combined Se and Cr is probably more effective than Se and Cr alone during this growing phase.

Blood glucose levels across treatments in this study were 4.99, 4.99 and 5.12 respectively for d0, d14 and d42. These values were within the normal range in cattle (Kaneko, 1997) and were not different at the beginning of the trial (P=0.08). However on d14, concentrations were highly significantly lowered (P≤0.01) except for treatment SEL. On d42, blood glucose values were again higher and tended to differ between treatments (P=0.07) with treatments SEL and SEL/CHR having the lowest values.

The differences in glucose concentrations from d0 to d14 differed very significantly between treatments (P≤0.05). From d14 to d42, the differences were even higher (P≤0.01) with biggest differences with treatment SEL/CHR. There was a positive correlation between cortisol and glucose on d14 (r = 0.56, p = 0.016) and a tendency on d42 (r = 0.39, p = 0.105) for SEL/CHR. This suggests that Se - Cr combination has acted on blood glucose concentration through simultaneous effects of both elements on blood cortisol and glucose.

It was also proposed that by lowering the cortisol levels in the animals, the dietary supplements of Se and Cr could improve the animal performance. The Pearson's correlation coefficient was used to analyse the relationships between the independent variables (cortisol and glucose concentrations) and the dependent variables (performance parameters). There were no correlations between most of the independent and dependent variables on d0, d14 and d42 except for treatment SEL/CHR.

Selenium and chromium provided alone have proven to be ineffective in alleviating stress and subsequently improving animal performance in much research. Lawlet et al. (2004) fed supranutritional and organically bound selenium to finishing beef steers and found no effect on production and carcass quality. Kegley and Spears (1995) fed 0.4 mg/kg of $CrCl_3$, high-Cr yeast and Cr nicotinic but there was no positive response. Others (Mowat et al., 1993) who investigated the finishing-

growing phase reported similar results. In contrast, Moonsie-Shageer and Mowat (1993) reported the carryover effect of the decreased cortisol concentrations on the performance during the adaptation period. Although it is accepted that the content of total diet Cr in a diet bear a little relationship to its effectiveness as biologically active Cr (NRC, 1997), it was reported that the highest dosage (1 mg of Cr per kg DM of feed) was more effective than the adequate levels.

Growth performance

The data related to growth rate of the steers are presented in Table 3. Repeated measures analysis of variance showed that the effect of supplemental Se and Cr were affected by days on feed (P≤0.01) and by the interaction days x treatment (P≤0.05). It was clear that all the animals were in a sustained phase of growth, which started to flatten around d112 (Figure 1). The contrast between LWT120 and LWT112 was not significantly different (P=0.10). The decision to stop the trial on d120 was thus justifiable because there was need to analysis all the growth parameters at a similar maturity.

The animals attained in average 345.461 kg live weight. The mean final ADG across treatments was 1.300 kg per day. These values are lower than the standards in the South African feedlot industry estimated to be around 420 kg for the live weight and an ADG of 1.700 to 1.800 kg (or even 2.000 kg) per day (Henning et al. 1999).

Although it was not statistically significant, the treatment SEL/CHR tended (P=0.197) to have high ADG (1.393 kg) as compared to treatments CON and CHR that gained 1.321 kg per day (5.37% less) and treatment SEL with 1.311 kg per day (6.26% less).

According to Slabbert and Swart (1989) the mature size and genetic potential of the animal establish the patterns of daily lean (protein) accretion whilst the nutritional level with other factors such as sex and the use of metabolic modifiers determine the extent to which these potentials are achieved. The steers used in the present trial were Bonsmara x Brahman x Nguni crosses. The Brahman is a medium frame and early maturing type of cattle. The Nguni cattle are of small frame and early maturing type while the Bonsmara is of medium frame and medium maturity type. The Brahman cattle were used in the breeding process for their harshness and resistance to tropical diseases. The Nguni and Bonsmara cattle are very close breeds; the heterosis effect is therefore small on growth since only the Bonsmara contributes to the improvement of growth rate. The weak growth potential of animals used in some trials such by Ammerman et al (1980) and Kincaid et al. (1999) have been also indicated as the cause of the lack of improvement in growth rate in different studies using Se in young animals. Moreover, because the genotypes that have excellent adaptability

Table 3. Mean values for performance of feedlot steers fed a supplement of Se and Cr (± SD).

Treatment	LWTD0	LWTD120	ADG	P-FCR
CON	185.297±19.813	343.888±33.631	1.321±0.203	6.66±0.84
SEL	187.352±21.295	344.706±27.184	1.311±0.125	6.68±0.49
CHR	185.352±21.578	343.529±23.436	1.321±0.141	6.65±0.56
SEL/CHR	182.5±21.777	349.722±37.079	1.393±0.185	6.37±0.70
AVERAGE	185±20.729	345.50±30.373	1.337±0.167	6.59±0.66
P-value	0.981	0.264	0.197	0.148

SD = standard deviation; LWTD1 = live weight on d1; LWTD120 = live weight on d120; ADMI = average daily gain; FCR = feed conversion ratio.

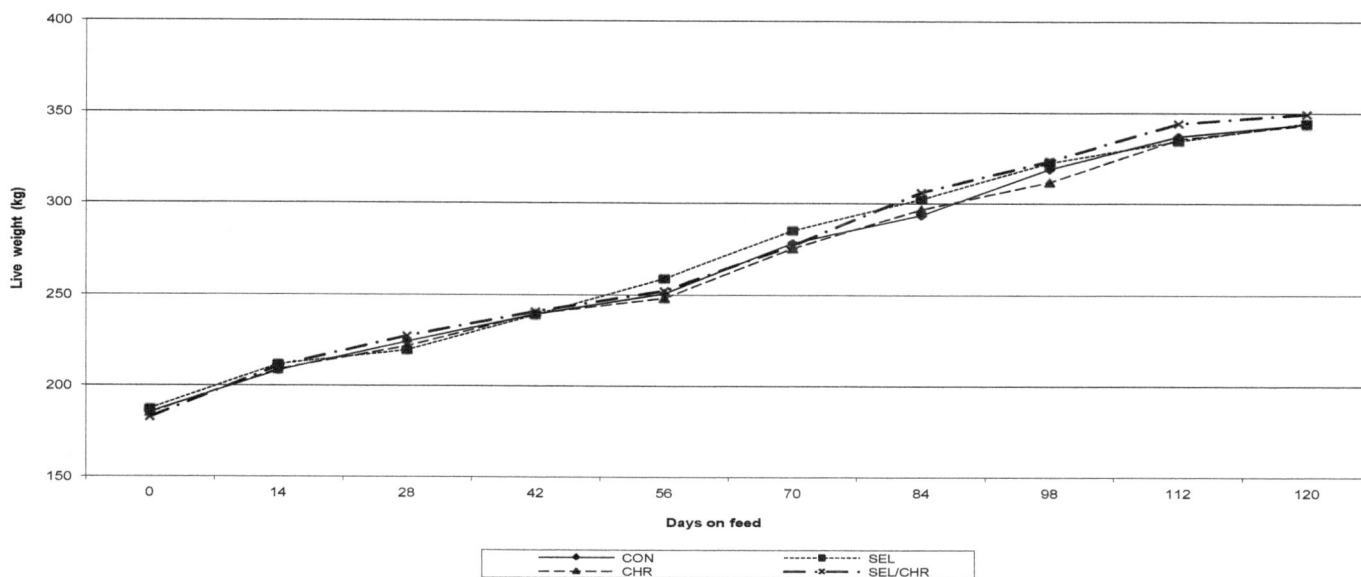

Figure 1. Growth performance of feedlot steers fed a supplement of Se and Cr.

traits for the tropical and sub tropical climates do not perform well under feedlot conditions (Bosman, 1998) it can be assumed that these animals achieved adequate growth potential for the given feeding period.

A functional selenium deficiency has been suggested as the major factor affecting the response to supplementation and the major challenge to Se activity is the level of oxidant stress presented by the diet or metabolism (Van Ryssen et al., 1992a). But as shown earlier, animals in this trial had a high Se diet therefore eliminating any possibility of Se deficiency. It has been also shown that the feed intake and subsequent growth rate of the animals can be depressed by a high intake of Se in pigs (Meyer et al., 1981) and in sheep (Echevarria et al., 1989). It has been agreed with Smolders et al. (1993) that the management and animal husbandry have generally a greater effect on growth rate and development than the mineral supply.

The difference in species and breeds in the absorption of Se was reported by Van Niekerk et al. (1990). Gerloff (1992) demonstrated that variations in genetic ability to absorb or retain Se may be present in cattle and variations of genetic lines on different farms may result in differences in response to Se supplementation. And more importantly, growth rate as an indicator of Se supplementation is an inaccurate measure because it is easily masked by such factors as sex and breeds (Van Ryssen et al., 1992b). Direct improvement of animal growth can be easily attained by improving the digestibility of the nutrients. However the dietary supplement of Se does not affect the digestibility of the nutrients (Nicholson et al., 1991a).

Treating steers with Cr (CHR) alone did not affect the performance as well in the present study. According to Pollard and Richardson (1999) research pertaining to chromium in the diets of beef cattle has been less consistent in its effect on performance and growth parameters. The high-Cr yeast and $CrCl_3$ dietary supplements did not alter the ADG and feed efficiency of weaned stressed calves. Feeding weaned stressed calves with a supplement of chelated Cr was ineffective in experiments by Wright et al. (1994) and Mathison and Engstrom (1995). Bunting et al. (1994) suggested that the lack of physiological stress during growing phase can explain the absence of dietary chromium effect. However, this is not in accordance with the present study which was conducted during the South African winter which is harsh and stressful weather and feedlot diet is also suspected to bring about a physiological stress.

Many other factors can also be considered when interpreting the results but the age of the animal is a very important factor. Kegley et al. (1996) used calves of less than 7 days. Arthington et al. (1997) used calves aged 3 to 4 weeks. Pollard and Richardson (1999) used suckling calves. Kegley and Spears (1995) used young calves on milk replacer supplemented with Cr-Nicotinic acid complex (Cr-NAC) or Cr-Cl$_3$ that failed to improve the performance. Here, the authors argued that the lack of stress, the stage of the digestive system development and the use of the inorganic form (Cr-Cl$_3$) known to be less absorbable (Merck and Co, 1996) could be considered. The growth pattern of any animal as seen in the well-known growth curve is minimal at this stage. The exponential growth pattern is seen in the feedlot after seven to nine weeks on feed when the cattle are brought in after weaning.

In another research, Kegley et al. (1996) pointed out that the cattle may have not been stressed because they were only shipped from 100 km as compared to studies showing Cr effects where cattle were shipped much greater distances and were severely stressed before the initiation of the study. In the present study, cattle were shipped over much longer distance (approximately 200 km). The performance was not improved nevertheless.

The main action of Cr is thought to be its involvement in glucose metabolism (Burton, 1995; NRC, 1997). However, it seems that adult ruminants derive a major portion of their glucose requirement from hepatic gluconeogenesis (Samsell and Spears, 1989). Therefore, there is no high expectation of Cr effect on the digestibility of the nutrients and subsequent improvement of feed intake.

Feed intake and feed conversion rate

The data related to feed intake was recorded on a pen basis. However, in order to obtain meaningful statistics, predicted feed conversion rate (Meissner, 1998) was instead used. The animals consumed in average 8.85 kg of concentrate feed DM per day, which represents approximately 2.6% of body mass. Muir et al. (1992) reported similar value. Given the mean ADG of 1.300 kg, the feed to gain ratio was 6.6:1. This figure is high compared with the average feed conversion ratio of 5:1

reported by Henning et al. (1999). It should be noted again here that the diet used was not a typical feedlot diet used in South Africa since known metabolic modifiers, e.g. monensin, where not used and the effect of breed as noted early could be prominent.

Because Se is also known to be ineffective in the improvement of the digestibility of the nutrients (Nicholson et al., 1991a) it is thus accepted that it would not improve the intake. Likewise, others (Nicholson et al., 1991b) did not find a positive effect of dietary Se supplement on feed intake and feed conversion. According to Perry et al. (1976) higher levels of Se than 0.1 mg/kg DM can lead to depression in rate of gain in cattle. Diets in this study had Se levels on average of 1.48 mg/kg DM.

Mineral composition of liver samples

The univariate procedures of the GLM (SAS, 1999) were used for each mineral under investigation in order to test the distribution. Data was analysed with or without the outliers defined here as the 5 highest and the 5 lowest values. The results were the same and therefore all variables were included in the interpretations of the results. Pearson's correlation coefficients were used to study the interactions between different elements.

Table 4 presents the liver mineral and trace elements concentrations. With the exception of Se and Co that had high concentration and Cr with low concentrations, all the values recorded in this experiment were within the normal range for cattle (Puls, 1994). The average Se concentrations recorded for the feed samples were higher than the normal ranges of 0.1 to 0.3 ppm (NRC, 1980) and 0.3 to 1.0 ppm (Puls, 1994) except for the control diet that was Se adequate. This was however a certain overall higher Se status for all the treatments.

Although Se feed content was in average 5 to 8-fold higher than the normal range, the liver content was only 3-fold higher than the normal liver concentration. This confirms findings from Van Ryssen (1996), that is, it seems that a proportion of Se consumed by the ruminant is reduced in the rumen to unavailable forms depending on the rumen degradability of the protein sources. Net Se absorption was reported to be about 35% in sheep (NRC, 1980). There is, therefore, a limit to the body Se absorption making Se intoxication rare in cattle. Probably that is why Se deficiency or toxicity symptoms are rarely reported in South Africa (Van Ryssen, 1996) though losses as a result of marginal deficiency could occur unnoticed.

An inverse relationship between Se uptake and the levels of dietary Se and the Se status was also reported by Jelinek et al. (1985). In contrast Zachara et al. (1993) showed that Se concentration in the liver increased linearly with Se level in the diet of goat. Others reported that higher dietary Se resulted in increased Se

Table 4. Liver mineral concentration (ppm) of feedlot cattle fed a supplement of Se and Cr (± S.E.).

Mineral	Treatment				Average	Significant p-value
	CON	SEL	CHR	SEL/CHR		
Ca	85.05 ± 5.48[a]	153.00 ± 5.66[b]	49.08 ± 5.66[c]	43.21 ± 5.48[c]	82.15	0.001
P	3580.56±03.34	3036.96±210.01	3091.00±210.01	2921.06±3.34	3159.44	
Mg	149.54 ± 8.46[a]	220.92 ± 8.74[b]	111.62 ± 8.74[a]	100.28 ± 8.46[a]	144.4	0.001
Co	1.79 ± 0.48[a]	8.38 ± 0.49[b]	1.69 ± 0.49[a]	0.93 ± 0.48[a]	3.161	0.0001
Cu	88.91 ± 7.21[a]	67.07 ± 7.44[b]	80.03 ± 7.44[ab]	95.50 ± 7.21[a]	82.87	0.0475
Fe	81.85 ± 5.43	77.96 ± 5.61	87.29 ± 5.61	81.64 ± 5.43	81.9	
Mn	4.33 ± 0.26[a]	5.66 ± 0.26[b]	4.38 ± 0.26[a]	5.35 ± 0.26[b]	4.934	0.001
Zn	50.12 ± 3.40	48.68 ± 3.51	41.85 ± 3.51	44.94 ± 3.40	46.33	
Se	1.74 ± 0.16	1.48 ± 0.17	1.44 ± 0.17	1.70 ± 0.16	1.597	
Cr	1.70 ± 0.14[a]	1.31 ± 0.15[a]	2.79 ± 0.15[b]	1.73 ± 0.14[a]	1.87	0.0001

Values with different superscripts differ (P≤0.05) within a row.

concentrations in all pig tissues (Lowry et al., 1985) and cows (Ammerman et al., 1980). Due to the liability of tissue Se, its losses from the body are rapid initially and then slower latter (Mahan, 1985). It is therefore possible that in 120 days on feed in this trial, the Se losses that could have been enhanced by stress were recovered and its status stabilised.

As compared to CHR and SEL/CHR, treatment SEL increased significantly (P≤0.05) the liver Ca, Co and Mn contents. These results would suggest that dietary supplement of Se is more effective than supplemental Cr in the improvement of the general status of minerals and trace elements in feedlot cattle. The low liver concentrations of Ca and Mg might be due to a negative correlation between liver Cr and some minerals. Others also (Chang and Mowat, 1992) demonstrated that the dietary Cr did not augment the tissue concentrations of Zn, Fe, Mg, Ca and P. This was further assessed by correlations between Se and Cr and other minerals.

The correlation coefficients and statistical significance of the minerals are described subsequently. Mineral interrelationships are complex as described by Puls (1994). The interrelating and antagonistic effects of the elements have to be considered if more than one element is added to the diet (AFRC, 1988). In the present study, liver Se was positively correlated with Cu (r=0.30, P=0.01) and phosphorus (r=0.44, P=0.0001). Cu had positive relationship with Fe (r=0.31, P=0.009) and Mn (r=0.263, P=0.027). Phosphorus had also positive correlations with Fe (r=0.30, P=0.01) and Zn (r=0.30, 0.01). Cr was negatively correlated with Ca (r=-0.54, P=0001) and Mg (r=-0.29, P=0.012). Calcium was strongly and negatively correlated with Mg (r=-0.75, P=0.001).

These results are not in consonance with those of Cloete et al. (1994) who did not find a positive relationship between Se and Cu. They supplemented sodium selenite to sheep and suggested that there was

no interaction between Se and Cu because the combination Se-Cu did not affect performance whilst Cu or Se supplemented alone affected the animal performance. In contrast Se and P were decreased in Se deficient cows (Salewski and Seegers, 1994; Klawoun and Landfried, 1996) showing that a positive correlation was present as demonstrated in the present study.

The regressions between Se and Cr, and different minerals are presented as follows:

1. $y^P = 2302.194 + 537.22a$, where y^P is phosphorus regression given 'a' concentration of Se;
2. $y^{Cu} = 62.395 + 12.832a$, where y^{Cu} is copper regression given 'a' concentration of Se, and
3. $y^{Ca} = 146.730 - 34.525b$, where y^{Ca} is calcium regression given 'b' concentration of Cr

The correlation procedures and stepwise and non-stepwise regression were used to study the relationships between the liver minerals in one hand, and the growth parameters in finishing steers on the other hand. It was assumed that a significant positive correlation between gain and mineral concentration in liver indicated a beneficial effect of a trace element. A significant negative correlation was an evidence of a detrimental or non-beneficial effect of the supplementation of a mineral. No direct and significant effect of minerals was noticed. Particularly for Se, Juniper et al. (2008) did not find neither any effects on the performance of ruminant animals fed high doses of the trace element (10 times the European Union or near 20 times the US Food and Drug Administration requirements).

Conclusion

Feedlot cattle are subjected to stress during transportation, handling and feeding of high energy diets. Local cattle breeds are thought to adapt less in feedlot environment. Blood cortisol and glucose measurements were lowered in Bonsmara-cross steers fed a combination of Se and Cr (SEL/CHR) during the period of high dietary stress. There were no correlations between blood cortisol and glucose levels and growth performance parameters on d0, d14 except a tendency for treatment SEL/CHR on d42. The overall live weight and average daily gain were lower than the standards in the South African feedlot industry but treatment SEL/CHR tended to have high ADG as compared to other treatments. Treatment SEL increased significantly liver Ca, Co and Mn and liver Se was positively correlated to Cu and phosphorus. Cr was negatively correlated to Ca and Mg. In the light of these results it could be suggested that further work with different sources of Se and or Cr and their combination and interactions with other minerals on stress alleviation and carry-over on performance and carcass characteristics should be undertaken.

ACKNOWLEDGEMENTS

The author is grateful to the Professor Emeritus PA Boyazoglu who was the promoter during the time the project was conducted respectively at the Medical University of Southern Africa and University of Pretoria; both institutions contributed financially. BASF chemicals (Ltd, Pty, South Africa) provided the feed premixes and the former Gillimberg Boerdery (Limpopo Province, South Africa) provided the experimental animals.

REFERENCES

AFRC (1988). AFRC Technical committee on responses to nutrients, report number 3, characterisation of feedstuffs: other nutrients. Nutr. Abstr. Rev. 58:549-571.

ALASA (1998). Handbook of feeds and plants analysis. ARC – Animal Nutrition and Animal Products Institute, South Africa.

Ammerman CB, Chapman HL, Bouwman GW, Fontenot JP, Bagley CP, Moxon AL (1980). Effect of supplemental selenium for beef cows on the performance and tissue selenium concentrations of cows and suckling calves. J. Anim. Sci. 51:1381-1386.

Anderson RA, Bryden NA, Evock-Clover M, Steele NC (1997). Beneficial effects of chromium on glucose and lipid variables in control and somatotropin-treated pigs are associated with increased tissue chromium and altered tissue copper, iron, and Zinc. J. Anim. Sci. 75:657-661.

Arthington JD, Corah LR, Minton JE, Elasser TH, Blecha F (1997). Supplemental dietary chromium does not influence ACTH, cortisol, or immune responses in young calves inoculated with Bovine Herpesvirus-1. J. Anim. Sci. 75:217-223.

Bosman DJ (1998). Cattle breeds and types for the feedlot. In: Henning, P.H. and Osler, E.H., (Eds.) Feedlot management – A course on the production of beef in intensive feedlot systems: 45 p. ARC - Animal Nutrition and Animal Products Institute. South Africa.

Bunting LD, Fernandez JM, Thompson Jr DL, Southern LL (1994). Influence of chromium picolinate on glucose usage and metabolic criteria in growing Holstein calves. J. Anim. Sci. 72:1591-1599.

Burton JL (1995). Supplemental chromium: its benefits to the bovine immune system. Elsevier Sci. 53:117-133.

Chang X, Mowat DN (1992). Supplemental chromium for stressed and growing feeder calves. J. Anim. Sci. 70:559-565.

Cloete SWP, Van Niekerk FE, Kritzinger NM, Van Der Merwe GD, Heine EWP, Scholtz AJ (1994). Production responses of sheep supplemented with Copper, Cobalt and Selenium on Kikuyu ryegrass pastures. J. South Afr. Vet. Assoc. 65:52-58.

Cole NA, Camp TH, Rowe LD, Stevens DG, Hutcheson DP (1988). Effect of transport on feeder calves. Am. J. Vet. Res. 49:178-183.

Domínguez-Vara IA, González-Muˇnozb SS, Pinos-Rodríguezc JM, Bórquez-Gastelum JL, Bárcena-Gamab R, Mendoza-Martíneze G, Zapataf LE, Landois-Palenciag LL (2009). Effects of feeding selenium-yeast and chromium-yeast to finishing lambs on growth, carcass characteristics, and blood hormones and metabolites. Anim. Feed Sci. Technol. 152:42–49.

Echevarria MG, Henry PR, Ammerman CB, Rao PV (1989). Effects of time and dietary selenium concentration as sodium selenite on tissue selenium uptake by sheep. J. Anim. Sci. 66:2299-2305.

Ford D (1998). South African feedlot industry and economics of beef production. In: Henning, P.H. and Osler, E.H. (Eds) Feedlot management – A course on the production of beef in intensive feedlot systems, ARC - Animal Nutrition and Animal Products Institute. South Africa, p. 16.

Gerloff BJ (1992). Effect of selenium supplementation on dairy cattle. J. Anim. Sci. 70:3934-3940.

Grandin T (1997). Assessment of stress during handling and transport. J. Anim. Sci. 75:249-257.

Henning PH, Steyn DG, Leeuw KJ (1999). Nutrition of feedlot cattle: current issues and science's answers. Sixth Biennal Symposium on Ruminant Nutrition. ARC – Animal Nutrition and Animal Products Institute, p. 29.

Hutcheson DP (1992). Stress influences nutritional requirements of receiving cattle. Feedstuffs 27:13-17.

Jelinek PD, Steele P, Masters HG, Allen JG, Copland MD, Petterson DS (1985). Erythrocyte selenium-75 uptake as a measure of selenium status in weaner sheep and its relationship to erythrocyte glutathione peroxidase activity. Austr. Vet. J. 62:327-331.

Juniper DT, Phipps RH, Ramos-Morales E, Bertin G (2008). Effect of dietary supplementation with selenium-enriched yeast or sodium selenite on selenium tissue distribution and meat quality in beef cattle. J. Anim. Sci. 86:3100-3109

Kaneko JJ (1997). Carbohydrates metabolism and its diseases. In: Kaneko, J.J., Harvey, J.W. and Bruss, M.L., (Eds.) Clinical biochemistry of domestic animals, 5th edn. Academic Press, pp. 45-81.

Kegley EB, Spears JW (1995). Immune response, glucose metabolism and performance of stressed feeder calves fed inorganic or organic chromium. J. Anim. Sci. 73:2721-2726.

Kegley EB, Spears JW, Brown TT Jr (1996). Immune response and disease resistance of calves fed chromium nicotinic acid complex or chromium chloride. J. Dairy Sci. 79:1278-1283

Kincaid RL, Rock M, Awadeh F (1999). Selenium for ruminants: comparing organic and inorganic selenium for cattle and sheep. In: Lyons, T.P. and Jacques, K.A., (Eds.) Biotechnology in the feed industry, 15th edn.: Nottingham: University Press.

Lowry KR, Mahan DC, Corley JR (1985). Effect of dietary phosphorus on selenium retention in postweaning swine. J. Anim. Sci. 60:1438-1446.

Mahan DC (1985). Effect of inorganic selenium supplementation on selenium retention in postweaning swine. J. Anim. Sci., 61: 173 -178

Mathison GW, Engstrom DF (1995). Chromium and protein supplements for growing-finishing beef steers fed barley-based diets. Can. J. Anim. Sci. 75:549-558. (Abstract)

Meissner HH (1998). The feedlot: Introduction and overview. In: Henning, P.H. and Osler, E.H. (Eds.). Feedlot Management: A course on the production of beef intensive feedlot systems. ARC - Irene, South Africa, pp. 31-35.

Merck Co I (1996). An encyclopedia of chemicals, drugs and biologicals, 12th edn. Whitehouse Station, N.J.: Merck Research Laboratories.

Meyer WR, Mahan DC, Moxon AL (1981). Value of dietary selenium and vitamin E for weanling swine as measured by performance and tissue selenium and glutathione peroxidase activities. J. Anim. Sci. 52:302-311.

Moonsie-Shageer S, Mowat DN (1993). Effect of level of supplemental chromium on performance, serum constituents, and immune status of stressed feeder calves. J. Anim. Sci. 71:232-238.

Mowat DN, Chang X, Yang WZ (1993). Chelated chromium for stressed feeder calves. Can. J. Anim. Sci. 73:49-55.

Muir PD, Cruickshank GJ, Smith NB, MacLean KS, Wallace GJ (1992). A comparison of grain and pasture finishing of heavyweight cattle. Proc. New Zealand Soc. Anim. Prod. 52:93-95.

Nicholson JWG, McQueen RE, Bush RS (1991a). Response of growing cattle to supplementation with organically bound or inorganic sources of selenium or yeast cultures. Can. J. Anim. Sci. 71:803-811.

Nicholson JWG, St-Laurent AM, McQueen RE, Charmley E (1991b). The effect of feeding organically bound selenium and α–tocopherol to dairy cows on susceptibility of milk to oxidation. Can. J. Anim. Sci. 71:135-143.

NRC (1980). Selenium. 393-420. Washington, D.C.: National Academy of Science.

NRC (1997). The role of chromium in animal nutrition, Washington, D.C.: National Academy Press.

Perry TW, Beeson WM, Smith WH, Mohler MT (1976). Effect of supplemental selenium of performance and deposit of selenium in blood and hair of finishing beef cattle. J. Anim. Sci. 42:192-195.

Pollard GV, Richardson CR (1999). Effects of organic chromium (Bio-Chrome) on growth, efficiency and carcass characteristics of feedlot steers. In: Lyons, T.P. and Jacques, K. (Eds) Biotechnology in Feed Industry, 15th edn. Nottingham: University Press, 103 p.

Puls R (1994). Mineral levels in animal health, 2nd edn. Clearbrook: Sherpa International.

Salewski A, Seegers N (1994). Effect of a selenium supplement on milk yield, health and fertility. Milchpraxis (Abstract). 32:196-197.

Samsell LJ, Spears JW (1989). Chromium supplementation effects on blood constituents in lambs fed high or low fiber diets. Nutr. Res., 9:889-899.

SAS (1999). SAS/SAT User's guide (Release 8.3). SAS Inst. Inc. Cary, NC.

Slabbert N, Swart D (Ed.) (1989). Nutritional manipulation of growth to produce beef carcasses of a desired composition. In: Proceedings of an Information Day Held at the Animal and Dairy Science Research Institute, pp. 71-83. Irene, South Africa.

Smolders EAA, Boxem T, Kalis C, Jorna T, van Houwelingen K, Zonderland J (1993). Copper, magnesium and selenium in young cattle on Finland pasture. Schapenhouderij en Paardenhouderij (Abstract), pp. 19-22.

Van Niekerk CH, Van Niekerk FE, Heine EWP, Coetzee J (1990). Concentrations of plasma copper and zinc and blood selenium in ewes and lambs of Merino and SA Mutton Merino sheep. South Afr. J. Anim. Sci. 20:21-26.

Van Ryssen JBJ (1996). the selenium concentration of animal feedstuffs in South Africa. AFMA MATRIX. pp. 3-5.

Van Ryssen JBJ, Bradfield GD (1992a). An assessment of the selenium, copper and zinc status of sheep on cultivated pastures in the Natal Midlands. J. South Afr. Vet. Assoc. 63:156-161.

Van Ryssen JBJ, Bradfield GD, van Malsen S, de Villiers JF (1992b). Response to selenium supplementation of sheep grazing cultivated pastures in the Natal Midlands. J. South Afr. Vet. Assoc. 63:148-155.

Wright AJ, Mowat DN, Mallard BA (1994). Supplemental chromium and bovine respiratory disease vaccines for stressed feeder calves. Can. J. Anim. Sci. 74:287-295.

Zachara BA, Mikolajczak J, Trafikowska U (1993). Effect of various dietary selenium (Se) intakes on tissue Se levels and glutathione peroxidase activities in lambs. J. Vet. Med. 40:310-318.

16

The effect of castration age on the productivity of meat from lamb hogs

Amanzhol Kusaynovich Dnekeshev and Abzal Kenesovich Kereyev*

Faculty of Veterinary Medicine and Biotechnology, West-Kazakhstan Agrarian Technical University Named after Zhangir Khan, West Kazakhstan Region, Uralsk, Kazakhstan.

This article contains information about influence of castration of lamb hogs at the age of 4, 5 and 6 months with Burdizzo forceps on some parameters of meat productivity. This study included 200 lamb hogs divided into 4 groups. The following criteria were studied: Pre-slaughter weight, carcase weight and fatness category of animals, which were monitored by periodical weighting at the age of 4 months before castration, and subsequently at 8, 11 and 14 months prior to control slaughter. At weighting before the slaughter at the age of 14 months animals which were castrated at the aged of 5 and 6 months had the best meat productivity – 45.1 ± 0.18 and 46 ± 0.16 kg, which, in percentage ratio in comparison to the first weighting increased by 36.4 and 34.6%, in comparison to the second weighting by 15.0 and 16.0%, and in comparison to the third weighting by 8.3 and 9.6%. It appeared that upon the attainment of the age of 11 and 14 months, lamb hogs castrated at the age of 6 months have better meat productivity than lamb hogs castrated at the age of 4 and 5 months or uncastrated animals.

Key words: Castration, live weight, carcass weight, Burdizzo forceps, lamb hogs.

INTRODUCTION

Castration of lamb hogs, alongside with good animal nutrition, is one of essential mass measures in veterinary surgery applied in sheep breeding for increase and improvement of meat productivity (Young, 2006). Many sheep breeding farms have to keep culled lamb hogs at autumn after the scheduled livestock valuation for stud purpose, and put them on feed till the age of one year or more (Nikitenko et al., 2007).

In many sheep farms, there is an opinion that it is necessary to castrate lamb hogs at the early age from 1 to 2 months. This is based on the fact that castration, which is mainly conducted in spring, is performed by open bloody method. By this method, testes of an animal are completely removed. So, at this age, lamb hogs are more tolerate to surgery interventions and reduce post-castration-complications.

In Kazakhstan, scientists and animal technicians, who work at breeding farms engaged into keeping of meat-wool breeds, start weaning and valuation of animals for stud purpose at the end of August, when the lambs attain the age of 3 to 4 months. The weather is hot at this period and it is the time of active reproduction of Wohlfahrtia magnifica, therefore application of any bloody methods of castration is not practical.

Many studies and observations show that the most effective ways of lamb hogs castration at this season are bloodless (percutaneous) methods (Kent et al., 1998; Wolfe et al., 1998; Stafford, 2007), which are implemented through violation of integrity of spermatic cord mainly by Burdizzo or Telyatnikov castration forceps (Bonelli et al., 2008).

Acute circulatory disorder in gonads, which occurs after this procedure, leads to atrophy and resorption of spermatic cords. After castration, sexual reflexes do not develop in males, so animals are calm (Farm Animal Welfare Council, 1994; Hosie et al., 1996) and well-fattened by feeding. Due to insufficient blood supply to

*Corresponding author. E-mail: Kereev-vet@mail.ru

tissues of testicles and accumulation of nutrients that cause activation of growth in percutaneously castrated animals. Under similar management conditions, such animals show higher weight gain than animals castrated by bloody methods (Brazle, 1992; ZoBell et al., 1993; Knight et al., 1999) or uncastrated animals. Remaining internal secretion in gonads has physiologically beneficial effects on growth and meat productivity of lamb hogs (Wellington et al., 2003). Therefore, the rational method of castration at this season should stimulate growth and meat productivity and destroy spermatogenesis in lamb hogs.

Taking into account the above listed difficulties and relevance of castration of defective lamb hogs in autumn, we tested the widely known and proven medium modified Italian Burdizzo castration forceps on lamb hogs of different age in breeding farm "Izdenis" located in West Kazakhstan region.

The aim of our study was to investigate the influence of castration performed by subcutaneous percutaneous method with the use of medium modified Burdizzo forceps on some parameters of meat productivity of lamb hogs of akzhaikskiy meat-wool breed born in 2009 and castrated at the age of 4, 5 and 6 months.

MATERIALS AND METHODS

This study was carried out in Taskalinskiy district of West Kazakhstan region of the Republic of Kazakhstan, in breeding farm "Izdenis" The study included 200 lamb hogs of akzhaikskiy breed culled after judging. These animals were weighted and divided into 4 groups of 50 animals in each by analogous method. All animals were subcutaneously percutaneously castrated by medium Burdizzo forceps. Study lasted for 10 months during grazing (feeding) and feedlots fattening of animals.

Lamb hogs in first group were castrated in late August at the age of 4 months. Animals in second group were castrated at the end of September at age of 5 months. Animals in third group were castrated in late October at the age of 6 months, and fourth group of animals was left uncastrated for control.

A small corral was built for 20 to 30 animals inside the main enclosure at the beginning of mass castration. Animals were placed into this corral by groups from the main herd. Before castration, the animals were fixed in side or in dorsal position. Veterinary specialist assured appropriate fixation of vascular cone and spermatic cord by fingers of left and right hands so that the vascular cone covered by skin near scrotum neck is laterally as possible.

After moving palpable excretory duct of seminal vesicle to vascular cone, forceps jaws were placed on vascular cone closely to the head of epididymis and at last impose forceps jaws were closed to the head of the epididymis. Spermatic cord was captured by middle part of forceps jaws and operator hold handles of medium Burdizzo forceps with both hands and keeps them in such state for at least 5 to 10 s.

Burdizzo forceps were designed so that after compression of forceps, operator needs both hands in order to return handles into the primary position. This operation is performed with the second testis in the same manner.

The growth of test lamb hogs was monitored by periodical weighing at the age of 4 months before castration and subsequently at 8, 11 and 14 months prior to slaughter.

In order to determine meat productivity, 5 test animals had been taken from each group every 2 to 6 months. Last slaughters were performed in all groups when lamb hogs were at the age of 14 months. Live-animal estimate of meet productivity of slaughter lambs was carried out by evaluation of pre-slaughter weight of animal according to the average life weight after 24 h period. Carcase weight was evaluated immediately after slaughter. The quality of the carcase was determined on the basis of the development of lean tissue and extent of fat deposition.

Most important meat productivity parameters including, pre-slaughter weight, carcass weight and fatness category of animals were recorded as mentioned by Aripov et al. (1990).

Statistical analysis

All obtained results were calculated by analysis of variance (Lakin, 1990) and analyzed by SPSS, 12.0 release for OC Windows.

RESULTS AND DISCUSSION

Pre-slaughter live weight data of the experimental animal was presented in Table 1. The result clearly indicated that the highest live body weight at 8 months was in group with uncastrated lamb hogs (39.9 ± 0.07 kg). In percentage ratio, this index was increased by 26.9% in comparison with the beginning of the test. Low level of live body weight was in group with animals, which were castrated at the age of 4 months where average live body weight was 36.8 ± 0.04 kg. In comparison with primary weight of animals, it increased by 21%. This gap can be explained by the fact that during this season, castrated animals grow slowly due to the poor pasture and watering. Spermatic cords of 30% of lamb hogs, which were castrated at this time, can not be crushed by forceps because testes of animals have not been completely formed. Therefore one (commonly) or two (less commonly) testes, in case of poor fixation of spermatic cords, recover again after castration.

In compliance with the results of control weighting at 8 months, live body weight of animals castrated at the age of 5 and 6 months were 37.9 ± 0.10 and 39.1 ± 0.12 kg, respectively. So, live body weight increased by 22.2 and 25.1% in comparison with primary data.

Table 2 shows that carcass weight of lamb hogs at the age of 4 month and similar record was obtained, 11.6 ± 0.20, 11.7 ± 0.13, 11.5 ± 0.14 and 11.5 ± 0.10 kg from first to forth group respectively. During the control slaughter at the age of 8 months the highest carcass weight was in the control group with uncastrated lamb hogs. It was 19.7 ± 0.30 kg, which in percentage ratio was 41.1% higher than carcass weight at the slaughter at 4 months.

At the winter-spring season during slaughter at the age of 11 months the highest carcass weight was in group with animals percutaneously castrated at 6 months: 22.9 ± 0.22 kg, which were 49.8% higher in comparison with first weighting and 16.6% higher in comparison with second weighting.

Last slaughters were conducted in all groups during the

Table 1. Live body weight of percutaneously castrated lamb hogs divided into age of castration.

Age of castration (months)	Live body weight at the age of 4 month before the test (kg)			Live body weight at the age of 8 month before the slaughter (kg)			Live body weight at the age of 11 month before the slaughter (kg)			Live body weight at the age of 14 month before the slaughter (kg)		
	n	Lim	$\bar{x} \pm S\bar{x}$	n	Lim	$\bar{x} \pm S\bar{x}$	n	Lim	$\bar{x} \pm S\bar{x}$	n	Lim	$\bar{x} \pm S\bar{x}$
4	50	28.2 - 30	29.1 ± 0.01	42	36 - 37.2	36.8 ± 0.04	37	38.6 - 40.3	39.8 ± 0.06	32	41.5 - 43.1	42.8 ± 0.05
5	50	29 - 30.5	29.5 ± 0.11	45	37 - 38.7	37.9 ± 0.10	40	40 - 41.2	40.8 ± 0.20	35	43 - 45.7	45.1 ± 0.18
6	50	28 - 30	29.3 ± 0.24	45	38 - 40.1	39.1 ± 0.12	40	41 - 42.5	42.2 ± 0.12	35	44 - 46.8	46 ± 0.16
Control	50	28.5 - 31	29.2 ± 0.06	43	38.5 - 40.5	39.9 ± 0.07	38	40 - 41.5	41.1 ± 0.03	33	42.6 - 43.5	43.1 ± 0.06

Table 2. Carcass weight of percutaneously castrated lamb hogs divided into age of castration.

Age of castration (months)	Carcass weight of lamb hogs slaughtered at the age of 4 month before the test (kg)			Carcass weight of lamb hogs slaughtered at the age of 8 month (kg)			Carcass weight of lamb hogs slaughtered at the age of 11 month (kg)			Carcass weight of lamb hogs slaughtered at the age of 14 month (kg)		
	n	Lim	$\bar{x} \pm S\bar{x}$	n	Lim	$\bar{x} \pm S\bar{x}$	n	Lim	$\bar{x} \pm S\bar{x}$	n	Lim	$\bar{x} \pm S\bar{x}$
4	5	10.3 - 12.9	11.6 ± 0.20	5	16 - 17.3	16.9 ± 0.12	5	18.9 - 20.1	19.6 ± 0.23	5	22.2 - 23.4	22.9 ± 0.32
5	5	10.2 - 13	11.7 ± 0.13	5	17 - 18.8	18.2 ± 0.15	5	20 - 22.3	21.8 ± 0.10	5	22.8 - 24.7	24.1 ± 0.03
6	5	10.4 - 12.8	11.5 ± 0.14	5	18 - 19.7	19.1 ± 0.02	5	21.5 - 23.8	22.9 ± 0.22	5	23.7 - 26.0	25.4 ± 0.08
Control	5	10.1 - 12.8	11.5 ± 0.10	5	18.5 - 20.2	19.7 ± 0.30	5	20.1 - 21.9	21.4 ± 0.03	5	22.4 - 23.5	22.1 ± 0.01

summer when age of all animals reached 14 months.

This season is characterized by excellent pasture feeding. Weighing at 14 months showed that the highest live body weight was animals castrated in 5 and 6 months with 45.1 ± 0.18 and 46 ± 0.16 kg, which was 34.6 and 36.4% higher in comparison with first weighting, 15.0 and 16.0% in comparison with second weighting and 8.3 and 9.6% in comparison with third weighting respectively.

During the control slaughter, the highest carcass weight was in the third group. It was 25.4 ± 0.08 kg on average. Carcasses of castrated animals killed at the age of 8 and 11 months were classified as medium category of fatness. These animals had satisfactory musculature with moderately performed hip. Spine of dorsal was slightly well defined and formed. Type of subcutis was the same in all groups of animals. Subcutis on the back and loin was in the form of layer. Its thickness was 0.2 to 0.3 cm, at the edges it was in the form of narrow and very thin strips. Meat carcasses of the third group of animals, classified to the highest category of fatness, had well developed muscles and roundish shape of hip. Subcutis covered the whole carcass from the root of tail to back edges of blade. It was in the form of layer. Its thickness on crupper was 0.8 to 1.0 cm, thickness on back was 0.5 to 0.6 cm and thickness on back and edges was 0.2 to 0.4 cm. Carcasses received from uncastrated lamb hogs (control) and classified to the highest category of fatness were organoleptically more compact with well-developed muscles, short and thick neck. But the extent of fat deposits was lower in comparison with castrated animals, that is, subcutis covered carcass as a layer, thickness of which did not exceed 0.5 cm.

Conclusion

This study indicated that lamb hogs castrated by subcutaneous percutaneous method at the age of 6 months have higher live weight and slaughter weight than lamb hogs castrated at the age of 4 months. Whereas noncastrated lamb hogs have higher live weight and slaughter weight than lamb hogs castrated at the age of 4 and 5 months, but their fat yield is less. Besides, according to the results of the second and third control slaughter, lamb hogs castrated by subcutaneous percutaneous method at the age of 6 months have better meat productivity than those castrated at the age of 4 and 5 months upon the attainment of the age of 11 and 14 months. Thus, it may be said that it is economically sound to castrate meat-wool lamb hogs kept after scheduled breed valuation for store till the age of one year and more, for improvement of meet productivity, by subcutaneous percutaneous method with medium modified Burdizzo forceps at the age of 6 months.

REFERENCES

Aripov UKh, Vinogradova VM, Vorobev PA (1990). Sheep breeding and goat breeding: directory. Moscow. Agropromizdat. pp. 146-150.

Bonelli P, Dimauro C, Pau S, Dattena M, Mollica A, Nicolussi PS (2008). Stress responses in lambs castrated with three different methods. Ital. J. Anim. Sci. 7:207-217.

Brazle FK (1992). Banding as a castration method in relationship to its effect on health and gain. J. Anim. Sci. 70:48.

Farm Animal Welfare Council (1994). Report on the Welfare of Sheep.

Hosie BD, Carruthers J, Sheppard BW (1996). Bloodless castration of lambs: results of a questionnaire. Br. Vet. J. 152:47-55.

Kent JE, Molony V, Graham MJ (1998). Comparison of methods for the reduction of acute pain produced by rubber ring castration or tail docking of week-old lambs. Vet. J. 155:39-51.

Knight TW, Cosgrove GP, Death AF, Anderson CB (1999). Effect of interval from castration of bulls to slaughter on carcass characteristics and meat quality. N. Z. J. Agric. Res. 42:269-277.

Lakin GF (1990). Biometry. Moscow. High school. p. 37.

Nikitenko DV, Nikitenko VE, Magomedov TA (2007). Dynamics of growth of castrates fabrics at different levels of feeding. Meat Prod. 3:62-63.

Stafford K (2007). Castrating older lambs: what are the issues? Vet. J. 173:477.

Wellington GH, Hogue DE, Foote RH (2003). Growth, carcass characteristics and androgen concentrations of gonad-altered ram lambs. Small Ruminant Res. 48:51-59.

Wolfe DF, Holl H (1998). Castration of the normal male: bulls, rams, and bucks. Large Animal Urogenital Surgery. Williams & Wilkins. Baltimore. MD. USA.

Young OA (2006). Changes in composition and quality characteristics of ovine meat and fat from castrates and rams aged to 2 years. N. Z. J. Agric. Res. 49:419-430.

ZoBell DR, Goonewardene LA, Ziegler K (1993). Evaluation of the bloodless castration procedure for feedlot bulls. Can. J. Anim. Sci. 73:967-970.

17

Phytotherapy in the control of helminthiasis in animal production

Erika Cosendey Toledo de Mello Peixoto[1] ,Andressa de Andrade[2], Fillipi Valadares[2], Luciana Pereira da Silva[3] and Regildo Márcio Gonçalves da Silva[3]

[1]Universidade Estadual do Norte do Paraná (UENP/Bandeirantes), BR-369, Km 54, Villa Maria, Caixa Postal 261, CEP 86360-000, Bandeirantes, Paraná, Brasil.
[2]Zootecnista, Rua Acapulco, n 293, Bairro Vila Industrial, CDEP – 85904-120, Toledo, Paraná, Brazil.
[3]Faculdade de Ciências e Letras de Assis, Universidade Estadual Paulista "Júlio de Mesquita Filho", Laboratório de Fisiologia Vegetal e Fitoterápicos, Avenida Dom Antônio, 2100, CEP: 19806-900, Assis-SP, Brasil.

Parasitic diseases constitute one of the main problems affecting livestock; however the use of chemical medicaments provides drug resistance residues in animal and environmental contamination. Changes in production concepts require that food must be produced in hygienic conditions, per healthy animals and that are not eliminating antibiotic residues, pesticides or other drugs. This scenario has favored organic production and the use of medicinal plants. For the control of endoparasites, several studies have published the benefits of *Azadirachta indica* A. Juss., *Punica granatum* Linn., *Musa* sp., *Operculina hamiltonii* G. DON., propolis, among others. However, despite the existence *in- vitro* studies that demonstrate the pharmacological properties of phytotherapics, there is still need for clinical trials to determine dosage and its effects *in- vivo*. Investigations of new bioactive natural substances can be of great value for the control of animal health and food safety, which is particularly important for organic production systems in which the use of chemical drugs is a limiting factor for certification.

Key words: Agroecology, biological agriculture, medicinal plants, parasitic worms.

INTRODUCTION

Parasitic diseases are one of the main problems affecting livestock. Losses with lower weight gain, increased mortality, lower carcass yield, lower production of meat and milk, and drug spending are factors that harm the profitability and animal welfare. These losses can be reduced through changes in pasture management, animal nutrition, and the use of chemicals anti-parasitic. Often anthelmintic chemicals, despite showing weak binding ability to the tissues, have high binding to plasma proteins. This results in longer retention of medicaments residue in the serum. Traditionally single doses of anthelmintics are administered, but the use of slow release drugs, also have result in increased risk of residues in food. Moreover, inadequate administration,

use of lower doses, misdiagnosis and lack of rotation of pharmacological bases, have caused drug resistance (Souza et al., 2008; Lima et al., 2010).

Consumers in different countries require more natural foods and higher quality (Casemiro and Trevizan, 2009). Changes in production concepts require that food must be produced in hygienic conditions, by healthy animals and which are not eliminating antibiotic residues, pesticides or other drugs. The use of growth promoters as antibiotics has been banned by the European common market, in poultry and pigs, since January 2006 (Regulation of European Commission no. 1881/2006).

The policy approach of the European Union, in relation to food safety aims assure a high level of safety, health

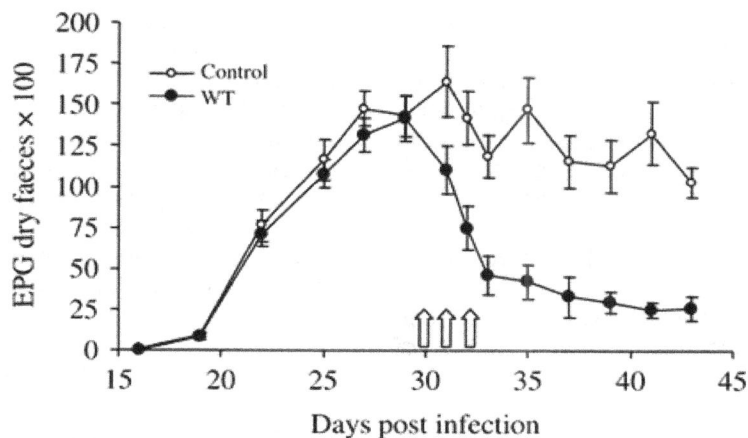

Figure 1. FEC (mean and standard deviation 0.2) of BHP sheep receiving WT drench, during experimental infection with mixed nematodes (Expt 2). The three arrows indicate the days of drench administration (Max et al., J. Agric. Sci. (2009), 147:211–218).

and welfare of animals in all its member States. For this, each member introduced control measures from farm to the consumer's table, focusing on proper surveillance (Regulation n° 178/2002 of the European Council, 2002). The protection of consumers to residues of pharmacologically active substances in foodstuffs of animal origin is guaranteed by the National Plan for Waste Management, which is based on Directive n° 96/23/EC of 1996 of the European Council. Thus, all countries wishing to export food products to the European Union must offer products which residue concentration does not exceed the limits, or does not present residues of substances classified as hazardous to consumers' health (Regulation n° 2377/90 of the European Council, 1990).

This scenario has favored the agroecologic and organic products' market. Organic agriculture is practiced over 35 million hectares in 154 countries. Despite the global recession, demand keeps growing. The European market was the most affected, but countries like Portugal and France, showed growth rates above 15% (Sahota, 2012). The North American market surpassed Europe in 2010, although the supply of organic foods continues short of demand. Thus, Latin America has become a major supplier of organic foods.

However, these systems require that the food must be produced without any harmful residues. Thereby advances in the search for natural methods, have been et obtained from the exploitation of herbal medicines (Nery et al., 2009). Searches related to natural products are justified by the possibility of avoiding and or minimizing the use of chemicals, reducing the occurrence of drug resistance, and the presence of residues in animal products. However, although there is a great variability of medicinal plants, their use in veterinary medicine is still not disseminated. Thus, the present study summarize the

main results obtained from the use of medicinal plants, for the control of helminth infections in production animals.

Tanniferous plants

Several authors reported the promising potential of plants rich in tannin on nematode control (Oliveira et al., 2011a). The anthelmintic activity of tannins in -vitro was characterized by decrease in hatching, reduced of development and motility of larvae and of the adults (Joshi et al., 2011). In- vivo, they caused reductions of eggs per gram of faeces (EPG) and the level of parasitic infestation (Max et al., 2009) (Figure 1). It was postulated that tannins can still exert the action as anthelmintic indirectly by enhancing the immune response. Its protection on the ingested proteins, against rumen degradation, would facilitate greater availability, of these proteins, in the lower gastrointestinal tract (Otero and Hidalgo, 2004). The tannins are not absorbed by the gastrointestinal tract of small ruminants, occurs disposal in feces and it reduces the pasture contamination (Mupeyo et al., 2011).

Oliveira et al. (2011b) evaluated the effects of extracts from the leaf and stem of Anadenanthera colubrine (Vell.) Brenan, Leucaena leucocephala (Lam.) de Wit and Mimosa tenuiflora (Willd.) Poir. On larval exsheathment of Haemonchus contortus in-vitro and verified the role of tannins in this process. Third-stage larvae of H. contortus were incubated with extracts for3 h and were exposed to sodium hypochlorite solution. The extracts were tested at 300 µg.ml^{-1} and accompanied by controls: phosphate buffer solution (PBS) and polyvinyl polypyrrolidone (PVPP). The larval exsheathment was evaluated for 60 min. The 6 extracts blocked larval exsheathment. After

PVPP addition, a tannin inhibitor,the exsheathment percentage was similar to the PBS (p > 0.05), except for *L. leucocephala* and *M. tenuiflora* leaf extracts. However, pre-incubation with PVPP of these 2 extracts significantly changed larval exsheathment when compared to the non-treated extracts (p < 0.05). These results suggest that *A. colubrina*, *L. leucocephala* and *M. tenuiflora* could be useful in gastrointestinal nematode control and that tannins are probably the main compounds involved in the observed effects.

Thus, consumption of tannipherous plants can determine important perspectives for the control of nematodes.

Propolis

Throughout history, man has learned to use natural products and one of the most used has been the propolis. Propolis is characterized resinous and balsamic material, collected by bees from the branches, flowers, pollen, buds and exudates of trees. It has been used in the treatment of various diseases in animals (Coelho et al., 2010). Its antibacterial activity (Orsi et al., 2012), anti-inflammatory and immunostimulant (Fischer, 2008) are particularly important for the supportive therapy of parasitosis.

Ghanem et al. (2009) investigated the gastroenteritis caused by coccidial infection in domestic goats. *Eimeria* oocysts were found in 66% of the fecal samples examined, and 6 species of *Eimeria* were morphologically identified. 4 groups of kids were used: healthy kids used as control (CH), naturally infected kids and left untreated, kids treated with Toltrazuril (20 mg/kg body weight, orally for 2 weeks) and kids treated with Propolis (1 ml of 3% aqueous solution/liter of drinking water for 7 days). The results showed that Toltrazuril was highly effective as an anticoccidial drug (94.33% reduction of OPG) and was more effective than propolis which moderately reduced OPG (54.66%). The efficacy of both drugs was further compared based on antioxidant assays, serum biochemical analysis and histopathological changes. There was a significant reduction (p < 0.05) in erythrocyte reduced glutathione, glutathione reductase, superoxide dismutase, catalase activities, serum albumin, calcium, sodium and potassium in *Eimeria*-infected kids compared to control. On the other hand, significant elevations (P < 0.05) in serum malondialdehyde, iron and nitrate were recorded in infected animals compared to control. The total protein and phosphorus showed non-significant decrease compared to control. Propolis and Toltrazuril treatment significantly alter the serum biochemical aberrances toward the control values. Heinzen et al. (2012) evaluated holstein calves and their crosses, with 90 days old, with *Trichostrongylus* sp. and *Strongyloides* sp. It was directly administered orally, 10 ml of alcoholic extract of propolis 30%, every 8 h for 4 consecutive days.

It was observed an average reduction of 48.48% of egg fecal output in 83% of animals. These values are not considered effective (Wood et al., 1995), but it is important to emphasize that they represent initial results. Perhaps, from additional studies by assessing different doses, dosage and frequency of administration, best results may be achieved. Preliminary tests *in-vitro* are important early step to validation of medicinal plants. However, in veterinary medicine, unlike human medicine, studies in animals are still scarce (Nery et al., 2009).

Azadirachta indica A. Juss

The Neem (*Azadirachta indica A. Juss.*) is a tree of the family Meliaceae, originally from India, that can reach up to 30m tall and live up to 200 years. The inhabitants of India and Asian countries use the extract of the leaves and the oil extracted from the seeds for more than 2,000 years as fertilizer of soil and pest control in agricultural and livestock. Its main active principle, azadirachtin, is responsible for pesticidal activity (Forim et al., 2010). It has over 50 terpenoids compounds and most of them, act on insects (Deleito and Borja, 2008; Maciel et al., 2010), mites (Broglio-Micheletti et al., 2010) and helminths (Chagas and Vieira, 2007; Lipinski et al., 2011).

Chagas and Vieira (2007) evaluated aqueous extract of dried leaves of Neem in sheep. The concentration of 240.000 part per million (ppm) *in-vitro* has inhibited eclosion of larvae of gastrointestinal nematodes in 89%. However, tests *in-vivo* with smaller doses (30g of dried leaves per animal per day for five days) did not promote reduction of egg fecal output. Chagas and Vieira (2007) recommended that research studies should be directed to the biological action of the oil extracted directly from the seed, which has a higher amount of azadirachtina in comparison with the leaves. Moreover, the administration of seed oil is also easier, since the leaves have bitter taste and are avoided by the animals.

The anthelmintic effect of Neem and garlic was evaluated by administering in buffaloes of 2 g of Neem and 2 g of dehydrated garlic (Lipinski et al., 2011). The animals were treated daily during 6 months. Feces and blood were collected in the beginning of the test and 2 more times later. Helminth eggs per gram of feces counting technique (EPG), blood chemistry and hematologic tests were utilized. The animals had a significant reduction of EPG (p < 0.01) in comparison with the beginning of the test. Furthermore, there were no hematological and biochemical indicatives of toxicity to the animals.

Punica granatum Linn.

Punica granatum Linn. belongs to the Punicaceae family, is a woody shrub that has small leaves, orange-red

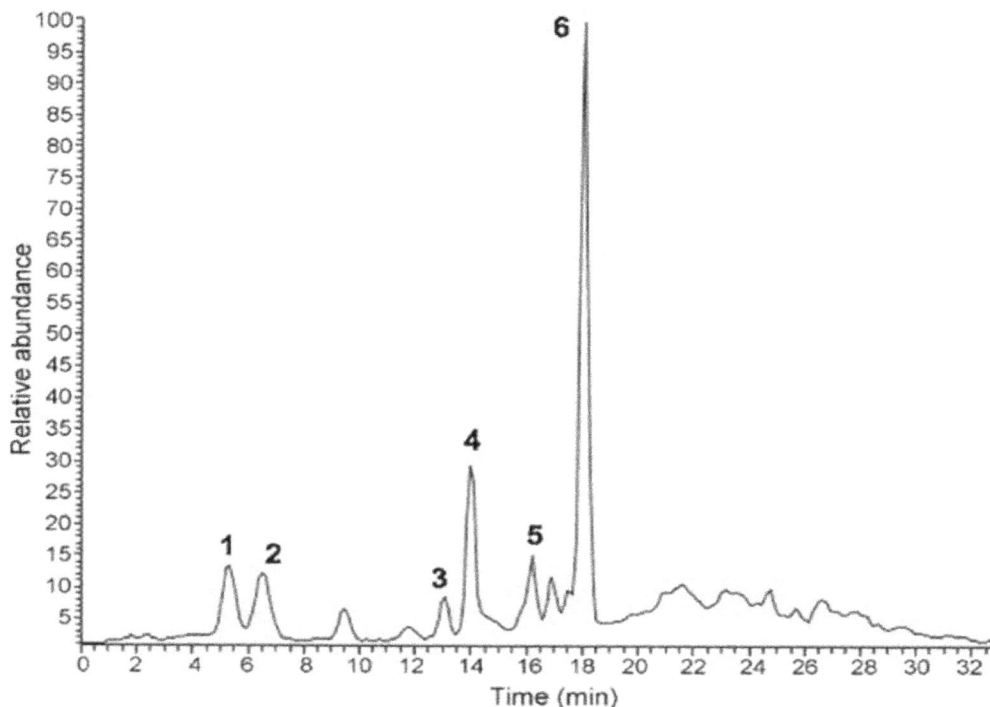

Figure 2. TIC chromatographic profile of *Pg*-FET. Peaks 1 to 2: punicalin isomers; peaks 3 to 4: punicalagin isomers; peak 5: ellagic acid exoside; Peak 6: ellagic acid. (Dell'Agli et al., J. Ethnopharmacol. 125 (2009) 279–285).

flowers, and fruits spherical at the end of its branches. Its main constituents are alkaloids (pelletierine, isopeletierine, metilpeletierine), tannins, phenolic compounds (anthocyanins, quercetin, phenolic acids) and flavonoids. Among its chemical constituents, it was observed that the punicalgina is probably a major contributor to the antimicrobial effect observed (Werkman et al., 2008).

In Brazil, *Punica granatum* Linn. is popularly used after boiling the peel of the fruit in water, against pharyngitis, tonsillitis and other infections (Silva et al., 2008). By studying the therapeutic applications of pomegranate, Werkman et al. (2008) stressed the possibility of employment in nosocomial infections due to bacterial resistance to conventional antibiotics. They still highlighted that the use of pomegranate can be easily prepared without compromising biological properties.

Pomegranate shows properties as anthelmintic, antibacterial, anti-inflammatory and antioxidant (Menezes et al., 2008). Its antioxidant action is relevant by the sequestration of free radicals and other oxidants, responsible for several degenerative cellular changes, often associated with chronic conditions such as the case of prasitoses. Flavonoids exhibit ideal conformation structural for scavenging free radicals, they are found in all plant organs, mainly in the fruits. Because it is rich in tannins, the *P. granatum* Linn. has been considered for the control of helminths in livestock. Dell'Agli et al. (2009)

studied *in vitro* the antiplasmodial activity of the methanolic extract of *P. granatum* Linn. by determining the activity of crude extracts, purified fractions and isolated compounds on asexual blood stages of *Plasmodium falciparum*. Additionally these authors assessed *in-vivo* the activity of the asexual and sexual stages developing in the vertebrate and mosquito host, using the rodent malaria model Plasmodium berghei. Urolithins, the ellagitannin metabolites, were also investigated for antiplasmodial activity. Chloroquine susceptible and resistant strains of *Plasmodium falciparum* were used for *in vitro* studies and the rodent malaria model *Plasmodium berghei*—BALB/c mice was used for *in vivo* assessments. Recombinant plasmepsins 2 and 4 were used to investigate the interference of *Punica granatum* Linn. compounds with the metabolism of haemoglobin by malaria parasites. The The Punica granatum Linn. methanolic extract inhibited parasite growth *in vitro* with a IC_{50} of 4.5 and 2.8 g/ml, for Chloroquine susceptible and resistant strains, respectively. The activity was found to be associated to the fraction enriched with tannins (*P. granatum*-FET, IC_{50} 2.9 and 1.5 g/ml) in which punicalagins (29.1%), punicalins, ellagic acid (13.4%) and its glycoside could be identified (Figure 2). Plasmepsin 2 was inhibited by *P. granatum* - extract and by *P. granatum* -FET (IC_{50} 7.3 and 3.0 g/ml), which could partly explain the antiparasitic effect. On the contrary, urolithins were inactive. Both

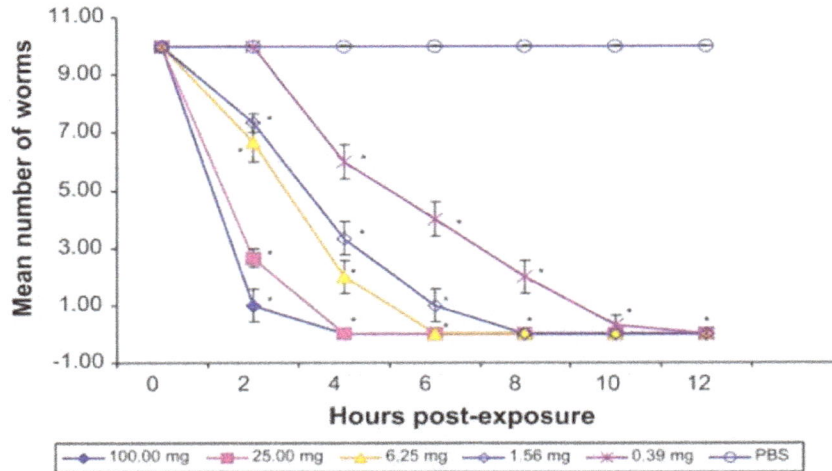

Figure 3. Graph showing the time and dose-dependent *in vitro* anthelmintic activity of *Musa paradisiaca* L. crude aqueous methanol extracts at 100.0–0.39 mg/ml concentrations in comparison with positive control levamisole (0.5 mg/ml), on mature live *Haemonchus contortus* of sheep. The inhibition of motility and/or mortality of the worms were used as the criterion for anthelmintic activity. Values shown are means, asterisk (*) indicates significantly different from previous value at P < 0.05 (Hussain et al., Veterinary Parasitology 179 (2011) 92–99).

P. granatum-MeOH extract and *P. granatum* -FET did not show any *in-vivo* efficacy in the murine model. The *in-vitro* studies support the use of *P. granatum Linn.* as antimalarial remedy.

Anjos et al. (2012) evaluated the anthelmintic effect of extract of the stalk bark of *Punica granatum* Linn. and the extract of *Musa paradisíaca* Linn. leaves (*Musa paradisíaca*). Using feces of naturally infected cattle. Anjos et al. (2012) evaluated aqueous extracts made from dried plant material and fresh. Better results were obtained by the extract of the dried bark of the stem of the pomegranate, which showed efficacy of 92.29% for haemonchus, and 96.97% for *Cooperia*. The aqueous extracts of banana leaves, were not effective, although several species of banana leaves exhibit tannins in its constitution.

Musa sp.

The banana may represent an important perspective to the natural control of helminths, not only due to the tannins in its constitution, but also for being one of the most consumed fruits. It was reported that its world´s production exceeded 91 million tones (FAO, 2012).

Batatinha et al. (2004) had already examined *in-vitro* the anthelmintic efficacy of the aqueous extract of the banana leaves (*Musa cavendishii* Lin.), and extract of papaya (*Carica papaya* Lin.). Higher efficacy was observed for treatment derived from banana leaves. These tests were performed *in vitro*, and they detected 95% reduction of the superfamily Strongyloidea, using the concentration of 130.6 mg/ml of extract of banana

leaves. Regarding the papaya seed extract, this result was only achieved using concentrations between 464 and 290 mg/ml. They concluded that 97.9% inhibition of larval development of *Haemonchus* spp. may represent an important perspective for the use of extract of banana leaves *in-vivo* tests.

Oliveira et al. (2010) evaluated the *in-vitro* efficacy of extracts of banana crop residues on the inhibition of larval gastrointestinal nematodes in sheep, among them *Haemonchus* spp. They have rated aqueous extracts of leaves, hearts and pseudostems of the banana Silver dwarf cultivar. They observed that all 3 extracts in concentrations equal to or greater than 75 mg/ml, reduced larval development with effectively above 96.9%.

Altaf et al. (2011) evaluated of anthelmintic effects of *Trianthema portulacastrum* Linn. whole plant and *Musa paradisiaca* Linn. leaves against prevalent gastrointestinal worms of sheep. *In-vitro*, it was determined using mature female *H. contortus* and their eggs in adult motility assay and egg hatch test, respectively. *In vivo* anthelmintic activity of crude powder and crude aqueous methanolic extract in increasing doses (1.0-8.0 g kg^{-1}) was determined in sheep naturally infected with mixed species of nematodes using fecal egg count reduction test and larval counts from coprocultures. These test were performed pre and post-treatments. Crude aqueous methanolic extracts showed a strong *in-vitro* anthelmintic activity and pronounced inhibitory effects on *H. contortus* egg hatching. Both plants exhibited dose and time dependent anthelmintic effects on live worms as well as egg hatching. *M. paradisiaca* (LC50 = 2.13 µg ml^{-1}) was found to be more potent than *T. portulacastrum* (LC50 = 2.41 µg ml^{-1}) in egg hatch test (Figure 3). However, *in-*

vivo, maximum reduction in eggs per gram of feces was recorded as 85.6 and 80.7% with crude aqueous methanolic extract of T. *portulacastrum* and M. *paradisiaca* at 8.0 g kg^{-1} on 15th day post-treatment, respectively as compared to that of Levamisole (7.5 mg kg^{-1}) that caused 97.0% reduction in eggs per gram. All the species of gastrointestinal nematodes, H. *contortus*, *Trichostronglyus spp.*, *Oesophagostomum columbianum* and *Trichuris ovis* which were prevalent, found susceptible (P < 0.01) to the different doses of both plants.

Operculina hamiltonii (G. Don) D.F. Austin & Staples (1983)

Operculina hamiltonii G. Don. *is* is a well-known plant in northeastern of Brazil. Its anthelmintic activity was evaluated on gastrointestinal nematodes of goats naturally infected. Additionally, the effects were observed associated or not by extract *Typha domingensis* Pers. (Silva et al. 2010). 30 caprine of Moxotó breed were submitted to the rhizome of *Typha domingensis* Pers. and to the root of *Operculina hamiltonii* G Don. The treatments were performed orally for 3 consecutive days, at a dose of 10 g of *Typha domingensis* pers. and 9 g of *Operculina hamiltonii* G Don. for 20 kg body weight. O. *hamiltonii* G Don. has determined percentage of reduction of egg fecal output corresponding at 84% and 70% on the seventh day and twenty-fifth day after treatment, respectively. Regarding the *Typha domingensis Per*, this plant showed reduction of 48% and 46% for the same evaluated periods.

Another experiment *in-vitro* was realized to evaluate the action of the ethanolic extract of O. *hamiltonii* G Don. on eggs and larvae of gastrointestinal nematodes of goats (Araujo et al., 2011). The extract was used at concentrations of 50, 25, 12.5, 6.25 and 3.12 mg.ml^{-1} in both tests. The percentage of viable eggs decreased as the concentration of the extract increased. When extract concentration reached 25%, the percentage of viable eggs decreased to 53.07%, and decreased further to 29.57% at concentration of 50%. The concentration of the extract of O. *hamiltonii* G Don. was responsible for the reduction of the percentage of viable larvae, where it was observed that the concentration of 50% affected negatively 66.87% of the larvae.

Brito-Junior et al. (2011) evaluated alcoholic extract of *Operculina hamiltonii* G Don. and *Marmodica charantia* Linn. against helminths of caprine naturally infected. The treatments were performed orally for 3 consecutive days, and the results were evaluated by the egg fecal output and larviculture. The extract *Operculina hamiltonii* G Don. has determined average reduction 63% of egg fecal output in 30 days, and 90% at 60 days after treatment, while *Marmodica charantia* Lin. did not decreased egg fecal output significantly.

However, larvicidal and ovicidal effect of the ethanol extract of leaves of *Marmodica charantia* Linn. were demonstrated by Cordeiro et al. (2010). This extract was evaluated at the concentrations of 50, 25, 12.5, 6.25 and 3.12% on feces of caprine naturally infected with gastrointestinal nematodes. The plates were examined under optical microscope to count the eggs in development and mobile larvae after 24, 48 and 72 h of incubation. The ovicidal and larvicidal activities, for larval motility test were observed in the concentrations above 12%.

Prosopis sp.

Prosopis juliflora Sw. was also assessed on larvae of gastrointestinal nematodes of caprine animals (Batatinha et al., 2011). Using their leaves, these researchers conducted 2 types of extracts methanol and aqueous. 6 different concentrations of the methanol extract (724.5; 557.3; 428.7; 329.8; 253.7 and 195.1 mg/ml) and 1 of the aqueous extracts (110.0 mg/ml) were used for the treatment of larvae cultures, in triple assays. Destilled water and doramectin were used to treat cultures considered to be negative and positive control, respectively. The results revealed a reduction of more than 90% of the infective larvae between the concentrations of 724.5 up to 253.7 mg/ml for the methanol extract and a low percentage of reduction (59.87%) for the aqueous extract. Only the methanol extract of *Prosopis juliflora* Sw. was effective in the *in-vitro* treatment of gastrointestinal nematodes of goat.

The anthelmintic effect of *Prosopis laevigata* (Willd.) MC Johnst (mezquite) n-hexanic extract was evaluated against H. *contortus* endoparasitic stages in artificially infected gerbils (De Jesu's-Gabino et al., 2010). The *in-vivo* effect of the plant extract was evaluated in gerbils artificially infected with H. *contortus*. Plant extract concentration was 40 mg/ml. 3 groups of gerbils were as follows: group 1, P. laevigata extract at 100 ml intraperitoneally; group 2, control – Tween 20 in water at a single dose of 100 ml IP; group 3 also served as a control, receiving water only, to determine the mortality due to causes other than the plant extract. An additional group (group 4) was administered fenbendazole, as a positive control. 5 days later the animals were euthanized and stomach and mucosa removed to quantify the nematodes. The parasite population in the plant extract treated group 1 was reduced by 42.5% (P < 0.05) with respect to the control group 2; and when control group 3 was used for comparison the parasitic reduction was estimated as 53.11%. This study shows the *in-vivo* anthelmintic effect of P. *laevigata* n-hexane extract for the first time, using gerbils as an *in-vivo* model, with potential use in sheep.

Lippia sidoides Cham.

Lippia sidoides Cham. is a plant that has been often used

Table 1. Plant species evaluated for antiparasitic activity for production animals.

Family	Plant	Action	References
Aizoaceae	*Trianthema portulacastrum* L.	Inhibitory effects on egg hatching; reduction EPG	Altaf et al. (2011)
Convolvulaceae	*Operculina hamiltonii*	Reduction EPG; percentage of viable eggs decreased	Silva et al. (2010), Araujo et al. (2011), Brito-Junior et al. (2011)
Cucurbitacea	*Marmodica charantia* L.	Did not decreased EPG; larvicidal and ovicidal	Brito-Junior et al. (2011), Cordeiro et al. (2010)
	Anadenanthera colubrina		
Fabaceae	*Leucaena leucocephala*	Larval exsheathment	Oliveira et al. (2011b)
	Mimosa tenuiflora		
	Prosopis juliflora Sw	Reduction of the infective larvae	Batatinha et al. (2011)
	Prosopis laevigata (mesquite)	Stomach and mucosa removed to quantify the nematodes; The parasite population in the plant extract treated group 1 was reduced	De Jesu's-Gabino et al. (2010)
Meliaceae	*Azadirachta indica*	Inhibited eclosion of larvae	Chagas and Vieira (2007)
		Reduction of EPG	Lipinski et al. (2011)
	Carapa guianensis	Reduction of the total number of larvae	Farias et al. (2010)
Musaceae	*Musa cavendishii* L.	Inhibition of larval development	Batatinha et al. (2004)
	Musa paradisiaca	Inhibition of larval development; larval exsheathment	Anjos et al. (2012)
		Reduced larval development	Oliveira et al. (2010)
		Inhibitory effects on egg hatching; reduction EPG	Altaf et al. (2011)
Myrtaceae	*Eucalyptus globulus*	Ovicidal and larvicidal activity	Macedo et al. (2009)
Punicaceae	*Punica granatum* L.	Antiplasmodial activity	Dell'Agli et al. (2009)
		Decrease of larval viability	Anjos et al. (2012)
Typhaceae	*Typha domingensis*	Reduction EPG	Silva et al. (2010)
Verbenaceae	*Lippia sidoides*	Lower egg development	Souza et al. (2010)

Eggs per gram of faeces (EPG).

as herbal medicine in northeastern of Brazil. The hydroalcoholic extract of *Lippia sidoides Cham.* was evaluated on embryonated eggs from feces of naturally infected goats (Souza et al., 2010). Embryonated eggs were obtained from feces of goats naturally infected with Trichostrongylidae nematodes and the fecal egg count was determined by using the modified McMaster technique. 50 ml of the suspension containing 40 eggs were transferred to polystyrene plates and incubated with 12 different concentrations, and evaluations were performed during 72 h at room temperature. The results demonstrated different efficacy of extracts, with lower egg development rate at 500 mg ml^{-1}. In conclusion, the hydroalcoholic extract of *L. sidoides* may play an important role on the *in-vitro* development of gastrointestinal nematode eggs, indicating ovicidal activity.

Eucalyptus globules Wood.

E. globulus Wood. belongs to the family Myrtaceae. Despite being from Australia, this plant is now spread worldwide. The essential oil from the leaves presents important pharmaceutical activity due to properties anti-inflammatory, analgesic, antioxidant, antibacterial, antifungal, insecticide and miticide against *Boohilus microplus*. Its effect on the emergence and development of larvae of *H. contortus* has been evaluated by using different concentrations (Macedo et al., 2009). The chemical composition determination of *Eucalyptus* Wood. essential oil (EGEO) was through gas chromatography and mass spectrometry. Egg hatch test (EHT) was *globulus* performed in concentrations 21.75; 17.4; 8.7; 5.43 e 2.71 mg ml^{-1}. In larval development test (LDT) they used the concentrations 43.5; 21.75; 10.87; 5.43 e 2.71 mg ml^{-1}. Each trial was conducted by negative control with Tween 80 (3%) and positive control, 0.02 mg ml^{-1} of thiabendazole in EHT and 0.008 mg ml^{-1} of ivermectin in LDT. The maximum effectiveness of eucalyptus essential oil on eggs was 99.3% in concentration of 21.75 mg.ml^{-1} and on larvae was 98.7% in concentration 43.5 mg ml^{-1}. The concentration of eucalyptus essential oil that inhibits 50% of the eggs and larvae was 8.3 and 6.92 mg ml^{-1}, respectively. The oil chemical analysis identified as main component of the monoterpen 1,8-cineol. Thus, eucalyptus oil showed ovicidal and larvicidal activity *in vitro* over *H. contortus*, determining its possible use for the control of gastrointestinal nematodes in sheep and goats.

Carapa guianensis Aubl.

The *Carapa guianensis* Aubl. family Meliaceae, is economically important due to the value of its timber. Its seed can determine a yield of 70% of medicinal oil, which is used as antirheumatic, antibacterial and insect repellent. In order to evaluate the action of the seed oil of *C. guianensis* Aubl. in larvae of gastrointestinal nematodes of goats and sheep. Farias et al. (2010) examined different concentrations: 100, 50, 30, 25 and 10%. Regarding at ewes, the fecal culture identified, the presence of larvae of *Haemonchus, Oesophagostomum, Strongyloides* and *Trichostrongylus,* noting predominance of *Haemonchus.* The percentage of the reduction of the total number of larvae was highly effective, except for the *Strongyloides.* Farias et al. (2010) observed that the oil of *C. guianensis* Aubl. at concentrations of 100, 50 and 30%, was similar to the action of Doramectin. Table 1

presents the plant species evaluated for antiparasitic activity for production animals.

CONCLUSION

Investigations of new bioactive natural substance may be of great value for the control of animal health and food safety due to the possibility to decrease the quantity and frequency of use chemicals. This consideration is particularly important for agroecological production systems, organic or biological-dynamical, which use of chemical drugs is considered a limiting factor.

REFERENCES

Anjos C., Matsumoto LS, Fertonani LH. S, Silva RMG, Silva CS, Silva BT, Mello-Peixoto ECT (2012). Extracts of *Punica granatum* L. and *Musa paradisiaca* on larval inhibition of Haemonchus spp. and Cooperia spp. from bovines. VI Symposium Iberoamerican on Medicinal Plants. Proceedings... Ponta Grossa. 15-19 Jun. CD-ROM.

Altaf H, Nisar KM, Iqbal Z, Sohail Sajid M, Kasib Khan M (2011). Anthelmintic activity of *Trianthema portulacastrum* 1 L. and *Musa paradisiacal* L. against gastrointestinal nematodes of sheep. Vet. Parasitol. 179:92-99.

Araujo MM, Vilela VLR, Silva WA, Sousa RVR, Feitosa TF, Athayde ACR (2011). *In vitro* anthelmintic effectiveness of ethanolic extracts of Operculina hamiltonii (G. DON) D.F. Austin and Staples (1983) - Batata de Purga. Ars. Vet. 27(3):192-196.

Brito-Junior L, Silva MLCR, Lima FH, Athayde ACR, Silva WW, Rodrigues RG (2011). Comparative study of anthelmintic action of purge potato (*Operculina hamiltonii*) and São Caetano Melon (*Mormodica charantia*) in goats (*Capra hircus*) naturally infected. Sci. Agrotec. 35(4):797-802.

Batatinha MJM, Santos MM, Botura MB, Almeida GM, Domingues LF, Almeida MAO (2004). *In vitro* effects of extracts of leaves of *Musa cavendishii* Linn. and seed of *Carica papaya* Linn. on cultures of larvae of gastrointestinal nematodes of goats. Braz. J. Med. Plants. 7:11-15.

Batatinha MJM, Almeida GN, Domingues LF, Simas MMS, Botura MB, Cruz ACFG, Almeida MAO (2011). Effects of aqueous and methanol extracts of algarroba on cultures of larvae of gastrointestinal nematodes in Goats. Braz. Anim. Sci. 12(3):514-519.

Brito-Junior L, Silva MLCR, Lima FH, Athayde ACR, Silva WW, Rodrigues RG (2011). Comparative study of the anti-helminthic action of the potato of purges (*Operculina hamiltonii*) and the cantaloups of São Caetano (*Mormodica charatia*) in naturally infected goats (*Capra hircus*). Ciênc. Agrotec. 35(4):797-802.

Broglio-Micheletti SMF, Dias NS, Valente ECN, Souza LA, Lopes DOP, Santos JM (2010). Action of extract and oil neem in the control of Rhipicephalus (Boophilus) microplus (Canestrini, 1887) (Acari: Ixodidae) in laboratory. Braz. J. Vet. Parasitol. 19:44-48.

Casemiro AD, Trevizan SDP (2009). Organic Food: Challenges for the Public Domain of a Concept International [Workshop advances in cleaner production]. São Paulo; pp. 1-9. São Paulo – Brazil – May 20th-22nd – 2009.

Chagas ACS, Vieira LS (2007). Ação de *Azadirachta indica* (Neem) em nematódeos gastrintestinais de caprinos. Braz. J. Vet. Res. Anim. Sci. 4(1):49-5.

Coelho MS, Silva JhV, Oliveira ERA, Amâncio ALL, Silva NV, Lima RMB (2010). Propolis and its use in production animals. Arch. Anim. Sci. 59:95-112.

Cordeiro LN, Athayde ACR, Vilela VLR, Costa JGM, Silva WA, Araujo MM, Rodrigues OG (2010). *In vitro* effect of ethanol extract of leaves São Caetano Melon (*Momordica charantia* L.) on eggs and larvae of gastrointestinal nematodes of goats. Braz. J. Medic. Plants 12(4):421-426.

Council Regulation nº 2377/90 of European Council, 26 of June of 1990, laying down a Community procedure for the establishment of maximum residue limits of veterinary medicinal products in foodstuffs of animal origin. Official J. L 224:1-8.

Council Regulation nº 178/2002 of Parliamentand and European Council, 28 of January of 2002, laying down the general principles and requirements of food law, establishing the European Food Safety Authority and laying down procedures in matters of food safety. Official J. Eur. Comm. 31:1-9.

Council Regulation nº1881/2006 of European Council, de 19 of December of 2006, that setting maximum levels for certain contaminants in foodstuffs. Official J. Eur. Comm. L364:5-24.

Deleito CSR, Borja GEM (2008). Neem (Azadirachta indica): an alternative in the control of flies in livestock. Braz. J. Vet. Res. 28:293-298.

Dell'Agli M, Galli GV, Corbett Y, Taramellib D, Lucantonic L, Habluetzel A, Maschia O, Carusoa D, Giavarini F, Romeod S, Bhattacharyae D, Bosisioa E. Antiplasmodial activity of Punica granatum L. fruit rind. (2009). J. Ethnopharm. 125:279-285.

Directive 96/23/CE of European Council, de 29 de Abril de 1996, concerning measures to control certain substances and residues thereof in live animals and animal products and repealing Directives 85/358/CEE e 86/358/CEE and the decisions 89/187/CEE e 91/664/CEE. Offic. J. L302:1-17.

FAO (2012) Food and agriculture Organization of United Nations. Agriculture data base ProdSTAT. On line http://faostat.fao.org/site/567/default.aspx#ancor (August 07 2012).

Farias MPO, Teixeira WC, Wanderley AG, Alves LC, Faustino MAG (2010). In vitro evaluation of the effects of seed oil Carapa guianensis Aubl. on larvae of gastrointestinal nematodes of goats and sheep. Braz. J. Med. Plants 12(2):220-226.

Fischer G, Hübner SO, Vargas GD, Vidor T (2008). Immunomodulation induced by propolis. J. Biolog. Instit. 75(2):247-253.

Forim MR, Matos AP, Silvia MFGF, Cass QB, Vieira PC, Fernandes JB (2010). The use of HPLC in the control of neem commercial products quality: Reproduction of the insecticide action. Quim. Nova. 33:1082-1087.

De Jesu's-Gabino AF, Mendoza-de Give P, Salinas-Sa´nchez DO, Lo´pez-Arellano ME, Lie´bano-Herna´ndez E, Herna´ndez-Vela´zquez VM, Valladares-Cisneros G (2010). Anthelmintic effects of Prosopis laevigata n-hexanic extract against Haemonchus contortus in artificially infected gerbils (Meriones unguiculatus). J. Helminthol. 84:71-75.

Ghanem MM, Mohamed ADA, Ramadan MY (2009). Clinical, biochemical and histopathological study on parasitic gastroenteritis associated with caprine coccidiosis: comparative effect of Toltrazuril and Propolis. Ion Ionescu de la Brad Iaşi 11(1):565-580.

Heinzen EL, Mello-Peixoto ECT, Jardim JG, Garcia RC, Oliveira NTE, Orsi RO (2012). Extract of propolis in the control of helminthiasis in calves. Acta Vet. Brasil. 6(1):40-44.

Joshi BR, Kommuru DS, Terrill TH, Mosjidis JA, Burke JM, Shakya KP, Miller JE (2011). Effect of feeding sericea lespedeza leaf meal in goats experimentally infected with Haemonchus contortus. Vet. Parasitol. 178:192-197.

Lima WC, Athayde ACR, Medeiros, GR, Lima DASD, Borburema JB, Santos EM, Vilela VLR, Azevedo SS (2010). Nematode resistant to some anthelmintics in dairy goats in Cariri Paraibano, Brazil. Braz. Vet. Res. 12:1003-1009.

Lipinski LC, Martinez JL, Santos MVR, Ferreira JN, Pfau DR (2011). Evaluation of the anthelmintic effect and metabolic changes in buffaloes (Bubalus bubalis) with the administration of neem cake and dehydrated garlic in the Southern of Parana State Res. Braz. J. Agroecol. 6(3):168-175.

Macedo ITF, Bevilaqua CML, Oliveira LMB, Camurça-Vasconcelos ALF, Vieira LS, Oliveira FR, Queiroz-Junior EM, Portela BG, Barros RS, Chagas ACS (2009). Eucalyptus globulus sobre Haemonchus contortus Ovicidal and larvicidal activity in vitro of Eucalyptus globulus essential oils on Haemonchus contortus. Braz. J. Vet. Parasitol. 18(3):62-66.

Maciel MV, Morais SM, Bevilaqua CML, Amóra SSA (2010). In vitro insecticidal activity of seed neem oil on Lutzomyia longipalpis (Diptera: Psychodidae). Braz. J. Parasitol. 12:105-112.

Max RA, Kassuku AA, Kimambo AE, Mtenga LA, WAKELIN D, BUTTERY PJ (2009). The effect of wattle tannin drenches on gastrointestinal nematodes of tropical sheep and goats during experimental and natural infections. J. Agric. Sci. 147:211-218.

Menezes SMS, Pinto DN, Cordeiro LN (2008). Biologics activities in vitro and in vivo of Punica granatum L. (pomegranate). Braz. J. Med. 65(11):388-391.

Mupeyo B, Barrya TN, Pomroya WE, Ramírez-Restrepoa CA, López-VillalobosaN, Pernthanerc A (2011). Effects of feeding willow (Salix spp.) upon death of established parasites and parasite fecundity. Anim. Feed Sci. Technol. 164:8-20.

Nery PS, Duarte ER, Martins ER (2009). Effectiveness of plants for the control of gastrointestinal nematodes of small ruminants: A review of published studies. Braz. J. Med. Plants. 11(3):330-338.

Oliveira NL, Duarte ER, Nogueira FA, Silva RB, Filho, DEF, Geraseev LC (2010). Efficacy of banana crop residues on the inhibition of larval development in Haemonchus spp. from sheep. Rural Sci. 40(2):488-490.

Oliveira LMB, Bevilaquai CML, Morais S M, Camurça-Vasconcelos ALF, Macedo ITF (2011a). Plantas taniníferas e o controle de nematóides gastrintestinais de pequenos ruminantes. Rural Sci. 41(11):1967-1974.

Oliveira LMB, Bevilaqua CML, Macedo ITF, Morais SM, Monteiro MVB, Campello CC, Ribeiro WLC, Batista EKF (2011b). Effect of six tropical tanniferous plant extracts on larval exsheathment of Haemonchus contortus. Rev. Bras. Parasitol. Vet. 20(2):155-160.

Orsi RO, Fernandes A, Bankova V, Sforcin JM (2012). The effects of Brazilian and Bulgarian propolis in vitro against Salmonella typhi and their synergism with antibiotics acting on the ribosome. Natural Product Research: Formerly Natural Product Lett. 5:430-437.

Otero MJ, Hidalgo LG (2004). Tanniferous plants and control of gastrointestinal nematodes of small ruminants. Livest. Res. Rural Dev. 16(2):1-9.

Sahota A (2012). Global Organic Food and Drink Market Presented at: BioFach Congresss 2012, Messezentrum Nürnberg, Germany, Feburary 15.

Silva MAR, Higino JS, Pereira JV, Siqueira-Júnior JP, Pereira MSV (2008). Antibiotic activity of the extract of Punica granatum Linn. over bovine strains of Staphylococcus aureus. Braz. J. Pharmacogn. 18(2):209-212.

Silva CF, Athayde ACR, Silva WW, Rodrigues, OG, Vilela VLR, Marinho PVT (2010). Evaluation of effectiveness of cattail (Typha domingensis Pers.) and potato-de-purge [Operculina hamiltonii (G. Don) DF Austin and Staples] fresh on gastrointestinal nematodes of goats naturally infected in semi-arid climate. Braz. J. Med. Plants 12(4):466-47.

Souza AP, Ramos CI, Bellato V, Sartor AA, Schelbauer CA (2008). Resistance of bovine gastrointestinal helminths to anthelmintics in the Planalto Catarinense. Rural Sci. 38(5):1363-1367.

Souza WMA, Ramos RAN, Alves LC, Coelho MCOC, Maia, MBS (2010). In vitro evaluation of "alecrim pimenta" (Lippia sidoides Cham.) hydroalcoholic extract (HAE) on the development of gastrointestinal nematode (Trichostrongylidae) eggs. Braz. J. Med. Plants 12(3):278-281.

Werkman C, Granato DC, Kerbauy WD, Sampaio FC, Brandão AAH, Rode SM (2008). Therapeutic use of Punica granatum L. (pomegranate). Braz. J. Med. Plants 10(3):104-111.

Wood IB, Amaral NK, Bairden K, Duncan JL, Kassai T, Malone JB, Pankavich JÁ, Reinecke RK, Slocombe O, Taylor SM (1995). Vercruysse, J. World Assoc. Advan. Vet. Parasitol. (W.A.A.V.P.) second edition of guidelines for evaluating the efficacy of anthelmintics in ruminants (bovine, ovine, caprine). Vet. Parasitol. 58:181-213.

18

Colony status of Asian giant honeybee, *Apis dorsata* Fabricius in Southern Karnataka, India

Basavarajappa S.* and K. S. Raghunandan

Apidology Laboratory, Department of Studies in Zoology, University of Mysore, Mysore-570 006, India.

The field survey conducted to collect more than twenty parameters of *Apis dorsata* colonies during different seasons at different districts of southern Karnataka revealed interesting results. Altogether, 2,407 normal colonies (comb with live bees) were recorded at various habitats of southern Karnataka. Of all, Mysore district has recorded highest (1,560) colonies followed by Chamarajanagar (544) and Kodagu (303) districts. Among the seasons, winter has recorded highest (839) colonies followed by rainy (807) and summer (761) seasons. *Apis dorsata* selected 20 tree species to nest 1,646 colonies at an elevation ranged from five to 80 feet. About 580 colonies nested on human built structures at the height of eight to 75 feet and 181 colonies were found on rock cliffs at 10 to 30 m elevation. The colony density was more (3.64) in Mysore district followed by Chamarajanagar and Kodagu districts. Accordingly, the abundance and percent nesting frequency were also varied considerably. Further, the colony aggregates, aggregates density per 5^2 meter area, abundance and percent frequency was 1.57, 3.19 and 134.07 respectively on trees, rock cliffs and human built structures at Chamarajanagar, Kodagu and Mysore districts. The morphometrics indicated considerable variations. *A. dorsata* produced multifloral honey (10,831.5 kg) and good quantity of beeswax (1,444.2 Kg) every year. The floral source included 687 flowering plants which belong to 223 familes, supplied continuous nectar flow during most of the seasons with little variations. *Apis dorsata* honey quality was moderate with below detection limit of pesticide residues. Continuous interference from various predators, pests and parasites along with man-made activities have stressed severely *A. dorsata* colony population which resulted in the colony decline considerably. Thus, there is a dire need to conserve *A. dorsata* population under natural conditions in this part of the state.

Key words: Colony status, Apis dorsata, southern Karnataka, India.

INTRODUCTION

In India, different *Apis* species occur at various agro-ecosystems. Among them the Asian giant honeybee, *Apis dorsata* Fabricius (Hymenoptera: Apidae: Apoidea) pollinators, produce multifloral honey, which contributes more than 80% to the total honey production in India (Bradbear and Reddy, 1998). In Karnataka, it is locally called 'Hejjenu', which is known for its big-sized colonies, ferocious behaviour and venomous sting to man and is a large feral insect (Engel, 1999) distributed unevenly at diversified ecosystems in the wild. It is one of the major livestock at various farm lands, forests and human inhabited ecosystems. Although reports are available on *A.dorsata* hive products (ex. Honey and beeswax) at certain regions of Karnataka, information on beeswax production, pests, predators, pathogens, pesticide poisoning, honey hunting and other unwanted human activities are sparse. Moreover, the overall colony status is not properly explored. Since, *A. dorsata* is one of the largest bees in the genus *Apis* (Oldroyd et al., 2000), build vertical colonies, which often found in dense

*Corresponding author. E-mail: apiraj09@gmail.com

aggregations (Dyer and Seeley, 1994) under arboreal conditions such as on the eaves of tall tree limbs, rock cliffs and human built structures (Reddy, 2002; Seeley et al., 1982) at an inaccessible elevations (Hepburn and Hepburn, 2007; Tan, 2007). Recording A. dorsata colonies and revealing their status requires multifaceted approach. In this regard, reports on the colony distribution at various habitats, potential nest host trees, foraging source, production of multifloral honey, honey quality and contamination, honey harvest and prevailed biological constraints at nest sites of A. dorsata from southern Karnataka are scanty. There exists a lacuna of understanding the full potential of A.dorsata population in this part of the state.

Further, A. dorsata is known for its migratory behaviour, it gathers high amount of honey upto 45 kg from a single colony (Ruttner, 1988; Vinutha, 1998). Therefore, recording nest host trees, nesting sites at hilly areas and human inhabited ecosystems are essential to quantify A. dorsata hive products. Moreover, honey is harvested from the combs of A. dorsata during specific seasons and hence honey can be harvested seasonally (Paar et al., 2004) and that shows greater variation in composition and characteristics (Anklam, 1998; Anupama et al., 2003; Joseph et al., 2007). However, in Karnataka honey is harvested from A. dorsata combs by conventional methods (Setty and Bawa, 2002), honey hunters give least importance to clean and hygiene while honey preservation. Further, nest architecture matters a lot while yielding good amount of honey. Does A. dorsata construct different shaped combs? Neither an influence of celestial cues for specific orientation of combs nor to protect from biological and environmental factors, scientific data are required to reveal information on such facts. Moreover, reports are available on variety of both invertebrate and vertebrate predators or parasites or pests including 40 different speices of birds (Abrol, 2003;Nagaraja and Rajagopal, 2011; Morse and Laigo, 1969), which cause severe damage to A. dorsata colonies (Nagaraja and Rajagopal, 2011). Reports on the extent of colony decline and colony status at this part of the state during different seasons are not available. Thus, this has necessitated the scientific reports and impelled us to conduct the detailed study on the colony status of A. dorsata. By considering various parameters, multifaceted approach was made and the results of such findings are presented in this paper.

MATERIALS AND METHODS

Study area

The field survey was conducted during 2008 to 2011 by selecting13 taluks in Chamarajanagar, Kodagu and Mysore districts of southern Karnataka. These districts lies in between 11° 92' to 12° 52' N latitude and 72° 22' to 76° 95' E longitude at an elevation of more than 867.33 meter above msl and located amidst the vicinity of south-western parts of Western Ghats covered by rich floral source

and represents the high degree of biological activity (Figure 1) (Kamath, 2001). The physiographic and meteorological details of the study area are depicted in Table 1. During field study, pre-tested questionnaire was prepared by including more than twenty parameters, namely occurrence of A. dorsata normal colonies at various habitats during different seasons, normal colony density, abundance, percent frequency of colony establishment, number of colony aggregates per bee tree, comb morphology, hive products (ex. Honey and beeswax) potential nest host tree species, nesting habitat, nesting elevation, floral source, physico-chemical characters of multifloral honey, honey quality and contamination, methods of honey harvest, possible causes stressing on the survival of A. dorsata and reason for colony decline were considered. Each parameter was further divided into various sub groups depending on the need and importance. Several agricultural ecosystems, rock cliffs at Male Mahadeshwara Hills, Biligiri Ranganath Hills, Chamundi Hills and human inhabited places were periodically visited during different seasons. While collecting information, specific methods were followed as and when required and the images of certain parameters were taken with the help of Canon-Power Shot S21S, 8.0 Mega Pixels Digital Camera with 12X Optical Zoom.

The normal colony density, abundance and frequency of colony establishment on different trees, rock cliffs and human built structures (HBS) were calculated as per the method of Phillips (1959) after little modification. Normal colony density = Total no. of normal colonies recorded/total no. of study sites visited to record the normal colonies. The normal colony abundance = Total no. of normal colonies recorded/ number of study sites where normal colonies observed. The frequency of colony establishment = No. of times colonies found at a particular study site/total number of observations made at each study site × 100. The nesting elevation was measured as per Krishnamurthy (2001). To record the comb aggregates, 25 sampling sites were randomly selected in each districts and in each sampling site, various nest host trees located at garden, cultivable land, on either sides of the road, on rock cliffs and various HBS were observed as per Woyke (2008) from a distance of 25 to 50 m. As comb aggregations were confined to few limbs of bee trees, on some parts of rock cliffs and HBS, only 5^2 m imaginary area was considered and photographed. Number of colonies was counted by using adobe software version CS3 and digital video camera with 16X Optical Zoom and recorded noteworthy variations (ex. shape of comb, size, comb length, width, cell depth, cell area and honey storing capacity in honey chamber and brood chamber) from the normal colonies (Vinutha, 1998; Shukla and Upadhya, 2007).

The weight of abandoned comb was taken before boiling it in water at 60°C. After boiling, the molten wax was filtered, smeared on silver plate and again the weight of dried wax was taken. Moreover, the causes for the abandonment of A. dorsata colonies were recorded. Nest host trees were observed at 114 sampling sites both by naked eyes and using a binocular (10 × 50X) by selecting one kilometer length Variable Width Line Transects (VWLTs) (Burnham et al., 1980). The trees were photographed and identified by using both photographic pictures and the methods as described by Gamble (1967). Moreover, foraging plants were recorded with the help of information as described by Gamble (1967) and Rao (1973). The foraging plants were further grouped into various types so as to reveal their percent occurrence.

The predators or pests were recorded at the vicinity of A. dorsata colonies by considering different lengthen VWLT's with a width of 50 feet as devised by Abrol (2003), Nagaraja and Rajagopal (2011). Total 125 VWLT's were selected and predators, pests were identified by following standard methods as described by Nagaraja and Rajagopal (2011) and Hepburn and Radloff (2011). To record parasitic mite's infestation, Mysore city was selected randomly and field observations were made on weekly basis between 0800 and 1200 h during September, 2008 to June, 2010. The worker bees

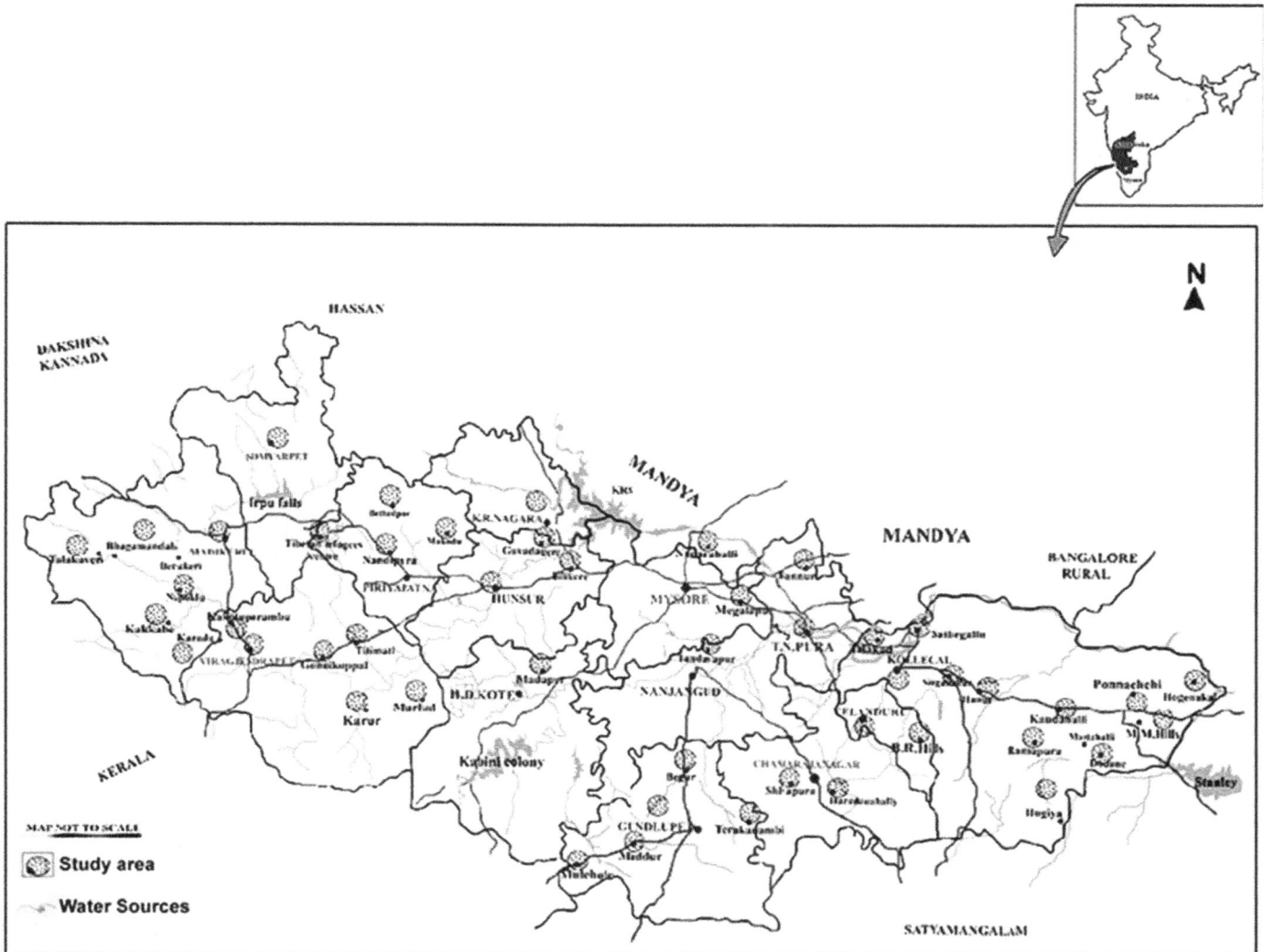

Figure 1. Study areas in southern Karnataka.

suspected to be infested with parasitic mites were collected from various places, that is, nearby nesting trees, trees in scrubby vegetation and at HBS by random sampling method. Total 80 observations were made periodically by using adult worker bees during their foraging in filed and in hive too. The infested bees were collected from honey hunters after the comb harvest and observed under Leica EZ4 Stereozoom microscope. The phoretic mites were recorded from different body parts of adult worker bees. Further, the parasitic wax moth infested normal and abandoned colonies were observed both by naked eyes and using a binocular (10 x 50X) and photographed by making imaginary quadrate (5^2 m) from 114 Line Transects. The incidence of predators, pests, parasites and pathogens of A. dorsata was made as per Abrol (2002, 2003) and Bailey and Ball (1991). As the vegetation distribution was not uniform at different districts of southern Karnataka, an all out search method (AOSM) was also adopted during the field survey. Moreover to reveal the quality and contamination of multifloral honey, the physico-biochemical parameters (ex. pH, specific gravity, electrical conductivity, absorbance and turbidity, glucose, fructose and invert sugar and total protein content) were estimated

as per Ouchemoukh et al. (2007). Further, sixteen A. dorsata honey samples were collected randomly from the study area and stored in airtight plastic containers until analysis for pesticide residues contamination as per Rissato et al. (2007), Jimenez et al. (2000) and Blascoc et al. (2004). To reveal overall colony status of A. dorsata standard methods as described by Savanurmath et al. (1993) were followed. The collected data was analyzed with the help of SPSS (ver.12.0, Chicago, Inc. USA).

RESULTS AND DISCUSSION

Occurrence, abundance and combs aggregates of A. dorsata normal colonies are predicted in Table 2. A. dorsata normal colonies were recorded highest (1,560) in Mysore district followed by Chamarajanagar and Kodagu districts respectively 544 and 303 colonies (Table 2). Accordingly, the normal colony density was more (3.64)

Table 1. Physiographic and meteorological details of different districts of southern Karnataka.

S/N	Districts	Longitude	Latitude	Elevation in ft (Height in above msl)	Temp (°C) (Min - Max)	RH (%) (Min - Max)	Average rainfall (in mm)
1	Chamarajanagar	11°55'27.22" N	76° 56' 18.10"E	2369	16.4 - 34C	21 - 85	731.80
2	Kodagu	12° 20' 14.98" N	75° 48' 24.87" E	3027	14.2 - 28.6	68 - 94.5	2800
3	Mysore	12°18'12.72"N	76°38' 46.00" E	2465	11 - 38	30 - 70	785
Range		11°55'27.22" - 12°20'14.98" N	75°48' 24.87"- 76° 56' 18.10"E	2369 - 3027	11 - 38	21 - 94.5	731.8 - 2800
Average		11.7°31'18.31"N	75.7° 47.3 29.66" E	2620.3	13.9 - 33.5	39.7 - 83.2	1202.3

Source: Karnataka State Gazetteer, Bangalore and India Meteorological Station, Bangalore.

Table 2. Occurrence, abundance, aggregates, comb morphometrics and hive products of *Apis dorsata*.

Asian giant honeybee, *A. dorsata*

S/N	Distribution of colonies in / at						Individual colony					Colony aggregates
	District	No.	Season	No.	Habitat	No.	Habitat	Density	Abundance	Frequency		
1	CN	544	R	807	T	1646	T	1.10	4.94	16.58		D:1.57
2	K	303	W	839	RC	181	RC	0.45	4.47	8.06		A:3.19
3	M	1560	S	761	HBS	580	HBS	3.64	4.78	71.48		F:134.07
Total		**2407**		**2407**		**2407**						

Comb morphology

Brood Comb	Size (cm)	Honey Comb	Size (cm)
Cell depth	1.74±0.07	Cell depth	2.03±0.03
Cell area	2.36±0.29	Cell area	2.70±0.14
Horizontal length	38.09±5.78	Horizontal length	15.89±2.70
Vertical length	28.63±2.12	Vertical length	6.13±1.01
Comb width	2.33±0.62	Comb width	12.66±0.51
		Honey storing capacity (ml)	1.03±0.84

Comb shape

Shape	%
'U'	53.2
'V'	31.3
Cone	6.2
Round	4.9
Uneven	2.6
Others	1.5

Hive products (kg)

Multifloral honey	10,831.5
Beeswax	1,444.2

CN = Chamarajanagar; K = Kodagu; M = mysore; R = rainy; W = winter; S = summer; T = trees; RC = rock cliff; HBS = human built structures.

in Mysore district while in Chamarajanagar district the colony density was 1.10, whereas the Kodagu district has scored only 0.45 colony densities (Table 2). Similarly, the abundance and percent frequency of colony establishment was varied considerably among these districts (Table 2).

Mysore district experiences moderate climate and floral source due to good water source available at many cultivable lands along with tall ramified trees for nesting. While Chamarajanagar district experiences dry climate, water ways are scanty and accordingly floral source was not good.

However, Kodagu district possess good forest coverage with congenial climate and many streams, canals and rivers provide good water source for the luxuriant growth of diversified flowering plants during major part of the year. Despite all these congenial conditions, tall and

Table 3. Nest host tree species and source of flora to A. dorsata.

S/N	District	Nesting — Nest host tree species	Nesting — Habitat	Nesting — Elevation	District	Species	Plant families	Floral source* — Plant types (%)		Forage source from (%)*	
1	Chamarajanagar	08	Trees	5-80 ft	Chamarajanagar	70	36	Herbs	32.0	Ornamental plants	31.0
2	Kodagu	10	Rock cliff	10-30 mt	Kodagu	249	91	Shrubs	26.0	Vegetables	15.0
3	Mysore	17	HBS	8-75 ft	Mysore	368	96	Trees	36.0	Fruit yielding plants	20.0
Total nest host tree species		20	-		Total	687	223	Others	6.0	Timber yielding plants	16.0
Total families		12								Weeds	18.0
								Total	100.0	Total	100.0

*Floral source: Continuous with little variation between different districts of southern Karnataka; HBS = Human built structures.

long branched tree limbs and rock cliffs are not ideal for nesting due to thickly covered epiphytic vegetation during most of the seasons. Perhaps, these features might have not encouraged A. dorsata swarms to establish more colonies at Chamarajanagar and Kodagu districts.

In general, winter scored highest (839) colonies followed by rainy (807) and summer (761) seasons. During winter, the climate is characterized by moderate temperature and humidity with fair floral source at many cultivable lands in southern Karnataka. However, during rainy and summer seasons, the temperature and relative humidity varied considerably along with variable rainfall and resulted with uneven distribution of foraging source. Perhaps during these seasons, A. dorsata might have undergone migration to some other places in search of suitable nest sites and forage. Hence, colonies were less during rainy and summer seasons. Our observations are in conformity with the observations of Dyer and Seeley (1994), Thapa et al. (2000) and Shrestha et al. (2002). A. dorsata established its colonies on various objects such as on the eaves of trees (1646) followed by HBS (580) and on rock cliffs (181) during different seasons. Altogether, 2,407 normal colonies were recorded at various habitats (Table 2). Further, the colony aggregates per 5² m

area was three to six combs with an average aggregate density 1.57, aggregates abundance 3.19 and the frequency of aggregations 134.07 found at different elevations (Table 3) respectively on tall tree limbs, high rock cliffs and certain HBS at Chamarajanagar, Kodagu and Mysore districts (Tables 2 and 3).

Solitary nests (only one or rarely more than two colonies on different branches of a bee tree) and colony aggregations (3 to 15 nests on different branches of a single bee tree) are common on many tall trees with broad limbs and giant rock cliffs amidst dry and wet deciduous forest vegetation, scattered tall trees at cultivable lands and on certain multistoried buildings (5 to 45 colonies on a single HBS) in southern Karnataka. Similar type of reports was made by Sahebzadeh et al. (2012) at Malaysia. However, data on solitary nests, colony aggregations and their abundance are beyond the scope of this paper and such observations along with the aggregation structure and their elevation at various bee trees and HBS in southern Karnataka will be published elsewhere.

Further, the brood, honey comb size and shape of colony indicated considerable variations. In brood chamber, the hexagonal shaped cell had 1.74 ± 0.07 cm depth with 2.36 ± 0.29 cm area.

Moreover, the horizontal and vertical length of brood comb was 38.09 ± 5.78 and 28.63 ± 2.12 with 2.33 ± 0.62 comb width. However, honey comb cells measured highest depth 2.03 ± 0.03 cm and 2.70 ± 0.14 cm area with 1.03 ± 0.84 ml honey storing capacity. But, the horizontal and vertical length of honey comb was less compared to brood comb. As brood part of the comb is meant for developing young ones, the width is normal when compared to honey storing area .

The comb width was very high (14.66 ± 0.51 cm) accordingly at honey storing part of comb. There were five comb shapes namely 'U', 'V', 'Cone', 'Round' and 'Uneven' recorded commonly at various habitats. Among them 'U' shaped comb was most predominant (53.2%) and hence considered as typical shape to assess the size and honey storing capacities. Results of such observations are depicted in Table 2.

Data from the Table 2 clearly demonstrated that, 10,831.5 kg multifloral honey and 1,444.2 kg of beeswax are produced from A. dorsata colonies per year (that is, three to four months of flowering during Kharif and Rabi seasons) indicated the hive products potential (that is, on an average honey yield and beeswax production respectively 4.5 and 0.6 kg per colony) in southern Karnataka. About 20 different tree species belong to twelve

Table 4. *A. dorsata* multifloral honey quality, contamination and harvest.

Multifloral honey quality and contamination

Physico-biochemical characters		Residual pesticides	
pH	3.68 ± 0.1	Organochlorine pesticides	
Specific gravity (g/cm^2)	1.350 ± 0.02	Organophosphate pesticides	
Electrical conductivity (µS/cm)	0.631 ± 0.03	Herbicides	BDL
Absorbance (OD at 359 nm)	2.224 ± 0.38	Pyrethroids	
Turbidity (%)	0.398 ± 0.30	Other pesticides	
		-	
Sugar content (g/100 g)			
Glucose (G)	51.68 ± 0.41		
Fructose (F)	57.36 ± 2.97		
Ratio of G:F	1.126 ± 0.07		
Invert sugar	108.841 ± 3.35		
Total protein content (g/100g)	2.205 ± 0.79		

Multifloral honey

Hunter's education level (%)		Hunting by (%)		Harvesting (%)	
Below primary	12.6	Locals	33.3	Traditional	66.5
Primary	57.8	Jenu Kurubas	67.7	Improved	33.5
High school	21.1	Total	100.0	Total	100.0
College	1.9	-			
Illiterate	6.6				
Total	100.0				

BDL = Below detection limit.

families were opted by *A. dorsata* to establish solitary nest or single-comb and high aggregates of open nests (Table 3). The nesting elevation ranged from five to 80 feet on tree species, 10 to 30 m at rock cliffs and eight to 75 feet at HBS (Table 3) and showed considerable variation in nesting. The single-comb was recorded at lower elevations and aggregates were found on tree limbs of tall bee trees along the rock cliffs. Consistent with the observations of Sahebzadeh et al. (2012), *A. dorsata* solitary nests and aggregations were recorded on specific trees that were selected often for nesting. Presumably, *A. dorsata* use certain criteria to select any particular site for nesting, that is, safety, bear the weight, size of nest, free from predators and enemies. It shows that single or numerous *A. dorsata* colonies can settle on various sites at different elevations which become suitable nesting niche to sustain the colony structure. The results agree with the explanation given by Sahebzadeh et al.

(2012).

Total 687 flowering plants which belong to 223 familes have extended forage interms of nectar and pollen to *A. dorsata* population. Amongst these, trees and herbs contributed respectively 36 and 32% floral source followed by shrubs (26%) and other flora (6%) to *A. dorsata* (Table 3). The existed forage source was further classified into ornamental plants, fruit yielding plants, timber yielding plants, vegetables and weeds, and their percent contribution is depicted in Table 3. Classifying the existing flowering plant source into various types is a common practice to understand the nectar and pollen source to honeybees during different seasons (Basavarajappa et al., 2010). Since, *A. dorsata* is a voracious forager, visited 687 flowering plants to collect nectar and pollen. These plants could bloom with characteristic apicultural values (that is, in terms of nectar, pollen and both nectar and pollen supply) during different seasons and extended continuous floral

source to *A. dorsata* at diversified habitats of southern Karnataka. Similar type of observations was reported by Rao (1973), Basavarajappa et al. (2010) at different habitats of Karnataka. The results agree with the explanation given by the aforementioned authors.

A. dorsata honey quality, its contamination and the process of honey harvesting is shown in Table 4. The multifloral honey had 3.68 ± 0.1 pH, 1.35 ± 0.02 g/cm^2 specific gravity, 0.631 ± 0.03 µS/cm electrical conductivity, 2.224 ± 0.38 optical densities with 0.398 ± 0.30% turbidity. The sugar glucose and fructose content was respectively 51.68 ± 0.41 and 57.36 ± 2.97 with a ratio between glucose and fructose 1.126 ± 0.07. Moreover, the invert sugar content was 108.841± 3.35 (Table 4). Further, the total protein content was 2.205 ± 0.79 g/100 g of honey. Since, honey is one of the internationally traded commodities (Nanda et al., 2003) regular analyses of physical and chemical constituents are essential to maintain internationally acceptable quality. In this regard, researchers from different parts of the world regularly analyze the physico-chemical parameters of honey before it is being marketed. Further, Organochlorine, organophosphate, herbicides and pyrethroid pesticide residues analysis in multifloral honey samples indicated below detection limit (BDL) of all these pesticide residues (Table 4). Thus, results clearly indicated that honey from the colonies of A. dorsata is appropriate to the quality standards of TSE and CODEX and good for human consumption. Our results agree with the earlier reports of Nanda et al. (2003), and Ouchemoukh et al. (2007).

In southern Karnataka, the Jenu Kurubas (67.7%) and local farmers (33.3%) harvest A. dorsata colonies for multifloral honey by employing traditional and improved methods respectively 66.5 and 33.5% (Table 4). The honey hunters are illiterates (6.6%) and not having good education to learn and adopt scientific harvesting methods. This becomes one of the major setbacks for scientific multifloral honey harvest and resulted conventional hunting. During conventional hunting, honey hunters destroy whole colony by using fire torches and burn the hive bees. This caused serious loss to the colony population and increase A. dorsata colony decline.

Although, the combs of A. dorsata in the wild are free from pesticide residues, various biological agents and man-made activities acted as stressors and enhanced the colony decline. Possible causes stressing on the survival of A. dorsata population are depicted in Table 5. The biological agents include predators (11.1%), pests (38.5%) and man-made activities namely pesticide poisoning (1.6%), uprooting of trees (8.2%), trimming of tree limbs (8.4%), cultural practices (8.9%) (Ex. Clearing of weeds by burning, removal of dead tree limbs, fronds etc, in cultivable lands) and conventional honey hunting (10%) have reduced A. dorsata colonies. Further, incidences of colony reduction due to biological agents and man-made activities during different seasons are shown in Table 5. Highest (38.7%) colony decline was recorded due to the infestation of pests (Ex. Galleria mellonella and Neocypholaelaps ampullula followed by predators (11.1%) (Ex. Monkey, Bat, Flying fox, Green bee-eater, Honey buzzard, Drango, Flycatcher, Tree snakes, Garden lizard, Spiders, Wasps etc.), and conventional hunting activities (10%). Uprooting of bee trees, trimming their limbs and cultural practices at rural and urban habitats have caused serious loss (around 8% each) to A. dorsata colony. Accordingly, every year A. dorsata colonies are declining considerably in southern Karnataka (Table 5).Interestingly, the phoretic mite, N. ampullula is first record on A. dorsata and kept in Colombia University, USA under the accession numbers OSAL0106881-886 (Kolmpen, 2011, Personal communication). Although, Neocypholaelaps species is not a major pest of A. dorsata, its frequent occurrence on the body of foraging worker bees has created nuisance in the hive and results in continuous disturbance in the

colony. Moreover, they interfere with normal activities and become troublesome to hive bees. This might have altered the working efficiency of forager bees and hive bees, finally encouraged the disintegration inside the colony. Perhaps, it could weaken the colony population and such colonies become prone to other problems. The greater wax moth, G. mellonella commonly infests weak colonies and abandoned combs found on the tree limbs, rock cliffs and HBS. However, the pathogenic diseases such as bacterial, viral, protozoan and fungal diseases were not identified during the present investigation. Thus, pathogenic diseases were not problematic to A. dorsata population during the present investigation.

A. dorsata cover larger area (ex. 10^2 km) to gather good amount of nectar and pollen to supplement its huge colony population. During its foraging, various flowering plant species which are partly or completely exposed to pesticide sprays during their bloom by farmers were visited. As a result, the nectar and pollen gets contaminated with residues of various pesticides (Bright et al., 1998). During foraging, several hundreds to thousands of forager bees become victims to pesticide poisoning. Moreover, hive bees along with developing brood also become victim to pesticide poisoning. Ultimately this could cause colony decline Bright et al., (1998).

Caron (1978), Seeley et al. (1982), Jadczak (1986), Novogrodzki (1990), Abrol (2003), Kastberger and Sharma (2000), Thapa et al. (2000), Thapa and Wongsiri (2003) and Nagaraja and Rajagopal (2011) have reported the vertebrates predation on honeybee colonies. The monkey (Macaca sp.), drongo (Dicrurus sp.), bee-eater (Merops and Nyctyornis sp.), common crow (Corvus sp.), oriental honey buzzard (Pernis sp.), rats, honey badgers, foxes, lizards and toads feed on honey, brood, pollen and hive bees. Mammals are major enemies of honeybee colonies (Jadczak, 1986). Species of bears are commonly known as mammalian predators to honeybee colonies. They usually dismantle the hives to feed on honey, pollen, brood and adult bees (Jadczak, 1986). They tear the hives into small pieces and carry off honey comb to escape from mass stinging of bees. They also knock down the colonies of A. dorsata. Thus, bears menace is more common during honey flow seasons. Many primates eat honey and brood from the beehive. Macaca sp. has been reported to damage brood of A. florea (Seeley et al., 1982). The monkeys remove adult bees from the comb and feed on honey and brood and it is very common in south India. The monkeys in troop jump on to the beehive open the top cover and shake it for the bees to fly away. Later they carry away with super and brood combs (Nagaraja and Rajgopal, 2011). Nyctyornis species creates nuisance at the hive by making a quick fly fast and disturb A. dorsata colonies (Kastberger and Sharma, 2000). For Merops orientalis, honeybees form a considerable part of its diet. It captures individual bees and rarely attacks A. dorsata hive by

Table 5. Possible causes stressing on the survival of *A. dorsata* population in southern Karnataka.

Possible causes						Incidence				Population decline		
Biological		Man-made		Stressing agents	%	Season				Year	%	Reference
Agents	%	Activities	%			Rainy	Winter	Summer	Mean			
Predators	11.1	Pesticide poisoning	1.6	Predators	1.6	8.8	11.5	13.0	11.1	2006-07	4.0	Basavarajappa et al., 2010
Pests	38.7	Uprooting of trees	8.2	Pests	8.2	32.1	25.0	59.1	38.7	2007-08	10.0	
Pathogens	-	Trimming tree limbs	-	Pathogens	8.4	-	-	-	-	2008-09	6.3	Present work
Mean	24.9	Cultural practices	8.9	Pesticide poisoning	8.9	8.3	13.2	11.8	7.1	2009-10	8.0	
		Conventional honey hunting	10.0	Uprooting of trees	10.0	11.6	10.5	2.6	8.2	2010-11	9.3	
		Others	6.5	Trimming tree limbs	6.5	11.9	11.3	2.0	8.4			
		Mean	7.2	Cultural practices	7.2	2.8	11.7	2.1	8.8	On an average 1.3 to 1.7% decline of *A. dorsata* population recorded every year		
				Conventional honey hunting		14.3	7.0	8.7	10.0			
				Others		9.6	3.9	6.0	6.5			
				Mean		12.2	10.5	11.7	6.7			

passing close to the colonies and feed on individuals that pursue them to a perch (Kastberger and Sharma, 2000). Moreover, flocks of *M. orientalis* were reported to launch coordinated attacks at *A. dorsata* nests at low ambient temperatures when the bees flying performance is impaired (Thapa et al., 2000). Similarly, *Dicrurus* species create nuisance to the hive bees (Nagaraja and Rajagopal, 2011). It is commonly seen at the vicinity of *A. dorsata* colonies and at various apiaries in Karnataka (Nagaraja and Rajagopal, 2011). Further, *Corvus splendens* also troublesome to honeybee colonies, its menace was recorded at different parts of India (Nagaraja and Rajagopal, 2011). *Pernis ptilorhycus* generally hunt *A. dorsata* colonies to collect brood and honey (Thapa and Wongsiri, 2003). In this way all these predators becomes troublesome to hive bees and their brood and stored hive products. Such troublesome activities perhaps weaken the colony gradually and finally initiate the process of colony desertification.

However, in the present investigation certain mammals (ex. Monkey, Bat and Flying fox), birds (ex. Green bee-eater, Honey buzzard and Drango) reptiles (ex. Tree snakes, Garden lizard), arachnids (ex. Spiders) and insects (Predatory wasp, *Vespa cincta*) have directly or indirectly interfered with the strong and weak colonies, collected the developing brood, honey, pollen, hive bees and individual worker bees. All these activities affected the colony integrity and altered the colony strength and influence the process of colony desertification that finally leads to colony abandonment. Thus, the observations are in conformity with the earlier reports of Van Lawick-Goodall (1968), Caron (1978), Seeley et al. (1982), Jadczak (1986), Novogrodzki (1990), Abrol and Kakroo (2000), Kastberger and Sharma (2000), Thapa et al. (2000), Thapa and Wongsiri (2003), Nagaraja and Rajagopal (2011).

Furthermore, other man-made activities namely uprooting of nesting trees, trimming of tree limbs,

cultural practices and conventional honey hunting at cultivable lands, bee trees or at nest sites perhaps developed the nuisance to hive bees in *A. dorsata* colonies. This might have gradually initiated the colony desertification and finally end up with colony abandonment and that results in decline. However, the intensity of all these activities varied considerably during dif-ferent seasons and accordingly *A. dorsata* colony decline has been recorded during different years (Table 5). Highest (10%) *A. dorsata* colony was recorded during 2007-2008 followed by 9.3% during 2010-2011. However, the percent decline was less that is, 4.0, 6.3 and 8 respectively during 2006-2007, 2008-2009 and 2009-2010 (Table 5).During 2007 and 2008, untold incidents such as pesticide poisoning, uprooting and trimming of trees accompanied with cultural practices and conventional honey hunting practices were enor-mous at various habitats of southern Karnataka. Perhaps, this could be one of the reasons for highest decline of *A. dorsata* colonies during

2007-2008. However, on an average 1.3 to 1.7% *A. dorsata* colonies depleting every year at various habitats of southern Karnataka. Thus, it clearly demonstrates that, *A.dorsata* colonies are declining under natural conditions due to various biological agents and unwanted man-made activities at the vicinity of nesting habitats. Hence, colony status of *A. dorsata* is at brink in Southern Karnataka.

SUMMARY AND CONCLUSION

A. dorsata is distributed at various ecosystems by establishing its big sized colonies. *A. dorsata* selected more than twenty tree species, several rock cliffs at hilly areas and HBS for nesting at southern Karnataka. Total 2,407 normal colonies and 2,028 abandoned combs were recorded during different seasons. The solitary nests or single comb and colony aggregates per 5^2 m area was three to six combs with an average aggregate density 1.57, aggregates abundance 3.19 and the frequency of aggregations 134.07 found at different elevations on tall tree limbs, high rock cliffs and at certain HBS in southern Karnataka. The hive products potential per *A. dorsata* colony was 4.5 kg honey and 0.6 kg beeswax and around 10,831.5 kg multifloral honey and 1,444.2 kg beeswax per year from the existing *A. dorsata* colonies. Multifloral honey contained a mixture of sugars namely glucose, fructose, which was in good ratio with required amount of total invert sugars. The total protein content was good with acidic pH.

The specific gravity and the electrical conductivity were in good range. Moreover, the honey had little turbidity with moderate absorbance. Interestingly, none of the honey samples were contaminated with detectable pesticides limit. Thus, the combs of *A. dorsata* in the wild were free from pesticide residues in southern Karnataka. The parasitic/or phoretic mite, *N. ampullula,* greater wax moth, *G. mellonella* infested the weak normal colonies and abandoned combs. Altogether, the developing brood, honey, worker bees and hive bees were predated by certain animals, which altered the live colony integrity and influenced the colony desertification. The pathogenic diseases were not much problematic to *A. dorsata* population and did not show any incidences of pathogenic diseases. Besides parasites and predators impact, the intensive man-made activities at different habitats have encouraged the colony decline 1.3 to 1.7% every year. This is how, *A. dorsata* population decreasing considerably. Although information on colony status could help reveal the current trends of *A. dorsata* population, in depth molecular approaches are essential to support the present work. However, it is beyond the scope of this paper and results of such studies shall be published elsewhere. Thus, multifaceted approach in the present investigations could help assess the colony status of *A. dorsata*, which is at threatened state and require conservation in its natural abode.

ACKNOWLEDGEMENT

The authors are very much grateful to the University Grants Commission and UGC-SAP-Phase III , New Delhi for the financial assistance. They are greatly indebted to Dr. Diana Sammataro, Research Entomologist, USDA-ARS, Carl Hayden Honey Bee Research Center, USA and Dr. Hans Kolmpen, Ohio State University, Columbus, USA for guidance and identification of ectoparasitic phoretic mite of *A. dorsata*. They sincerely thank The Principal Chief Conservator of Forests (Wildlife) & Chief Wildlife Warden, Bangalore for granting the permission to conduct field survey at certain forest areas of southern Karnataka. Thanks are also due to M/s. Shiva Analyticals (India) Limited, Bangalore for their co-operation during the conduct of pesticide analysis and the India Meteorological Station, Bangalore for the supply of meteorological details. Sincere thanks are also due to the Chairman, DOS in Zoology, University of Mysore, India for the necessary laboratory facilities.

REFERENCES

Abrol DP, Kakroo SK (2002). Pest, Predators and Pathogens of honeybee, Apis mellifera L. in Jammu, India. Proceedings of 6th AAA Int. Conf.. World Apiexpo. Bangalore, India. p. 112.
Abrol DP (2003). Honeybee Diseases and their Management. Kalyani Publishers, New Delhi. pp. 70-102.
Anklam E (1998). A review of the analytical methods to determine the geographical and botanical origin of honey. J. Food Chem. 63(4):549-562.
Anupama K, Bhat K, Sapna VK (2003). Sensory and physico-chemical properties of commercial samples of honey. Food Res. Int. 36(2):183-191.
Bailey L, Ball (1991). Honeybee Pathology. 2nd Edn. Academic Press, London. pp. 10-110.
Basavarajappa S, Raghunandan KS, Hegde SN (2010). Seasonal incidence of the parasitic mite (Arachnida: Acarinidae) on Apis dorsata F. (Hymenoptera: Apidae) in Mysore, Karnataka. Hexapoda 17(2):166-173.
Blascoc C, Linoo M, Pico Y, Pena Y, Font AG, Silverira MIN (2004). Determination of organochlorine pesticide residues in honey from the central zone of Portugal and Valencian community of Spain. J. Chromatogr. 1049:155-160.
Bradbear N, Reddy MS (1998). Existing apicultural practices within Karnataka.Mission Report, FAO and UN, Pune. pp. 1-20.
Bright AA, Chandrasekaran M, Muthuswami M (1998). Bee pollinators importance and preservation. Kissan World 25(4):61-63.
Burnham KP. Anderson DR, Laake JL (1980). Estimation of density from line transect sampling of biological populations. Wildlife Monograph 72:202.
Caron DM (1978). Marsupials and Mammals. In: Honeybees pests, predators and diseases. (Edn. Morse RA). Cornell Uni. Press, Ithaca and London. pp. 227-256.
Dyer FC, Seeley TD (1994). Colony migration in the tropical honey bee Apis dorsata F. (Hymenoptera: Apidae). Insectes Soc. 41:129-140.
Engel MS (1999). The taxonomy of recent and fossil honey bees (Hymenoptera: Apidae; Apis). J. Hym. Res. 8:165-196.
Hepburn R, Hepburn C (2007). Bibiliography of Apis cerana Fabricius (1793). Apidologie 37:651-652.
Hepburn Radloff. (2011). Honeybees of Asia. Springer-Veerlag Berlin

Hidelberg. New York.

Hepburn RH, Hepburn C (2007). Bibliography of the giant honeybee of Apis dorsata Fabricius (1793) and Apis laboriosa F. Smith (1871). Apidologie 38:219.

Jadczak AM (1986). Honeybee diseases and Pests. Maine Dept. Agric. Food and Rural Resources. USA. pp. 24.

Jimenez JJ, Bernal JL, Nozal MJ, Novo M, Higes M, Llorente J (2000). Determination of rotenone residues in raw honey by solid-phase extraction and high-performance liquid chromatography. J. Chromatogr. 871:67-73.

Kamath US (2001). Karanataka State Gazetter. Govt. Press. Bangalore. p. 1-49.

Kastberger G, Sharma DK (2000). The predator-prey interactions between blue bearded bee eaters (Nyctornis athertoni Jardine and Selby) and giant honeybees (Apis dorsata F.) Apidologie 31:727-736.

Kolmpen H (2011). Personal Communication.

Krishnamurthy HR (2001). How to measure a height of the tree. Vignana Sangaathi, Hampi Uni. Hampi. p. 11.

Morse RA, Laigo FM (1969). The mite Tropilaelaps clareae in Apis dorsata colonies in the Philippines. Bee World 49:116-118.

Nagaraja N, Rajagopal D (2011). Honeybees – Diseases, Parasites, Pests, Predators and their Management. MJP publishers, Chennai, India. pp. 74-90.

Nanda V, Sarkar BC, Sharma HK and Bawa AS (2003). Physico-chemical properties and estimation of mineral content in honey produced from different plants in Northern India. J. Food Composition and Analysis. 16:613-619.

Novogrodzki R (1990). Amphibians and reptiles. In: Honeybee pests, predators and diseases (Edn. Morse RA.), Cornell Uni. Press, Ithaca and London. pp. 227-256.

Oldroyd BP, Osborne KE, Mardan M (2000). Colony relatedness in aggregations of Apis dorsata Fabricius (Hymenoptera: Apidae). Insectes Soc. 47:94- 95.

Ouchemoukh S, Louaileche H, Schweitzer P (2007). Physicochemical Characteristics and pollen spectrum of some Algerian honeys. J. Food Control. 18:52-58.

Paar J, Oldroyd BP, Huettinger E, Kastberger G (2004). Genetic structure of an Apis dorsata population: The Significance of Migration and Colony Aggregation. J. Heredity 95(2):119-126.

Phillips J (1959). Succession, development, the climax and complex organism. An analysis of concepts. Part 1 and 2. J. Ecol. 22:559-571, 23:488-508.

Rao RR (1973). Studies on the flowering plants of Mysore district. Ph.D., thesis, Uni. Mysore, Mysore. pp. 428-1000.

Reddy CC (2002). Beekeeping policies and programs in India. 6th AAA Inter. Con. & World Apiexpo.2002. February 24- March 1, 2002 Bangalore- India. p. 03.

Rissato SR, Galhiane MS, Almeida MV, Gerenutti MB, Apon M (2007). Multiresidue determination of pesticides in honey samples by gas chromatography-mass spectrometery and application in environmental contamination. J. Food Chem. 101:1719-1726.

Ruttner F (1988). Biogeography and Taxonomy of honeybees. Springer-Verlag, Berlin. pp. xxii+284.

Sahebzadeh N, Mardan M, Ali AM, Tan SG, Adam NA, Lau WH (2012). Genetic Relatedness of Low Solitary Nests of Apis dorsata from Marang, Terengganu, Malaysia. PLoS ONE 7(7):1-9.

Savanurmath, CJ, Basavarajappa S, Ingalahalli SS, Sanakal RD, Singh KK, Hinchigeri SB. (1993). A Status Paper on: The Sericultural activities and the incidence of silkworm viral diseases in northern districts of Karnataka. J. Karnatak Univ. Sci. 36:197-211.

Seeley TD, Seeley RH, Akratanakul P (1982). Colony defense strategies of the honeybees in Thailand. Ecol. Monographs 52:43-63.

Shrestha JB, Mandal CK, Shrestha SM, Ahmad F (2002).The trend of the giant honeybee Apis dorsata Fab. colony migration in Chitwan, Nepal. Wildlife 7:16-20.

Shukla GS, Upadhyay VB (2007) Economic Zoology. pp. 188-201. Rastogi Publ. Meerut, India.

Setty SR, Bawa KS (2002). Characteristics of honey resources in a tropical forest:Productivity and extraction of Apis dorsata honey in Biligiri Rangan Hills Widlife sanctuary India. 6th AAA Inter. Con. & World Apiexpo.2002. February 24- March 1, 2002 Bangalore, India. p. 54.

Tan NQ (2007). Biology of Apis dorsata in Vietnam. Apidologie 38:221-229.

Thapa R, Wongsiri S (2003). Flying predators of the giant honeybees, Apis dorsata and Apis laboriosa in Nepal. Am. Bee J. 143:540-542.

Thapa R, Wongsiri S, Oldroyd BP, Prawan S (2000). Migration of Apis dorsata F. in Northern Thailand. In: Matsuka, L.R.Verma, S. Wongsiri, K.K.Shrestha and U.Prathap (Eds). Asian bees and beekeeping: Progress of research and development. Oxford and IBH Publishing Co. Pvt. Ltd. New Delhi, India. pp. 39-43.

Van Lawick-Goodall J (1968). The behavior of free living chimpanzees in the Gombe stream reserve. Anim. Behav. Mon. 1:159-131.

Vinutha S (1998). Morphometric studies on the rock bee, Apis dorsata Fabr. (Hymenoptera : Apidae). III Congress of IUSSI Indian Chap. Nat. Symp. Div. Soci. Insects & Arthropods and Func. of Eco. March 7th – 9th .p. 19.

Woyke J (2008). Why the eversion of endophallus of honeybee drone stops at the partly everted stage and significance of this. Apidologie 39:627-636.

19

Effects of age, breed and sex on the serum biochemical values of Turkeys (*Meleagridis gallopova*) in South-eastern Nigeria

Ogundu Uduak E, Okoro V. M. O., Okeke G. U., Durugo N., Mbaebie G. A. C. and Ezebuike C. I.

Department of Animal Science and Technology, Federal University of Technology, Owerri, Imo State, Nigeria.

This study was carried out to establish the serum biochemical values of turkey (*Meleagridis gallopova*) as well as determine the sex, age and breed effects on these characters. Blood samples collected in the morning from each turkey at 6th and 12th week of age, comprising of 60 black and 45 white breeds were used in this study. Total protein, albumin and alkaline phosphate means were 5.56±0.15, 2.5±0.1 and 26.44±1.55 g/dl respectively. Cholesterol, aspartate transaminase (AST) and alanine transaminase (ALT) means were 161±2.83 mg/dl, 74.87±1.87 U/L and 17.86±1.33 U/L respectively while globulin had a mean of 3.16±0.2 mg/dl. The mean total protein value (4.81±0.17 and 6.38±0.28 g/dl) of black and white breeds respectively indicated significant difference ($P<0.05$). Albumin, cholesterol and globulin values also indicated significant difference ($P<0.05$) between the breeds, while alkaline phosphate, AST and ALT showed no significant difference ($P<0.05$) between the breeds. The mean total protein, albumin, globulin and AST of the turkeys revealed significant difference ($P<0.05$) between ages 6 and 12 weeks while alkaline phosphate, cholesterol and ALT showed no significant ($P>0.05$) difference between the ages. Sex generally had no significant ($P>0.05$) effect on the characters.

Key words: Turkeys, serum, age, breed, sex, biochemical values.

INTRODUCTION

Publications for white blood cell counts and differentials, serum albumin, aspartate transaminase (AST) and alanine transaminase (ALT) have not been previously reported for domestic or wild turkeys (Bounous et al., 2000). Strong differences are reported (Sturkie and Textor, 1978; Warren, 1995) between serum chemistry characteristics of local breeds of turkey at different ages from different geographical and agricultural zones of the world. Serum biochemical values have been established in most domestic mammalian species (Jain, 1986; Adejumo et al., 2005). However, limited information is available for domestic avian species, (Ola et al., 2000; Oke et al., 2001) and even less has been established for

Turkey species more so in the local breeds (Zinki, 1986). Publications for parameters available in domestic turkeys are limited to packed cell volume, cholesterol, glucose, calcium and total protein (Rhian et al., 1944; Bell et al., 1957; Bell and Sturkie, 1965). These findings were aimed at establishing the diagnostic baseline of blood characteristics of farm animal (Orji et al., 1986b).

Research (Pagot, 1992; Cannon, 1992) has shown that biochemical characteristics of blood - albumins, transferine, alkaline phosphatase, immunoglobulin G_1 and G_2 are established genetic markers. These biochemical and molecular characters enable the study of the genetic structure or parentage situation of livestock population under small holder conditions like ours in the tropics. Phenotypic characterization could also be supported by genotype identification through laboratory studies of gene markers or biochemical polymorphisms (Gall et al., 1994).

This reference data presented for biochemical values could significantly boost the genetic improvement and breeding of the turkeys through the use of marker assisted selection by providing baseline information. There is need therefore to establish the reference data to characterize the blood serum biochemistry for the turkeys in Nigeria.

MATERIALS AND METHODS

Source of the birds

70 white and 70 black turkey poults were procured at day old from Gofons Hatcheries Nigeria Limited, Owerri, Imo State, Nigeria. They were raised at the poultry unit of Federal University of Technology Owerri Teaching and Research farm from day old to 12 weeks of age.

The experiment was conducted between May 1st and August 30th 2006. Prestarter diet of 29.76% crude protein, starter diet of 26.4% crude protein and finisher diet of 21.35% crude protein were offered ad-libitum from day old to 12 weeks of age. Clean water was also offered ad-libitum throughout the experiment. The birds were aged approximately 6 weeks at first bleeding and 12 weeks old at second bleeding. They had mean body weight of 2.5 and 4.6 kg at 6 weeks and 12 weeks old respectively. In addition, the breed weighed 3.4 and 4.2 kg respectively for black and white breeds. They were also certified clinically healthy at the time of bleeding.

Blood collection

Blood samples were collected from 60 black and 45 white turkeys randomly at 6 and 12 weeks of age. A 23-G needle was used to puncture the brachial vein after cleaning the area with methylated spirit, and blood sample collected. A total of 8 ml of blood was collected from each bird; 3 ml was transferred into ethylene diamine tetra acetic (EDTA) acid treated bottle, while the remaining 5 ml was left in the syringe to coagulate to produce sera for blood biochemistry analysis.

Blood biochemistry analysis

The blood sample in the syringes were spinned on a centrifuge at 3000 rpm and the serum collected with a Pasteur pipette, transferred into a serum bottle and then stored at 2 to 8°C in a refrigerator for serum analysis. Total serum protein was determined by using the quantitative in vitro procedure with the burette method according to Tietz (1995). The serum protein concentration (g/dl) for each blood sample was read off on spectrophotometer. The serum enzyme - AST, ALT and Alkaline phosphate were determined by using RANDOX SGOT, RANDOX SGPT and RANDOX ALP kits respectively. AST otherwise known as Glutamic-oxaloacetic transaminase was measured by monitoring the concentration of oxaloacetate hydrazones formed with 2, 4-dinitrophenyl hydrazine. ALT- Glutamic pyruvic transaminase was measured by monitoring the concentration of pyruvate hydrazones formed with 2, 4 - dinitrophenyl hydrazine. Alkaline phosphate was measured by the concentration of phosphate from P-nitrophenyl phosphate. Albumin and globulin were determined as component parts from which the

total protein was derived from. Cholesterol was determined spectrophotometrically in plasma.

Experimental design

A 2×2×2 factorial experiment in a completely randomized design model was used to estimate the treatments effects, that is, age, sex and breed effects. The model is as follows:

$$Y_{ij} = \mu + S_i + e_{ij}$$

Where Y_{ij} is the individual observation; μ = the population mean; S_i = the sex, age and breed effects on blood parameters, and e_{ij} = the error term.

Measurements were taken according to age, sex and breed with each treatment replicated three times. ANOVA procedure according to SAS (1999) was used to test for significant differences between the treatment means. All recorded weights and serum biochemical values were expressed as means and standard error of the means (Mean ± SEM). The means were then subjected to the Duncan new multiple range tests as outlined by SAS® software pack (SAS version 8) to separate the means that were significantly different.

RESULTS AND DISCUSSION

Table 1 shows the means for normal serum biochemical values for Turkeys. Table 2 shows the breed effect on serum biochemical values at 12 weeks of age. The white breeds showed significantly higher total protein and globulin while the black breeds showed significantly higher albumin and cholesterol. There was no significant breed effect on alkaline phosphate, aspartate transaminase (AST) and alanine transaminase (ALT).

Table 3 presents the age effect of serum biochemical values at 6 and 12 weeks of age. The total protein globulin and aspartate transaminase showed significantly higher values at 6 weeks, while albumin was significantly higher at 12 weeks. Table 4 presents the sex effect of serum biochemical values on turkey. The results observed showed that sex has no effect on the serum biochemical values of turkeys.

The females had higher mean values for alkaline phosphate, globulin and ALT, but were not statistically significant. The slightly higher mean values for total protein and albumin (5.56±0.15 2.51±0.1 g/d respectively) in this study were similar to the reported values (3.6-5.5 and 1.1-2.1 g/dl) by Bounous et al. (2000) as established reference interval for 4 months old turkey in temperate environment. Our values for total protein and albumin compared favorably with the findings of Verma et al. (1975) for adult domestic chicken, Orji et al. (1986a) and Ozbey et al. (2004) for the guinea fowl. This could be attributed to temperature fluctuations in the tropics which have effect on the blood biochemical parameters and subsequently on production. The ALP value observed in this study was higher when compared with values (225 499 iu/L) reported by Bounous et al. (2000). Quist et al (2000) reported similar low value (353.36 mg/dl) in the control experiment with quails. Its high value could be

Table 1. Means for normal serum biochemistry analytes for Turkeys.

Analyte	Mean ± SEM
Total protein (g/dl)	5.56±0.15
Albumin (g/dl)	2.51±0.1
Alkaline phosphate (g/dl)	26.44±1.55
Cholesterol (mg/dl)	161.42±2.83
Globulin (mg/dl)	3.16±0.20
Aspartate transaminase AST (U/L)	74.87±1.87
Alanine transaminase ALT (U/L)	17.86±1.33

Means with different superscripts are significantly (P<0.05)

Table 2. Breed Effect on Serum Biochemistry of Turkey at 12 weeks of age.

Analyte	Mean ±SEM of black breeds	Mean ±SEM of white breeds
Total protein (g/dl)	4.81±0.17[b]	6.38±0.23[a]
Albumin (g/dl)	2.67±0.10[a]	2.34±0.16[b]
Alkaline phosphate (g/dl)	24.70±2.06	28.12±2.31
Cholesterol (mg/dl)	171.46±4.11[a]	150.63±3.60[b]
Globulin (mg/dl)	2.10±0.18[b]	4.39±0.33[a]
Aspartate transaminase AST (U/L)	76.91±2.72	72.79±2.56
Alanine transaminase ALT (U/L)	18.42±1.38	17.30±2.26

Means with different superscripts are significantly different (P<0.05).

Table 3. Age effect on serum biochemistry analytes for 6weeks and 12weeks old turkeys.

Analyte	Mean ± SEM of 6 weeks old	Mean ± SEM of 12 weeks old
Total protein (g/dl)	6.22±0.33[a]	4.90±0.16[b]
Albumin (g/dl)	1.46±0.12[b]	3.48±0.12[a]
Alkaline phosphate (g/dl)	23.41±2.57	28.97±3.37
Cholesterol (mg/dl)	160.91±5.44	161.90±5.47
Globulin (mg/dl)	4.74±0.33[a]	1.39±0.11[b]
Aspartate transaminase AST (U/L)	78.95±2.67[a]	70.87±4.14[b]
Alanine transaminase ALT (U/L)	17.89±2.83	17.82±2.09

Means with different superscripts are significantly different (P<0.05).

Table 4. Sex effect on serum biochemistry analytes for turkeys.

Analyte	Mean ±SEM of males	Mean ±SEM of females
Total protein (g/dl)	5.66±0.58	5.53±0.63
Albumin (g/dl)	2.48±0.55	2.13±0.15
Alkaline phosphate (g/dl)	22.57±1.55	27.96±4.87
Cholesterol (mg/dl)	165.34±4.30	149.21±8.04
Globulin (mg/dl)	3.18±0.99	3.38±1.10
Aspartate transaminase AST (U/L)	73.36±4.19	70.63±4.97
Alanine transaminase ALT (U/L)	15.09±2.04	16.47±1.43

Means with no superscripts are not significantly different (P > 0.05).

related to the calcification of turkey bones. The serum cholesterol value was within the range for guinea fowls reported (Ozbey et al, 2004) giving the impression that the serum cholesterol levels is not affected by the temperature variation in the tropics.

The AST value observed in our study was low when compared with the reference interval observed by Bounous et al. (2000). The increase in temperature to 33°C reported (Ozbey et al., 2004) did not have a significant effect on AST value in quail and as such, the low value observed in this study could be attributed to specie difference. The ALT value observed in this study was slightly higher than reported in the literature (Ozbey et al., 2004).

The differences (P<0.05) observed in relation to breed revealed that the white breed possesses higher values for the total protein and globulin, while higher values for albumin and cholesterol were observed in the black breed. These findings were inline with the observations of Nwosu (1979) and Oluyemi (1998) who reported significant differences in serum biochemical values for chicken and duck due to breed and species. The apparently higher values for serum total protein and globulin observed in the white over the black breed might be due to inherent physiological traits in the breed. The higher globulin value also gives the impression that the white breed can resist infection better, recalling that globulin binds with hemoglobin and other niafers and fights infection.

The higher values obtained in the black breed for albumin and cholesterol suggests that the black breed can stand more stress. This could be attributed to the fact that album as a reservoir of protein contributes to the colloidal osmotic pressure, acid base balance and also carrier for small molecules like vitamins, minerals, hormone and fatty acids. While cholesterol being a precursor of steroid hormones, vitamin D and bile acids could suggest a better reproductive performance of the black breed.

Age difference observed for total protein, globulin and AST at 6 weeks could be attributed to the fact that the young animal whose feed intake comprises of higher proportion of protein for growth and muscle development will exhibit more in the serum. This will be subsequently used up for the set purpose. The superiority of the globulin at this age gives the impression of a better resistance and survival at this age. The apparent superiority of albumin in the 12 weeks old could be as a result of the accumulated total protein breakdown since it serves as the major reservoir of protein.

Conclusion

Turkey improvement over the years has been based primarily on selection programmes with considerations of the body traits as key breeding parameter. This study therefore has tried to establish baseline information on serum biochemical markers which can be used for characterization of turkey and for genetic evaluation of poultry as stated by Nwosu (2005) as vital breeding parameters for the prediction of these growth traits.

REFERENCES

Adejumo DO, Ladokun AO, Ososanya TO, Sokunbi OA, Akinyemi A (2005). Hematology and serum biochemical changes in castrated and intact male weaner pigs administered Testosterone Enanthate. Proceedings of 10th Annual Conf. Anim. Sci. Ass. Of Nig (ASAN), Univ. of Ado-Ekiti, Sept, 12-15 2005. pp. 93-95.

Bell DJ, Sturkie PD (1965). Chemical constituents of blood. In Avian Physiology, 2nd Edition. Comstock Publishing Associates, Ithaca, New York. pp. 32-84.

BELL DJ, MCINDOE WM, GROSS D (1957).Tissue components of the domestic fowl. 3. The non-protein nitrogen of plasma and erythrocytes.Biochem. J. 71:355–364

Bounous RD, Wyatt RD, Gibb PS, Kilburn JV, Quist CF (2000). Normal Hematologic and Serum Biochemical Reference Interval for Juvenile Wild Turkeys. J. Wildl. Dis. 36(2):393-396.

Cannon MS (1992). The morphology and cytochemistry of the blood leukocytes of Kemp's ridley sea turtle (Lepidochelys kempi). Canadian Journal of Zoology. 70:1336-1340.

Jain NC (1986). Schalms Veterinary Hematology, 4th Edition. Lea and Febiger, Philadelphia, Pennsylvania. P. 932.

LE-GALL O, TORREGROSA L, DANGLO T, CANDRESSE T, BOUQUET A (1994). Agrobacterium-mediated genetic transformation of grapevine somatic embryos and regeneration of transgenic plants expressing the coat protein of grapevine chrome mosaic nepovirus (GCMV). Plant Sci. 102:161-170.

Nwosu CC (1979). Characterization of the local chicken of Nigeria and its potential for egg and meat production, in: Proc. 1st National Seminar on Poultry Production, Ahmadu Bello University, Zaria. pp. 187-210.

Nwosu CC (2005). Strategies for the improvement of Animal Genetic Resources in developing Countries. An Invited paper on the plenary session of the Genetic Soc. Of Nig. Annual Conference 2005. pp. 2-14.

Oluyemi JA, Ologhobo AD (1998). The significance and the management of the local ducks in Nigeria. In: Sustainablility of the Nigerian Livestock Industry in 2000AD. Egbunike, G. N. and E. D. Iyayi (eds) pp. 96-103.

Orji BI, Okeke GC, Ojo OO (1986b). Hematological studies on the Guinea fowl (Numida meleagridis pallas): II. Effect of age, sex and time of bleeding on protein and electrolyte levels in blood serum of guinea fowls. Niger. J. Anim. Prod. 13:100-106.

Ozbey O, Yildiz N, Aysondu MH, Ozmen O (2004). The effect of high temperature on blood serum parameter and the egg productivity characteristics of Japanese Quails (Cortunix coturnix japonica). Int. J Poult. Sci. 3(7):485-489.

Pagot J (1992). Animal production in the tropics and sub-tropics Macmillan. pp. 239-310.

Quist CF, Bounous DI, Kilburn JV, Nettles VF, Wyatt RD (2000). The effect of dietary aflatoxin on wild turkey poults. J. Wildl. Dis. 36(3):436 444.

Rhian M, Wilson WO, Moxon AC (1944). Composition of blood of normal Turkey. Poult. Sci. 23:224-229.

SAS (1999). SAS Users Guide. Statistics released version 8.0 Statistical Analysis System Institute, Inc. Cary, N. C.

Sturkie PD, Textor K (1978). Sedimentation of erythrocytes in chickens as influenced by method and sex. Poult. Sci. 39:444-447.

Tietz NW (1995). Clinical guide to laboratory tests. 3rd Ed. WB Saunders Company. Philadelphia. P. A. pp. 518-519.

Verma PN, Rawat JS, Pandy MD (1975). Effect of age and sex on the serum proteins of the White Leghorn birds. Indian Vet. J. 52(7):544 546.

Warren AG (1995). Ducks and geese in the tropics. World Anim. Rev 3:35-36.

Zinki JG (1986). Avian hematology. In Schalm's Veterinary hematology 4th Edition. N. C. Jain (ed). Lea and Febiger, Philadelphia Pennsylvania. pp. 256-273.

20

Statutory regulations of dead animal carcass disposal in Nigeria: A case study of Enugu State

ONYIMONYI, Anselm Ego, MACHEBE, Ndubuisi Samuel and UGWUOKE, Jervas

Department of Animal Science, University of Nigeria, Nsukka, Enugu State, Nigeria.

The present study examined the statutory regulations governing the disposal of dead animal carcasses in Nigeria. A detailed literature review of the criminal code (Cap 77 Laws of the Federation of Nigeria, 1990), Animal Diseases (control) Act [Cap 18 LFN1990 and a structured interview of 120 livestock farmers in Enugu State of Nigeria was carried out. The Criminal Code and Animal Diseases (control) Act have numerous provisions that offer protection to live animals from being deliberately infested with disease and in the case of death strictly specifies the manner in which such dead animal carcass shall be disposed. Results obtained from the structured interview shows that 87% of farmers interviewed dispose dead young/immature animal carcasses by burning their carcass, whereas 13% resort to burying. The carcasses of matured dead animals are offered to the unsuspecting consumer as meat. Enforcement of the relevant provision of the statutes mentioned above is practically not in place. No prosecution of any offender of the provisions of these statues is known. It is concluded that whereas there are enabling statutory provisions that clearly stipulates the manner in which dead animal carcass shall be disposed in Nigeria, what is obtainable in practice is totally in contrast with the provisions of the statutes.

Key words: Statutory, dead, carcasses, Nigeria.

INTRODUCTION

Nigeria, the most populous black nation is still battling to meet the animal protein requirement of its citizens. Average animal protein consumption per caput per day is far below the recommendation of national and international organizations. There are clear indicators that average animal consumption may have declined to very low levels as evidenced by clear clinical manifestations of animal protein deprivation in the diets of majority of the population, especially children. The common animals slaughtered for meat in Nigeria are cattle, goats, sheep, pigs and poultry. Others include camel, buffaloes, donkeys, horses, rabbits and other games and forest animals that are edible (Addas et al., 2010).

The management of these animals is becoming increasingly difficult in the face of emerging climatic variables. The increasing temperature and relative humidity occasioned by changing climates have resulted in cases of death as a result of heat stress. There are also cases of emerging and re-emerging diseases of animals. Where this disease challenge is not properly managed, death may result. What then happens to the carcass of such a dead animal? Are there statutory provisions spelling out how such a dead carcass shall be disposed? Where such provisions exist, are the followed? These are the basic research questions this paper seeks to address.

MATERIALS AND METHODS

The study area

The study was carried out in Enugu State of Nigeria. Enugu State is one of the 36 states in Nigeria and is located between latitude 5°56'N and 7° 06'N, and longitude 6° 53'E and 7° 55'E (Ezike, 1998). The State has 17 Local Government Areas (LGA). NPC (2006) census report showed that the State has a population of 2,452,996. The vegetation of the State is mainly forest type but stretches out into derived savannah in the northern fringes. Enugu State experiences distinct wet and dry seasons with a total annual rainfall of about 1,700 mm (Enugu State Government Official Gazette, No. 25, 1997). Farming is the major occupation of people in the State. Major crops cultivated include, cassava, yam, cocoyam, vegetables, oil palm etc, while major livestock reared are poultry, goat, sheep and cattle. ENADEP (2007) report showed that Enugu State is made up of three agricultural zones namely: Enugu North zone, Enugu East zone and Enugu West zone.

Sampling technique and data analysis

The study was conducted using the following approach/methodology (i) review of existing regulations on dead animal disposal in Nigeria. This involved a detailed literature review of the Criminal Code (Cap 77 laws of the Federation of Nigeria, 1990), Animal Diseases (control) Act cap 18 of the LFN 1990 and (ii) interview of livestock farmers in Enugu State of Nigeria. The interview was conducted using a structured questionnaire designed to capture the background information of the farmers, farmer's awareness on existing statutory regulations on dead carcass disposal and methods of carcass disposal systems practiced by the farmers.

A total of 120 livestock farmers in the State constituted the population for the study. A multi-stage random sampling technique as described by Ozor and Nnaji (2011) was used to select livestock farmers in the State. Thus, two (Enugu North and Enugu West) out of the three agricultural zones in the State were randomly selected and two LGAs were also randomly drawn from each of the two zones. For Enugu North zone, Nsukka and Udenu LGAs were randomly selected, while Udi and Oji River LGAs were selected from Enugu West zone. Within a LGA, two communities were also randomly drawn to give a total of eight communities. They include Ibagwa-Ani and Okpuje (Nsukka), Obollo Eke and Amala (Udenu), Agbala-Enyi and Ugwuoba (Oji River), and Awhum and Nsude (Udi). Fifteen (15) livestock farmers were selected from each of the town community making a total of 120 respondents. Variables investigated include demographic characteristics of livestock farmers, farmers' awareness on existing regulatory control on dead carcass disposal, and methods of carcass disposal practiced by farmers in the State. Data were analyzed and presented using frequencies and percentages.

RESULTS AND DISCUSSION

Demographic characteristics of livestock farmers in Enugu State

Demographic characteristics of livestock farmers in Enugu State are presented in Table 1. Most of the livestock owners in Enugu State were men (81.7%), while 22% were women. Thus, livestock enterprise in the state is a male dominated enterprise. This agrees with the findings of Oni and Yusuf (1999). From the analysis, it

can be adduced that the number of livestock farmers of 52 within age bracket <30 years is six times those in age bracket >50 years with a frequency of 8%. According to Ajala et al. (2007), farmers within age bracket (<30 years) are young and can easily take the risk of accepting new innovations aimed at improving livestock production. These young farmers can still face the challenges of livestock enterprise despite the huge labour demand.

The results also revealed that majority of the respondents (55%) were married, whereas 45% were single. Only about 6.7% of respondents were widow/widower. The high number of married men in livestock enterprise in the State may not be surprising because farmers rely very heavily on family labour which is usually not accounted for in the entire cost of production. Livestock farming in the State is the major occupation of majority (58.3%) of the respondents. Only about 30.8% of livestock farmers combine livestock farming with other businesses like trading. Few civil servants (6.7%) and retirees (3.3%) are into livestock enterprise in the State. The involvement of traders, civil servants and retirees in livestock production could be a way of diversifying their income base (Nwanta et al., 2011). Recently, the Federal Government of Nigeria reiterated the need for Nigerians to go into farming as an alternative source of income for the country outside crude oil. This may account for the increase in the number of farmers that are engaged solely in livestock farming.

The result also showed that majority of livestock farmers were holders of First School Leaving Certificates, while 30% had West Africa School Certificate. Farmers with Post Secondary Educational Certificates were about 20%. Very few livestock farmers had no formal education (5%). The high number of livestock farmers who had formal education compare to the low number in those who had no formal education is a welcomed development. Similar findings have been documented (Adesehinwa et al., 2003; Nwanta et al., 2011). It is obvious that the level of educational development of livestock farmers in the study area would help in bridging communication gap between extension officers and farmers especially in the adoption and application of new technologies in livestock production and management. This concurs with the report by Nwanta et al. (2011) and Mishra et al. (2009). Mishra et al. (2009) reported that education and training improves business performance and returns of farmers through the adoption of better technology and management practices. The study revealed that majority of the farmers (41.7%) had 6 to 10 years farming experience, while about 9.2% had 1 year farming experience. Sixteen (16) farmers (13.3%) had above 10 years farming experience.

Statutory regulations on disposal of dead animal carcass

The statutes examined seem not to provide for the

Table 1. Demographic characteristics of livestock farmers in Enugu State of Nigeria.

Characteristic	No. of respondents	Percent
Sex		
Male	98	81.7
Female	22	22
Occupation		
Sole farming	70	58.3
Trading and farming	37	30.8
Civil servant	8	6.7
Others (Retiree and farming)	4	3.3
No response	1	0.9
Level of education		
No Formal Education	6	5
First School Leaving Certificate	49	40.8
West African School Certificate	36	30
Post Secondary Education (HND, B.Sc, M,Sc, PhD)	24	20
No response	5	4.2
Farmers' experience		
1 year	11	9.2
1 - 5 years	40	33.3
6 - 10 years	50	41.7
>10 years	16	13.3
No response	3	2.5
Age (years)		
<30	31	25.8
30 - 40	52	43.3
40 - 50	29	24.2
>50	8	6.7
Marital status		
Married	60	50
Single	54	45
Widow/widower	6	5

disposal of dead animal carcasses. Available provisions are concerned with the disposal of carcasses of diseased animals. Does it mean that the legislators never contemplated that there could be other causes of death apart from disease? Or is that the principle that any livestock found dead from unknown cause is presumed to have died because of disease applies? (agr.wa.gov/FoodAnimal/AnimalHealth/docs/LivestockDisposalManual10709.pdf) S.244 of the Criminal Code Act Cap 77 LFN 1990 provides that: Any person who; (1) knowingly takes into a slaughter house for the slaughter of any animal intended for the food of man, the whole or any part of the carcass of any animal which has died of any diseases or (2) knowingly sells the whole or part of the carcass of any animal which has died of any disease or which was diseased when slaughtered is guilty of a misdemeanor and is liable to imprisonment for 2 years. Paragraph (b) of Section 247 of the same code provides that any person who does any act which is, and which he

knows or has reason to believe to be likely to spread the infection of any diseases dangerous to life, whether human or animal is guilty of a misdemeanor and is liable to imprisonment to 6 months.

Sections 8 of the Animal Diseases (Control) Act Cap 18 LFN 1990 also have a provision that further protects the unsuspecting consumer from access to diseased meat. Subsection (1) of that section provides that: any person having in his charge or under his control any animal infected or suspected to be infected with any of the diseases listed in the first schedule to this Act shall keep such animal separate from other animals not so infested or suspected to be infected and shall forthwith give notice of the fact of the animal being so infected or suspected to be infected to a veterinary officer or the nearest veterinary surgeon or the prescribed officer in the LGA. Subsection (4) further provides that a veterinary officer, if he is of the opinion that any animal is infected with any disease, or if he has reason to believe that any animal

Table 2. Farmers' awareness of existing regulatory control on dead carcass disposal.

Activity	No. of respondents	Percent
Knowledge of regulatory control on dead carcass disposal		
Yes	19	15.8
No	101	84.2
If yes, what was the source of information		
Farmers forum	11	57.9
Public health workshop	2	10.5
Radio/TV	5	26.3
Internet	1	5.3
Have you heard of any farmer prosecuted for improper carcass disposal?		
Yes	0	0
No	120	100
Improper carcass disposal could be source of environmental health hazard		
Yes	74	61.7
No	40	33.3
Don't know	6	5.0

has been exposed to infection shall administer veterinary vaccines or biological or issue such orders, directions or prohibitions as he may consider necessary or advisable to prevent the spread of the disease and may cause any such animal to be slaughtered if he considers that the slaughter of such animal is necessary for the prevention of the spread of the disease and shall inform the police forthwith.

Statutory provisions for the disposal of dead animal carcasses

According to the final report of the Avian Influenza Control and Human Pandemic Preparedness and Response Project (2007) [www.jhuccp.org/whatwedo/projects/avian-influenza-control-and-human-pandemic-preparedness-and-response], the disposal of H5N1 infected bird carcasses is primarily determined by the volume of birds, logistics of disposal as well as environmental and economic factors. The report identified the following technologies as reliable for pathogen inactivation: rendering, incineration, compositing, burial, land filling and alkaline hydrolysis.

S.9 of the Animal Diseases Act provides that where any animal dies of a disease or is slaughtered in accordance with the provisions of this Act or is slaughtered otherwise than in accordance with the provisions of this Act and its carcass is in the opinion of the veterinary officer infected with disease, such carcass shall be disposed-off by burning or in such manner as the veterinary officer may direct [see S.9 (1) (a) and (b)].

S.10 of the Act provides for a punishment of 3 months imprisonment or a fine of ₦250 for any person who is guilty of an offence, non-compliance or contravention of this Act.

Farmers' response to awareness of statutory regulation on dead animal carcass in the study area

Table 2 shows the response of livestock farmers to their awareness of statutory regulation on disposal of dead animal carcass. Analysis of the result showed that 10% farmers representing 84.2% had no knowledge of the existence of regulatory control on dead carcass disposal. The remaining farmers, representing 15.8% acknowledge that they were aware of existing regulation on carcass disposal in the State. Of the 19 farmers that affirmed their knowledge of regulatory control on carcass disposal, 57.9% (11) got the information from attending Farmers forum. About 26.3% got the information from radio and television broadcast, while 10.5% of farmers got the information from attending public health workshop. Few farmers sourced the information from the internet. It is surprising that in a State where majority of the livestock farmers are educated, and coupled with the existence of agricultural extension services, most farmers are yet unaware that they can be sectioned for improper disposal of dead carcasses. All the 120 livestock farmers interviewed mentioned that they have not heard about the prosecution of any livestock farmer in the State by regulation agencies for improper disposal of carcass.

Disposal of dead animal carcasses as observed in the study area

Table 3 presents farmers response to methods of disposing dead animal carcass. Analysis of the results indicated that majority of livestock farmers in the state practice some form of waste disposal method or the other. On farm, burial of dead carcass (49.2%) was observed to be the most practiced method of dead

Table 3. Farmers response to method of dead carcass disposal in Enugu State of Nigeria.

Activity	No. of respondents	Percent
Which carcass disposal method do you mainly practice in your farm?		
Burial (on farm)	59	49.2
Incineration	45	37.5
Composting	10	8.3
Rendering	0	0
Landfill	6	5
Burial and incineration imposes serious environmental problems (e.g. pollution of ground waters etc.)		
Yes	46	38.3
No	74	61.7
Method of disposing dead young/immature animal		
Burial (on farm)	104	86.7
Burning	16	13.3
Composting	0	0
Rendering	0	0
Landfill	0	0
Others	0	0
Method of disposing mature dead animal carcass		
Burial (on farm)	56	46.7
Burning	15	12.5
Composting	1	0.8
Rendering	0	0
Landfill	0	0
Others (Slaughter and sent to market)	48	40

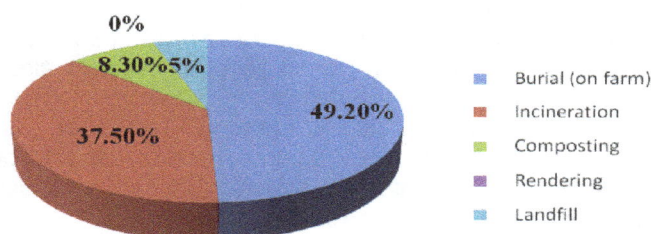

Figure 1. Which carcass disposal menthod do u practice mainly in your farm?.

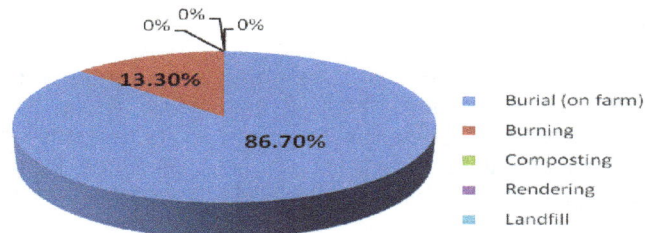

Figure 2. Methods of disposing dead young/immature animal carcass.

carcass disposal method applied by farmers in the State (Figure 1).

This was followed by incineration (37.5%), compositing (8.3%) and landfill (5.0%). Rendering as a means of dead carcass disposal was not observed in the study area. Rendering which is the practice of using heat to convert dead animal carcasses and animal by-products into marketable products such as meat and bone meal for animal feed, human food additives, or cosmetics (Livestock Disposal Manual, 2009) is not known in the area. The suitability of any disposal method is dependent on certain criteria among which are compliance with local, state and/or federal regulation. Unfortunately, these criteria among others are not being considered by most livestock farmers in the study area.

Majority of the farmers (61.7%) admitted not knowing that burial and incineration methods of carcass disposal impose serious environmental problems. Only about 38.3% acknowledged the fact that both methods when not properly done cause serious environmental problems.

On method of disposing dead young/immature animals, the study showed that 104 farmers representing 86.7% of the respondents bury dead young/immature animals on farm, while 13.3% resort to burning (Figure 2). None of the farmers practice other methods of disposal like compositing, rendering and landfill for dead young/immature animals. On the other hand, although majority of the farmers dispose mature dead animal carcass using on farm burial (46.7%), about 12.5 and

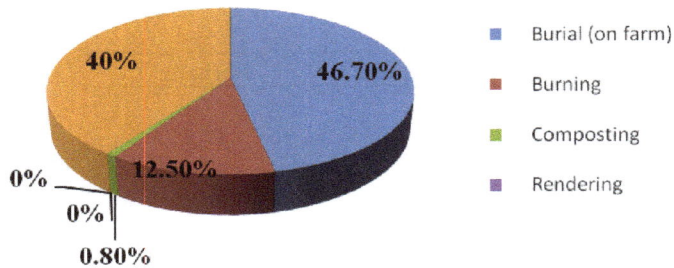

Figure 3. Methods of disposing mature dead animal carcass.

Figure 4. Incinerating area for dead young/immature animal carcass.

0.8% of them practice burning and compositing as means of disposing mature dead animal carcass (Figure 3). Of great concern is the number of farmers that slaughter and send the carcass to marketing. About 40% of dead mature animal carcass finds its way into the market paving way for serious health problems when consumed by unsuspecting consumers. This scenario is challenging because dead mature animal carcass is not meant for human consumption as this may lead to the spread of diseases, especially when the cause of the animal death is unknown.

This situation is a very serious and challenging one which must be urgently addressed by both the government and non-governmental organization/agencies.

Survey report showed that 87% of the respondents dispose dead animal carcasses by burning their carcasses, whereas only 13% resort to burying and this applies where the carcass is a very young or immature animal. Dead matured animal carcasses in most cases find their way into the meat table of unsuspecting consumer.

Figure 4 shows an incinerating area where immature/young carcasses are disposed-off in one of the sampled farms.

As earlier observed, the available statutes appear not to discuss the disposal of dead animal carcasses where the cause of death is not disease. This may be in agreement with the principle that any livestock found dead from unknown cause is presumed to have died of disease.

A very important observation in the statutes is the importance of the knowledge of the state of health of the animal by the farmer and meat dealer. S.22 of the Animal Diseases (Control) Act provides that,' where an owner or person in charge of any animal suffering from disease is charged with an offence against any of the provisions of this Act, he shall be presumed to have known of the existence of such disease in such animal unless he satisfies the court that he had no such knowledge and could not within reasonable time have obtained such a knowledge'. It is common for herdsmen and livestock owners to readily identify signs of diseases in their flock (Alawa et al., 2002). Such early identification of disease symptoms in a flock is very important as it will guide the farmer not to offer such an animal for sale as meat. Respectfully, it is our submission that the provisions of these sections ought to be a strict liability offence in cases where any reasonable man would have suspected such a carcass to have died of any form of disease and still proceeds to offer such a diseased carcass for meat.

More worrisome is the fact that the operators of the meat industry in the study area do not even know of the existence of these statutes. There is the need to advocate for adequate enforcement of legislations on routine veterinary examinations at the slaughter houses in Nigeria (Alhaji, 2011). This will checkmate the incidence of slaughtering disease animals as meat and thus, save the unsuspecting consumer from buying unwholesome meat. Agencies of government involved in animal production should incorporate educating farmers on the extant regulations governing all facets of animal production on their priority list. As already observed by Mafimisebi et al. (2012), most Nigerian farmers already have considerable knowledge on various aspects of animal production including treating various diseases. Regular trainings will help keep them abreast with best practices with the resultant effect of offering wholesome and safe meat to the consumers.

Conclusions

There are detailed statutory regulations of dead/diseased animal carcasses in Nigeria. Livestock farmers in the study area are not aware of the existence of such statues. Awareness has to be created among the various practitioners on the need to dispose-off dead animal carcasses as stipulated by legislation. There is the need to continually train farmers on best practices that will ensure that only wholesome and disease-free animals are offered to the Nigerian consumer as meat.

REFERENCES

Addas PA, Midaun A, Milka M, Tizhe MA (2010). Assessment of Abattoir Foetal Wastage of cattle, Sheep, and Goat in Mubi Main Abattoir Adamawa State, Nigeria. World J. Agric. Sci. 6(2):132-137.
Adesehinwa AOK, Makinde GEO, Oladele OI (2003). Socio- economic

characteristics of pig farmers as determinant of pig feeding pattern in Oyo state, Nigeria. *Livestock Research for Rural Development 15 (12)*. Retrieved March 30, 113, from http://www.lrrd.org/lrrd15/12/ades1512.htm

Ajala MK, Adesehinwa AOK, Mohammed AK (2007). Characteristics of smallholder pig production in Southern Kaduna area of Kaduna State, Nigeria. *Am-Eurasian* J. Agric. Environ. Sci. 2 (2):182-188.

Alawa JP, Jokthan GE, Akut K (2002). Ethnoveterinary medical practice for ruminants in the subhumid zone of northern Nigeria. Prev. Vet. Med. 54(1):79-90

Alhaji NB (2011).Prevalence and economic implications of calf foetal wastage in an abattoir in Northcentral Nigeria. Trop. Anim. Health. Prod. 43(3):587-90.

Avian Influenza Control and Human Pandemic Preparedness and Response Project (2007) [www.jhuccp.org/whatwedo/projects/avian-influenza-control-and-human-pandemic-preparedness-and-response]

ENADEP (2007). Enugu State Agricultural Development Programme (ENADEP) Annual Report, Enugu.

Enugu State Government Official Gazette. (1997). Enugu State Official Gazette. p. 25.

Ezike JO (1998). Delineation of old and new Enugu State. Bulletin of the Ministry of Works, Lands and Survey, Enugu State.

Mafimisebi TE Oguntade AE Fajemisin AN Aiyelari OP (2012). Local knowledge and socio-economic determinants of traditional medicines' utilization in livestock health management in Southwest Nigeria.

J. Ethnobiol. Ethnomed. 8:2

Mishra KA, Wilson CH, Williams RP (2009). Factors affecting the financial performance of new and beginning farmers. Agric. Fin. Rev. 69:160-179.

National Population Commission (NPC) (2006). Official Census Report. Abuja, Nigeria.

Nwanta JA, Shoyinka SVO, Chah KF (2011). Production characteristics, disease prevalence, and herd-health management of pigs in Southeast Nigeria. J. Swine. Health Prod. 19(6):331-339.

Oni OA, Yusuf SA (1999). The effects of farmers' socio-economic characteristics on livestock production in Ibadan Metropolis. In: Proceeding 4th Annual Conferene Animal Science Association of Nigeria, IITA, Ibadan. pp. 245-248.

Ozor N, Nnaji C (2011). The role of extension in agricultural adaptation to climate change in Enugu State, Nigeria. J. Agric. Exten. Rural Dev. 3(3):42-50.

21

Sheep and goats *Cysticercus tenuicollis* prevalence and associated risk factors

Endale Mekuria[1], Shihun Shimelis[1], Jemere Bekele[2] and Desie Sheferaw[2]

[1]College of Veterinary Medicine, Haramaya University, P. O. Box 05 Hawassa, Ethiopia.
[2]School of Veterinary Medicine, Hawassa University, P. O. Box 05 Hawassa, Ethiopia.

The purpose of this study was to estimate the prevalence of *Cysticercus tenuicollis*, identify factors that can influence its occurrence and to assess the distribution of the cyst in the visceral organs of sheep and goats slaughtered at Dire Dawa municipal abattoir. A total of 845 animals (425 sheep and 420 goats) were examined at the abattoir. The overall prevalence of *C. tenuicollis* was 24.6% (95% CI = 21.7 - 27.5) and 22.8 and 26.4% in sheep and goats respectively. Body condition of sheep was the only risk factor in which the prevalence of *C. tenuicollis* significantly varied (x^2 = 19.353, *P < 0.05*). Sheep with poor body condition (39.8%) were found most infected compared to medium (21.8%) and good (14.5%) body condition. There was no significant variation in the prevalence of *C. tenuicollis* between sheep and goats, because species mainly dependent on grazing, and hence, had equal exposure and opportunity to get infected. The cyst was found most frequently attached to liver, omentum and peritoneum both in sheep and goats. In conclusion, the presence of *C. tenuicollis* at a higher prevalence and the consequent effect on small ruminant signify the need for the control of stray dog population, deworming of dogs, and avoidance of backyard slaughter and proper disposal of infected viscera to curtail the problem.

Key words: Prevalence, *Cysticercus tenuicollis*, sheep and goats, Dire Dawa, Ethiopia.

INTRODUCTION

Cysticercus tenuicollis is the larval stage of *Taenia hydatigena*, which is a tapeworm of dogs, cats and wild canids (Kaufmann, 1996; Urquhart et al., 1996; Taylor et al., 2007). The intermediate hosts for the mature metacestode, *C. tenuicollis*, are sheep, goats, cattle, dromedaries, antelope, rarely pigs (Troncy, 1989) and also reported in deer and horse (Taylor et al., 2007) and monkey (Tsubota et al., 2009). Metacestode stage are frequently found attached to the omentum, mesentery and to the serosal surface of abdominal organs, especially liver (Taylor et al., 2007) and more frequently in the omentum of goats and sheep (EL-Azazy and Fayek, 1990; Radfar et al., 2005; Senlik, 2008; Samuel and Zewde, 2010). Normally, infection with *Taenia*

hydatigena is not very pathogenic in dog. The severity of infection depends on how many *Taenia* eggs the animal swallowed on a single occasion and the subject's age, young animals most susceptible. Severe infection of liver/tissues may result in liver/carcass condemnation at slaughter (Taylor et al., 2007). The prevalence of infection is considerably high in the world in some countries even more than 85% of sheep population was found to be infected with this metacestode (Garcia-Marin and Peris-Palau, 1987). Some studies indicated that the prevalence of *C. tenuicollis* increase with age (Togerson et al., 1998; Bhaskar et al., 2003) and other studies indicated that at the second age prevalence is increased and after that age the prevalence rates decreased

(Dajani and Khalaf, 1981; Senlik, 2008).

There was very few studies conducted, and hence, there is scarcity of information about *C. tenuicollis* prevalence or status in Ethiopia. Therefore, the aims of this study were to estimate the prevalence of *C. tenuicollis* and to identify factors that can influence the prevalence. Also to assess the organ distribution of the cysts both in sheep and goats slaughtered at Dire Dawa municipal abattoir.

MATERIALS AND METHODS

Study area

Dire-Dawa administrative city geographically lies within 9° 27' and 49°N latitude and 41°38' and 21°E longitude. The mean annual rainfall is 604 mm and the mean daily temperature is 25.4°C. It is generally characterized by semi-arid environmental condition, and has low amount of precipitation.

Study population

The study animals were sheep and goats slaughtered at Dire-Dawa municipal abattoir, and their sources were Dire-Dawa, Erer, Shinle and Eastern Hararghe areas. In the areas of their origin, the animals were owned by smallholder farmers under traditional management system. All selected animals were grouped into 2 age groups based on the number of pairs of incisors (Gatenby, 1991; Steele, 1996). Breeds of the study, sheep and goats were classified based on the phenotypic characteristics (Ayalew et al., 2004), and the body condition was scored following the guidelines set by Abebe (2007).

Study design and sample size

The study was cross-sectional study whereby the study animals were selected from the slaughter line using systematic random sampling technique. List of the animals to be slaughtered, from which study animals were selected, was prepared while the animals were kept in lairage. The required sample size was determined based on expected prevalence of 50% and the formula given by Thrusfield (2005). The study considered 95% confidence interval and 5% precision level. Accordingly a total of 845 animals (that is, 425 sheep and 420 goats) were selected and studied. For this study sex, age, species, breed, origin of animals and body condition were considered as risk factors.

Study methodology

The date and the species, origin, breed, sex, age and body condition of animals were recorded prior to slaughter. Then organs were thoroughly inspected by applying the routine meat inspection procedures during postmortem examination paying attention to the visceral organs and tissues in abdominal, thoracic and pelvic cavities (FAO, 1995; Gracey et al., 1999).

Data management and analysis

Collected data were stored in a Microsoft Excel spread sheet and analyzed with STATA version 11 (Stata Corp. College Station, TX) statistical software. Prevalence was calculated as percentage

value. Statistical association of *C. tenuicollis* prevalence with species, sex, age, body condition, origin and breeds of the animals was analyzed using χ^2 test.

RESULTS

Prevalence and associated risk factors

The overall prevalence of small ruminant *C. tenuicollis* was found to be 24.6% (95% CI = 21.7 - 27.5). From a total of 425 sheep and 420 goats examined by postmortem examination 22.8% sheep (95% CI= 18.9 - 27.1) and 26.4% goats (95% CI = 22.3 - 30.9) were found positive for *C. tenuicollis* infection. The association of the overall prevalence of *C. tenuicollis* with the considered risk factors was shown in Table 1.

The prevalence of *C. tenuicollis* in sheep and goats vs. the considered risk factors was shown in Tables 2 and 3.

Organ distribution of *C. tenuicollis*

In this study, the predominant predilection site for *C. tenuicollis* cyst was liver in both sheep and goats. Of the 97 positive sheep and 111 positive goats, liver accounts for 40.2% and 26.1% in sheep and goats respectively. The detailed for organ distribution of *C. tenuicollis* cyst was shown in Table 4.

DISCUSSION

Out of 425 sheep and 420 goats examined 22.8% (95% CI: 18.8-26.8%) and 26.4% (95% CI: 22.2-30.7%) were found to be positive for *C. tenuicollis*. This finding is comparable with the report of Senlik (2008) and Dada and Belino (1978) from Turkey and Nigeria respectively. The prevalence of *C. tenuicollis* in sheep and goats in this study is relatively lower than that reported from central Ethiopia (Samuel and Zewde, 2010) and in other countries (Radfar et al., 2005; Garcia-Marin and Peris-Palau, 1981). This variation in the prevalence mainly accounted to the grazing behaviour and management system prevailing in the local areas (Senlik, 2008). The study animals were selected from smallholder and backyard management system. In such areas dogs are kept by the animal owners, and believed that the dogs are useful for the community in preventing predators from their livestock. In the area, especially in rural, treating dogs for parasitic diseases is not practiced. Backyard slaughter of small ruminants and disposal of viscera and trimmings on open field is common. All these are very important for the life cycle to continue between the final and intermediate hosts. During this study there was no significant variation in the prevalence of *C. tenuicollis* between sheep and goats. This was mainly due to dependence of both species on grazing, and hence,

Table 1. Prevalence of *C. tenuicollis* at Dire-Dawa abattoir versus considered risk factors.

Factor	No. examined	Infected number	Prevalence (%)	95% CI	χ^2	P-value
Species						
Sheep	420	96	22.8	18.8 - 26.8		
Goat	425	112	26.4	22.2 - 30.7	1.480	0.224
Sex						
Female	406	96	23.6	19.5 - 27.8		
Male	439	112	25.5	21.4 - 29.6	0.396	0.529
Age						
Young	62	16	24.8	10.7 - 31.2	0.480	0.489
Adult	783	164	20.9	21.9 - 27.9		
Body condition						
Good	275	50	18.2	13.6 - 22.8		
Medium	421	110	26.1	21.9 - 30.3		
Poor	149	48	32.2	24.7 - 39.8	11.291	0.004
Animal origin						
East Hararge	267	65	24.3	19.2 - 29.5		
Erer	111	23	20.7	13.1 - 28.3		
Dire-Dawa	333	84	25.2	20.5 - 29.9		
Shinilie	134	36	26.9	19.3 - 34.4	1.350	0.717

Table 2. Prevalence of *C. tenuicollis* versus the considered risk factors in sheep.

Factor	No. examined	Infected number	Prevalence (%)	95% CI	χ^2	P-value
Sex						
Female	214	43	20. 1	14.7 - 25.5		
Male	211	54	25.6	19.7 - 31.5	1.824	0.177
Age						
Young	23	4	17.4	1.5 - 33.3		
Adult	402	93	23.1	19.0 - 27.3	0.407	0.523
Body condition						
Good	145	21	14.5	8.7 - 20.2		
Medium	197	43	21.8	16.0 - 27.6		
Poor	83	33	39.8	29.1 - 50.4	19.353	0.000
Animal origin						
East Hararge	116	25	21.6	14.0 - 29.1		
Erer	71	14	19.7	10.4 - 29.1		
Dire-Dawa	164	33	20.1	14.0 - 26.3		
Shinilie	74	25	33.8	22.9 - 44.7	6.221	0.101
Breed						
Afar	91	22	24.2	15.3 - 33.0	0.157	0.925
Harar	73	17	23.3	13.5 - 33.1		
Black Head Ogaden	261	58	22.2	17.2 - 27.3		

Table 3. Prevalence of *C. tenuicollis* versus the considered risk factors in goats.

Factor	No. examined	Infected number	Prevalence (%)	95% CI	χ^2	P-value
Sex						
Female	192	53	27.6	21.2 - 34.0	0.251	0.616
Male	228	58	25.4	19.8 - 31.1		
Age						
Young	39	9	23.1	9.6 - 36.5		
Adult	381	102	26.8	22.3 - 31.2	0.248	0.618
Body condition						
Good	130	29	22.3	15.1 - 29.5		
Medium	224	67	29.9	23.9 - 35.9	2.997	0.223
Poor	66	15	22.7	5.2 - 32.9		
Animal origin						
East Hararge	151	40	26.5	19.4 - 33.6		
Erer	40	9	22.5	9.4 - 35.6		
Dire-Dawa	169	51	30.2	23.2 - 37.1	3.562	0.313
Shinilie	60	11	18.3	8.4 - 28.2		
Breed						
Somali	184	47	25.5	19.2 - 31.9		
Afar	152	41	27.0	19.9 - 34.0		
Harar	84	23	27.4	17.8 - 37.0	0.137	0.934

Table 4. Organ distribution of *C. tenuicollis* in visceral organs of sheep and goats.

Location	Sheep		Goat	
	Frequency	Proportion	Frequency	Proportion
Diaphragm	6	6.19	5	4.5
Liver	39	40.2	29	26.1
Liver and lung	1	1.0	-	-
Liver and peritoneum	1	1.0	2	1.8
Liver and omentum	2	2.1	2	1.8
Liver, peritoneum and omentum	-	-	1	0.9
Liver and pelvic cavity	-	-	1	0.9
Lung	8	8.3	10	9.0
Lung and peritoneum	1	1.0	-	-
Lung and pelvic cavity	-	-	2	1.8
Omentum	18	18.6	22	19.8
Omentum and pelvic cavity	1	1.0	1	0.9
Pelvic cavity	6	6.2	12	10.8
Peritoneum	13	13.4	22	19.8
Peritoneum and pelvic cavity	1	1.0	-	-
Peritoneum and omentum	-	-	2	1.8

they had equal exposure and opportunity to get infected.

The analysis of risk factors considered for this study showed no significant effect on the preference of *C.*

tenuicollis. Body condition of sheep was the only risk factor in which the prevalence of *C. tenuicollis* poor body condition (39.8%) were found most infected significantly varied (χ^2 = 19.353, $P < 0.05$). Sheep with compared to medium (21.8%) and good (14.5%) body condition. But among all other risk factors considered both in sheep and goats except the slight difference in figures of the prevalence statistically no significant variation ($P > 0.05$) observed (Tables 2 and 3). This finding is in line with the report of Togerson et al. (1998), Senlik (2008) and Samuel and Zewde (2010) from Northern Jordan, Turkey and Central Ethiopia respectively. When all studied animals considered together, still the prevalence significantly varied (χ^2 = 11.291, $P < 0.05$) with body condition. Small ruminant with poor body conditions were the most affected compared to the medium and good body conditions. When animals suffer from shortage or scarcity of nutrition, and infected with gastrointestinal internal parasites their immunity compromised. Hence, possibly this can be accounted for the higher prevalence of the cyst in poor body condition animals.

Among the predilection sites observed during this study liver was found to be the predominant one, and it followed by omentum and peritoneum. Samuel and Zewde (2010) and Senlik (2008) reported that omentum is the predominant predilection sites for *C. tenuicollis*. The present study revealed that diaphragm and pelvic cavity was infected with *C. tenuicollis*..

CONCLUSION AND RECOMMENDATIONS

During this study high prevalence of *C. tenuicollis* was recorded both in sheep and goats slaughtered at Dire-Dawa municipal abattoir. Besides, the cyst was found distributed throughout the abdominal and pelvic cavities. It was found attached with many visceral organs and tissues, like liver, omentum, peritoneum, lung and diaphragm were the principal organ and tissues where the cyst was located. Hence, from the result of the current study the backyard slaughter of small ruminant should be discouraged, and the livestock health extension workers need to inform dog owners to deworm their dogs regularly. Controlling of stray dogs also play key role in the reduction of this high prevalence of *C. tenuicollis*.

REFERENCES

Abebe G (2007). Body Condition Score of Sheep and Goats, Ethiopian Sheep and Goats Production Improvement program (ESGPIP). P. 3.

Ayalew W, Geahn E, Tibbo M, Mamo Y, Rege EO (2004). Current state of knowledge characterization of farm animal genetic resource and breeding in Ethiopia. pp. 1-22.

Bhaskar RT, Vara PPV, Hafeez MD (2003). Prevalence of C. tenuicollis infection in slaughtered sheep and goats at Kakinda, Andhra Pradesh. J. Parasitic Dis. 27:126-127.

Dada BJ, Belino ED (1978). Prevalence of Hydatidosis C. tenuicollis in slaughtered livestock in Nigeria. Vet. Record. 103:311-312.

Dajani YF, Khalaf FH (1981). Hydatidosis and tenuicollis in sheep and goats of Jordan: A comparative study. Ann. Trop. Med. Parasito. 75:175-179.

El-Azazy OM, Fayek S (1990). Seasonal Pattern of *Fasciola gigantica* and *C. tenuicollis* in sheep and goats in Egypt. Bull. Anim. Health Prod. Afr. 38:369-373.

FAO (1995). Manual-On-Meat-Inspection for Developing Countries, Specific Diseases of Sheep and Goat, Rome, Italy.

Garcia -Marin JF, Peris-Palau B (1987). Visceral Cysticercosis in Lambs fattened in Zaragora Province, Spain, Incidence and nature of lesions. Rev, Iberica Parasitol. pp. 195-199.

Gatenby RM (1991). Sheep: The tropical agriculturalist. London and Basingstoke, MACMILLAN education Ltd. ACCT. pp. 6-10.

Gracey JF, Collins DS, Huey RJ (1999). Meat hygiene. 10th edition, London. W. B. Sounders Company Ltd. pp. 226-293.

Kaufmann J (1996). Parasitic Infection of Domestic Animals, A Diagnostic Manual. P. 303.

Radfar MH, Tajalli S, Jalalzadeh M (2005). Prevalence and Morphological Characterization of C. tenuicollis (*Taenia hydatigena*) from sheep and goats in Iran. Vet. Arh. 75:469-496.

Samuel G, Zewde GG (2010). Prevalence, risk factors, and distribution of C. tenuicollis in visceral organs of slaughtered sheep and goats in Central Ethiopia. Trop. Anim. Health Prod. 42(6):1049-1051.

Senlik B (2008). Infulence of Host Breed, Sex and Age on the Prevalence and Intensity of C. tenuicollis in sheep. J. Anim. Vet. Adv. 7(5):548-551.

Steele M (1996). Goats: The tropical agriculturist. London: MACMILAN education Ltd. ACCT. pp. 79-83.

Taylor MA, Coop RL, Wall RL (2007). Veterinary Parasitology, 3rd Ed. Black Well Science Ltd, Iowa, USA. pp. 210-211.

Thrusfield M (2005). Veterinary epidemiology, 3rd edition. Blackwell Science limited. Oxford, UK. pp. 229-246.

Togerson P, Williams R, Abo-Shehada MN (1998). Modelling the prevalence of *Echinococcus* and *Taenia* species in small ruminants of different ages in Northern Jordan. Vet. Parasitol. 79:35-51.

Troncy PM (1989). Manual of Tropical Veterinary Parasitology, 3rd Edition, Technical center for Agriculture Rural Co-operation, Word service to Agriculture and World Association for Advancement of Veterinary Parasitology (WAAVP), CAB International publisher, London. P. 111.

Tsubota K, Nakatsuji S, Matsumoto M, Fujihira S, Yoshizawa K, Okazaki Y, Murakami Y, Anagawa A, Yuzaburo Oku Y, Oishi Y (2009). Abdominal cysticercosis in a cynomolgus monkey, short communication. Vet. Parasitol. 161:339341

Urquhart GM, Armour J, Duncan L, Dunn AM, Jennings FW (1996). Veterinary Parasitology, 2nd edition, Longman Scientific and Technical publisher. Glasgow, Great Britain. P. 122.

22

Immunotoxic potential of sodium fluoride following subacute exposure in Wistar rats

D. K. Giri, R. C. Ghosh, M. Mondal, Govina Dewangan and Deepak Kumar Kashyap

Department of Veterinary Pathology, College of Veterinary Science and Animal Husbandry, Anjora, Durg, Chhattisgarh, India.

Fluoride pollution in drinking water is an international problem as the fluoride present is often at levels above acceptable limits. This study was done with an objective to determine the immunotoxic potential of sub acute exposure of sodium fluoride in Wistar rats with special reference to the Dinitroflurobenzene contact skin sensitivity test and pathological alterations in splenic histoarchitecture. The rats were intoxicated orally to sodium fluoride at dose levels of 5, 25 and 50 mg/kg body weight, respectively for 28 days. On day 21 the external surface of right ear pinna of rats were sensitized to 40 µl of 2% Dinitroflurobenzene in vehicle followed by challenge application of 40 µl of 1% Dinitroflurobenzene in vehicle on day 25. The left ear pinna served as control and 40 µl of vehicle (4:1 acetone- olive oil) was applied. The increase in ear thickness (in mm) was measured with Engineers micrometer at 0, 6, 12, 24 and 48 h post challenge. The results revealed a dose dependent decrease (p < 0.001) in mean ear thickness (in mm) of sodium fluoride intoxicated rats following challenge application of 1% Dinitroflurobenzene in vehicle. The blood picture revealed dose dependent leucocytosis associated with neutrophilia and lymphocytopaenia. There was also dose dependent decrease in organosomatic index of spleen and severe depletion of lymphocytes in white pulp of spleen. This report highlights the proposition that prolonged exposure to fluoride contaminated drinking water is likely to result in immunotoxicity.

Key words: Fluoride, immunotoxicity, rats, dinitroflurobenzene, spleen.

INTRODUCTION

Environmental pollution is a dangerous problem globally and one of the most complex health hazards results from fluorides. Fluorine being most electronegative element holds its ubiquitous presence. Fluorides are naturally occurring harmful contaminant in the environment (Raghuvansi et al., 2010). It is a cumulative poison and thus leads to fluorosis; a serious public health problem. Endemic fluorosis is prevalent in many parts of the world where drinking water contains more than 1 to 1.5 ppm of fluoride and causes damage not only to hard tissues of teeth and skeleton, but also to soft tissues, such as brain, liver, kidney, spleen and endocrine glands (WHO, 1984;

Santhakumari and Subramanian, 2007; Sharma et al., 2007; Ozsvath, 2009).

Fluoride can be an adjuvant for mucosal and systemic immunity and is reported to affect oral immunity in chickens. Studies with sodium fluoride (NaF), however, indicate that fluoride can damage human lymphocyte chromosomes, induce adverse effects in the spleen (Podder et al., 2010) inhibit growth and general health in rabbits and increase their nonspecific immune-related acid phosphatase and lysozyme activities (Liu et al., 2012).

Spleen is the largest secondary lymphoid organ

containing about one-fourth of the body's lymphocytes and initiates immune responses to blood-borne antigens (Nolte et al., 2002; Balogh et al., 2004). This function is charged to the white pulp which surrounds the central arterioles. The white pulp is composed of lymphocytes (B cells, CD4+ T cells and CD8+T cells), macrophages, dendritic cells, plasma cells, arterioles, and capillaries in a reticular framework (Cesta, 2006). Fluoride is reported to alter the histoarchitecture of spleen by causing reduction in the content of white pulp with concomitant increase in red pulp infiltrated by lymphocytes (Machalinska et al., 2002). Das et al. (2006) also observed disorganization in the histoarchitecture of the spleen and thymus of male albino rats after NaF treatment. Fluoride causes depletion of energy reserves due to its adverse effects on several enzymes and thus impairs the immune functions of white blood cells leading to reduction in the immune functions (Sharma et al., 2004).

The relationship between fluoride and immunity in animals is an ongoing topic of discussion and debate. The present communication is intended to put forth the immunotoxic potential of sodium fluoride on cell mediated immune response in Wistar rats being assessed by Dinitroflurobenzene contact skin sensitivity test together with the pathological alterations in blood picture pertaining to the leucocytes and histoarchitecture of spleen.

MATERIALS AND METHODS

Experimental design and test chemical

The experimental investigation was planned to adjudge the toxicopathological effects of sodium fluoride on cell mediated immune response in Wistar rats after obtaining approval from Institutional Animal Ethics Committee. The rats were maintained under regular supervision in controlled environment with 12 h light dark cycle and provided with standard feed and water *ad libitum* throughout the experimental period. The rats were acclimatized for 10 days before commencement of the experimental study. Twenty four healthy Wistar rats weighing 195 to 215 g (females) and 220 to 235 g (males) were randomly divided into four different groups having 6 (3 males + 3 Females) rats each. Rats of Group I served as control and were given only distilled water, orally. Rats of Groups II, III and IV were administered NaF orally, dissolved in distilled water, at the dose rate of 5.0, 25.0 and 50.0 mg/kg body weight for 28 days. These doses were selected based on the results of acute toxic dose of NaF determined by approximate lethal dose method. The test chemical Sodium fluoride (NaF) of 99% purity was obtained from Merck Limited, Mumbai- 400018.

Assessment of cell mediated immune response by dinitroflurobenzene (dnfb) contact skin sensitivity test

Cell mediated immune response; based on delayed type hypersensitivity reaction was measured by Dinitroflurobenzene (Sigma Chemical Co. Ltd., U.S.A.) contact skin sensitivity test as described by Phanuphak et al. (1974) and later slightly modified by Tamang et al. (1988). DNFB contact skin sensitivity test for

monitoring cell mediated immune response was studied on day 25 following primary sensitization at day 21.

Briefly, on day 21 of the experiment, 4 rats were randomly selected from each group and marked individually for this test. Dorsal aspects of ears (external pinna) of the selected rats were cleaned. Left ear was kept as control and only 40 µl of vehicle 4 acetone: 1 olive oil) was applied. The dorsal aspect of right ear of each rat was used to measure the contact sensitivity to DNFB. All the selected rats of each group were sensitized with 40 µl of 2% DNFB in 4: 1 acetone olive oil solution to the right ear. Four days following primary sensitization; on day 25, the sensitized rats were challenged with 40 µl of 1% DNFB in 4: 1 acetone olive oil solution to the right ear and the left ear was treated as respective control. Ear thickness (in mm) was measured with Engineer's micrometer at 0, 6, 12, 24 and 48 h post challenge. The clinical observations were recorded.

Haematology

At the end of the experiment on day 28 under diethyl ether anaesthesia blood samples were collected in heparinised vials from retro-orbital venous plexus of rats as described by Moore (2000). Thin blood smears were prepared for differential leukocyte count during blood collection. The total leucocyte count (TLC) was done as per Jain (1986), by using W.B.C. diluting fluid (Merck Limited, Mumbai - 400018) and Haemocytometer (Neubauer's chamber and WBC diluting pipette). Total leucocytes counts are expressed in thousands/cu.mm of blood. The differential leucocytic count (DLC) was done as per Coles (1986), using Leishman's stain (Merck Limited, Mumbai- 400018). The percentages of different leucocytes were determined by examining the stained blood smear under oil immersion objective lens of light microscope.

Organo-somatic index of spleen

Following humane sacrifice and after recording the gross lesions, the spleen were carefully removed, blotted free of blood and weighed (in grams) over electronic digital balance. The organo-somatic index (OSI) was calculated by using the following formula as per Chattopadhyay et al. (2011).

$$\text{Organo-somatic index (OSI)} = \frac{\text{Organ weight (g)}}{\text{Live body weight (g)}} \times 100$$

Histopathology of spleen

The tissue samples of spleen were collected in 10% neutral buffered formalin for histopathological studies. The tissues were thoroughly washed in running water; dehydrated in ascending grades of alcohol; cleared in benzene and embedded in paraffin at 58°C. The paraffin embedded tissue sections of 4 to 5 µm were obtained as described by Luna (1972) and stained with haematoxylin and eosin (H and E) as per the method described by Bancroft and Stevens (1990) with slight modifications. The stained sections were examined under light microscope and the lesions were recorded, if any.

Statistical analysis

Statistical analysis was done using complete randomized design (CRD) -single factor analysis of variance by Snedecor and Cochran (1968). The mean values between treatment and control groups

Table 1. Contact skin sensitivity response (Mean increase in ear thickness in mm) after challenge application of 1% Dinitroflurobenzene of different experimental groups during sub acute sodium fluoride toxicity (Left side served as vehicle control and right side treated with DNFB) n = 4.

Experimental groups	Ear side	Ear thickness before sensitization	Ear thickness after challenge with DNFB at different time interval (hours)			
			6	12	24	48
Group I	Left	0.80 ± 0.90	0.80 ± 0.29	0.80 ± 0.41	0.80 ± 0.22	0.80 ± 0.30
	Right	0.80 ± 0.86^a	1.14 ± 0.39^{ab}	1.15 ± 0.54^b	2.17 ± 0.19^a	1.97 ± 0.48^a
Group II	Left	0.80 ± 0.82	0.84 ± 0.36	0.87 ± 0.24	0.89 ± 0.31	0.88 ± 0.59
	Right	0.81 ± 0.65^a	1.06 ± 0.44^b	1.31 ± 0.51^{a**}	1.86 ± 0.43^{b**}	1.74 ± 0.65^{b**}
Group III	Left	0.81 ± 0.75	0.83 ± 0.43	0.87 ± 0.81	0.91 ± 0.55	0.89 ± 0.73
	Right	0.81 ± 0.73^a	1.16 ± 0.50^{a*}	1.27 ± 0.92^{a**}	1.62 ± 0.16^{c**}	1.50 ± 0.86^{c**}
Group IV	Left	0.80 ± 0.81	0.84 ± 0.22	0.89 ± 0.80	0.92 ± 0.31	0.88 ± 0.70
	Right	0.81 ± 0.65^a	1.13 ± 0.31^{ab}	1.25 ± 0.71^{a**}	1.48 ± 0.50^{d**}	1.38 ± 0.84^{d**}

Values indicate Mean ± S.E. Superscripts may read column wise for comparison of means. Mean values with similar superscripts do not differ significantly from each other. *significant at P ≤ 0.05 from control group (Group I) and **significant at P ≤ 0.01 from control group (Group I).

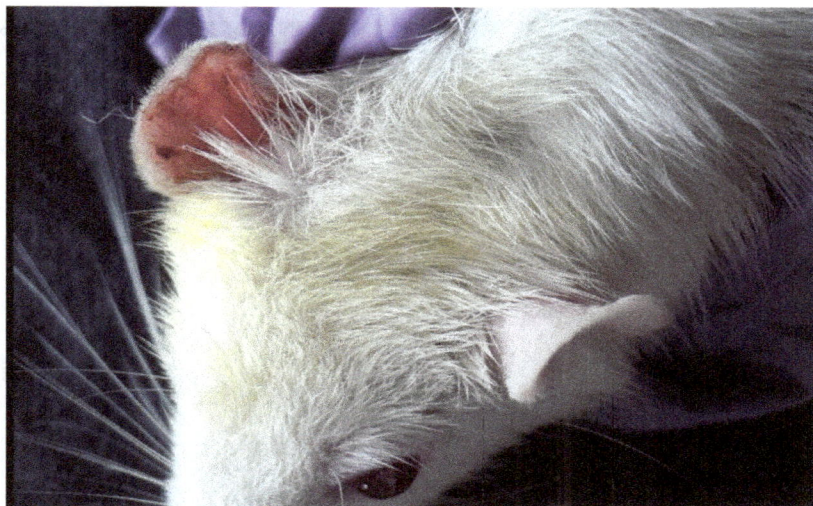

Figure 1. Ear of a control rat showing erythema 6 h post challenge application of 1% DNFB.

were tested for critical difference (CD), if any. The results are expressed as Mean ± S.E.

RESULTS

Cell mediated immune response to DNFB contact skin sensitization test

The observations of mean increase in ear thickness (in mm) of Wistar rats post challenge at different time interval is summarised and presented in Table 1.

Challenge application of 1% DNFB to the right ear caused erythema (Figure 1) followed by vesiculation and scabbing. Vesiculation was observed at 12 h which persisted up to 18 h in control rats. The ruptured vesicles were also noticed (Figure 2). The ear lesion turned into a scab between 20 and 24 h post challenge application (Figure 3) accounting for very intense increase in ear thickness. All these changes (erythema, vesicle and scabbing) were most prominent in rats of control group. However, there was moderate and mild erythema and vesiculation in rats of Groups II and III exposed to NaF at 5 and 25 mg/kg body weight, respectively which gradually

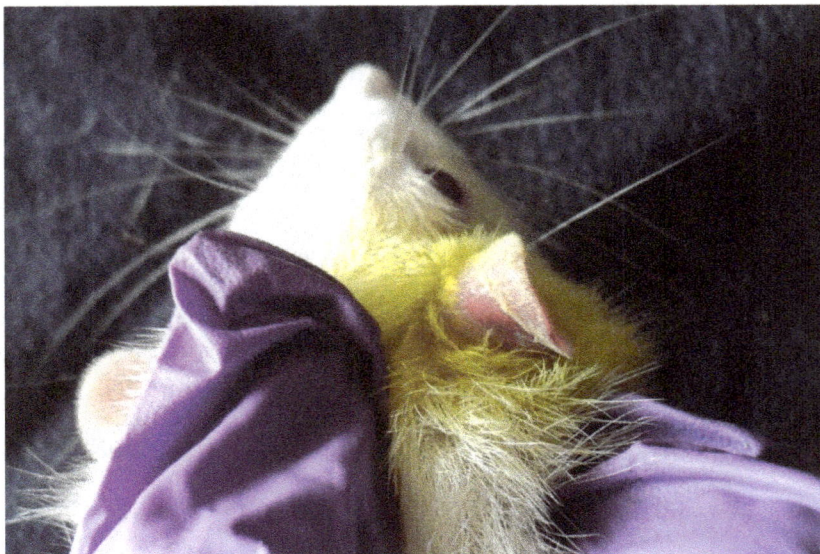

Figure 2. Ruptured vesicle on dorsal surface of right external pinna in a control rat 12 h post exposure to 1% DNFB challenge.

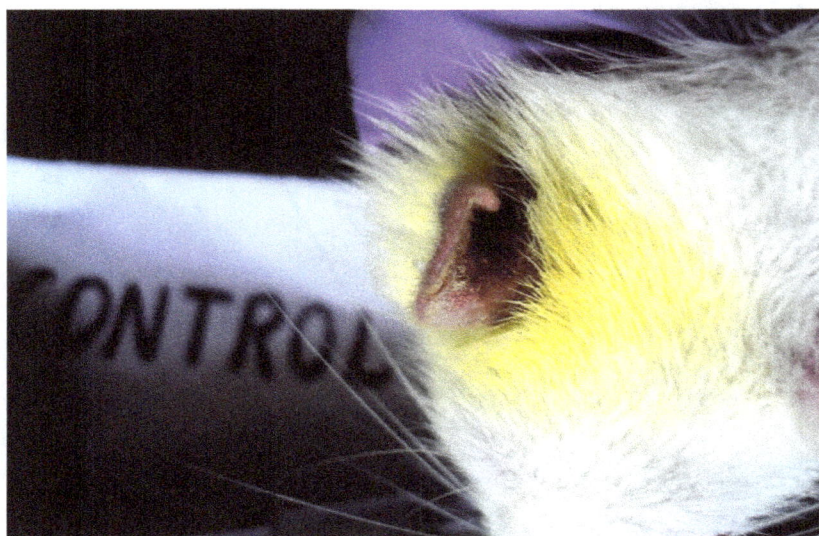

Figure 3. Profound increase in thickness of right external pinna in a control rat 24 h post exposure to 1% DNFB challenge due to scab formation.

became normal by 24 h. Furthermore, the intensity of ear lesion was very meagre in the rats of Group IV being intoxicated by the highest dose of NaF at 50 mg/kg body weight (Figure 4). A dose dependent decrease (P ≤ 0.01) in ear thickness of NaF intoxicated rats was observed at 12 to 48 h post challenge in comparison to the control rats. Moreover, the maximum immunological response was observed 24 h post challenge in the rats of control group (Group I). The scabs were found to be persisting even at 48 h in control group (Figure 5).

Haematology

The administration of NaF orally for 28 days to Wistar rats caused leucocytosis in rats of Group III (P ≤ 0.05) and Group IV (P ≤ 0.01) as compared to rats of Groups I and II. There were remarkable lymphocytopaenia (P ≤ 0.01) and neutrophilia (P ≤ 0.01) of all the rats belonging to Groups II, III and IV as compared to rats of control group (Group I). Non significant alterations were recorded for the monocyte percent and eosinophil percent in NaF

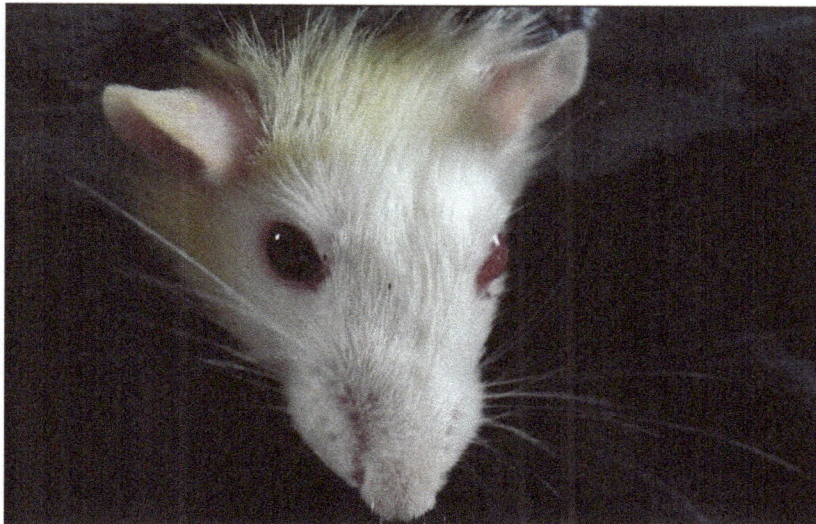

Figure 4. Very mild change in right ear thickness in a rat of Group IV (exposed to sodium fluoride at 50 mg/kg body weight, orally) 24 h post exposure to 1% DNFB challenge.

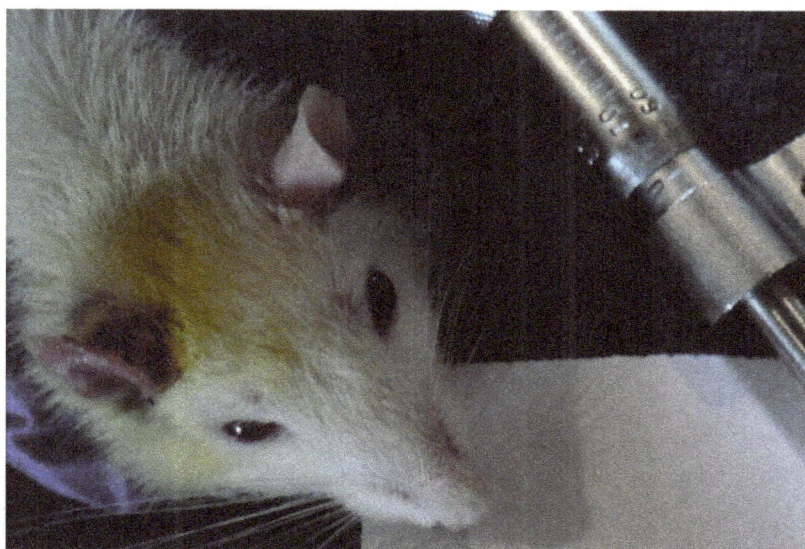

Figure 5. Persisting scab and increased thickness of right external pinna in a control rat 48 h post exposure to 1% DNFB challenge.

intoxicated groups with respect to the control rats. The results have been summarised and presented in Table 2.

Organosomatic index

The organosomatic index of spleen was found to be decreased (P ≤ 0.01) in comparison to the control rats following intoxication of rats with NaF. Further, the decrease was in a dose dependent manner. The results

have been represented in Figure 6.

Pathmorphology of spleen

At necropsy, spleen of control (Group I) rats appeared as normal. On the contrary, mild to moderate decrease in the size of spleen was observed in the rats of Group III and IV. The histopathological examination of spleen of Group IV represented several necrotising lymphocytes

Table 2. Effect of daily oral administration of sodium fluoride for 28 days on the leucocytic parameters of Wistar rats (n = 6).

Parameter	Groups			
	Group I	Group II	Group III	Group IV
TLC (Thousands/ cu.mm)	7.01 ± 0.01^c	7.02 ± 0.05^c	$7.20 \pm 0.04^{b*}$	$7.42 \pm 0.09^{a*}$
Lymphocyte (%)	66.75 ± 0.25^a	$58.00 \pm 0.82^{b**}$	$48.00 \pm 0.41^{c**}$	$41.25 \pm 0.25^{d**}$
Monocyte (%)	4.75 ± 0.25^a	4.25 ± 0.28^a	4.25 ± 0.24^a	4.25 ± 0.25^a
Neutrophil (%)	24.75 ± 0.48^d	$33.75 \pm 0.75^{c**}$	$44.25 \pm 0.48^{b**}$	$51.25 \pm 0.48^{a**}$
Eosinophil (%)	3.75 ± 0.25^a	4.00 ± 0.41^a	3.50 ± 0.29^a	3.25 ± 0.25^a
Basophil (%)	00.00	00.00	00.00	00.00

Values indicate Mean ± S.E. Superscripts may read row wise for comparison of means. Mean values with similar superscripts do not differ significantly from each other. *significant at P ≤ 0.05 and **significant at P ≤ 0.01 from control group (Group I).

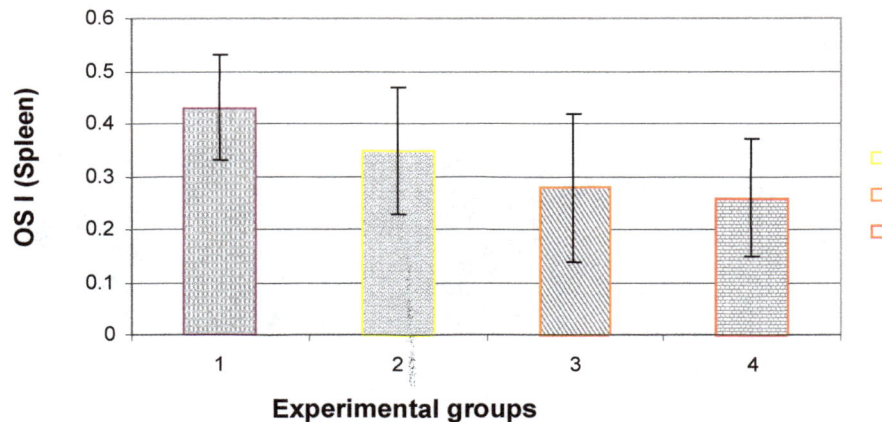

Figure 6. Organosomatic index of spleen in various experimental groups intoxicated orally with sub acute sodium fluoride toxicity.

and proliferation of reticuloendothelial cells. In addition to these changes, there was marked depletion of lymphocytes in the white pulp of spleen particularly in the periarteriolar lymphoid sheath (Figure 7). The splenic histoarchitecture in rats of Group III revealed a moderate depletion of lymphocytes in the white pulp (Figure 8) as compared to those from Group IV rats. On the other hand very mild depletion was noticed in splenic sections of Group II rats. No abnormalities could be identified in the splenic histological sections from control rats (Figure 9).

DISCUSSION

Contact hypersensitivity is a T- cell mediated cutaneous immune response to reactive haptens (Elmets and Bowen, 1986). After exposure of the skin to contact allergens, haptens covalently bind to discrete amino acid residues on carrier proteins. The epidermal Langerhans cell, a member of the dendritic-cell family, takes up haptenated proteins and processes them into antigenic peptides which are transported to the cell surface in association with major histocompatibility complex molecules (Griem et al., 1998; Wang et al., 2001). Matos et al. (2005) reported that DNFB induces the activation of the extra cellular signal-regulated kinases ERK1/2 and p38, and also up regulates CD40 expression.

The results of this study are in close agreement with Das et al. (2006) who reported that NaF treatment lowered cellular immunity in the rats. They observed a significant diminution in peripheral blood lymphocyte, monocyte and neutrophil counts in conjunction with a reduction in splenocyte counts. They had also reported the effects of NaF treatment on humoral immunity with lowered levels of plasma IgG specific to bovine serum albumin. On the other hand, De Vos et al. (2004) reported that mercurous chloride and NaF significantly suppressed concanavalin- A induced gamma-IFN

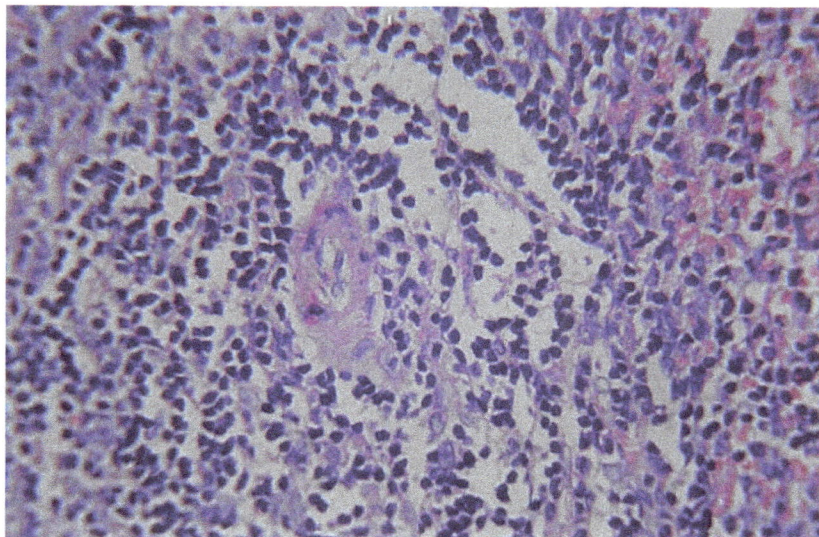

Figure 7. Microphotograph of spleen from Group IV rat (given NaF at 50 mg/kg body weight, orally) representing severe depletion of lymphocytes from white pulp, and proliferation of reticuloendothelial cells (H and E X 400).

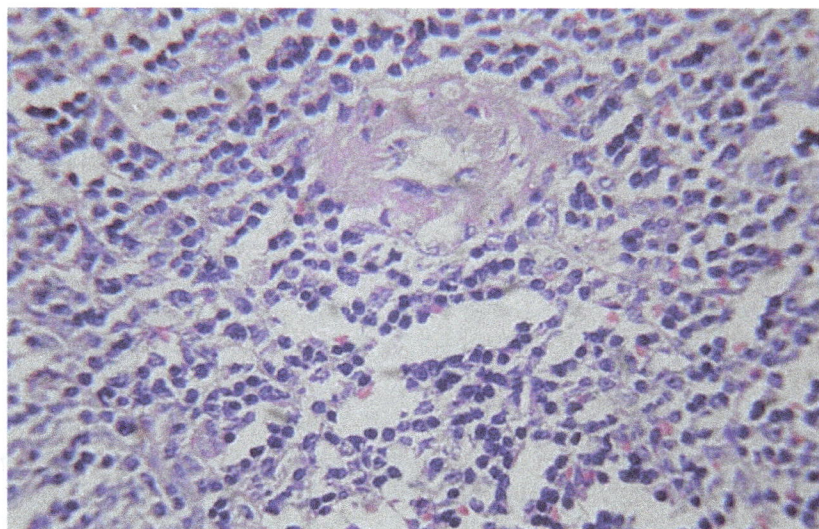

Figure 8. Microphotograph of spleen of a rat from Group III (given NaF at 25 mg/kg body weight, orally) showing moderate depletion of lymphocytes from white pulp (H and E X 400).

(interferon) production. They also reported suppressed Th_1 activity and stimulated Th_2 cytokine production in vitro on human peripheral blood mononuclear cells.

Our findings of significant increase in TLC is in accordance with the previous reports by Pillai et al. (1988) in Swiss albino mice exposed to 5.2 mg fluoride/kg body weight for 35 days and in rats treated with fluoride water containing1.5, 3, 4.5 and 6 ppm for 60 days (Sharma et al., 2004). Contrary to this Sharma et al.

(2007) reported no change in TLC in female rats exposed to water containing 6 ppm NaF for 15 and 30 days. Rao and Vidyunmala (2010) also documented dose and duration independent normal values in case of mice intoxicated with 5 and 10 mg fluoride/day for 30 and 60 days. Moreover, Santhakumari and Subramanian (2007) observed dose and duration dependent leucocytopaenia in Wistar rats exposed to 25 mg fluoride /l/rat/ day for 8 and 16 weeks. The decreased immune response in our

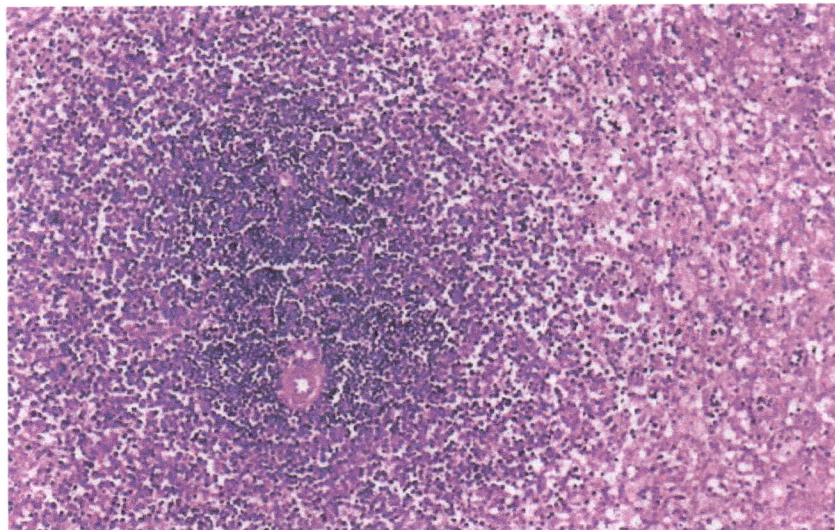

Figure 9. Normal histoarchitecture of spleen of a control rat (Group I) depicting the cell population in white pulp and adjoining red pulp.

study might be due to fluoride induced stress which further adds to immunosuppression. As fluoride alters cell membrane stability and integrity, it is also possible that membranous targets within immunocompetent cells which are essential for the induction of the contact hypersensitivity response are particularly vulnerable by fluoride. Immunosuppression could also have occurred due to complement-mediated damage to immune cells as opined by Elmets and Bowen (1986).

The reduction in organosomatic index of spleen are also in same line with those recorded by Podder et al. (2010) who observed that NaF when given at doses of 15 and 150 ppm through drinking water for 30 or 90 days in Swiss albino mice results in a decreased organo- somatic index of spleen without any change in body weight. However, in the present study a significant reduction in body weight of the rats was also recorded in NaF intoxicated groups. Fluoride causes a significant increase in the G0/G1 fraction and a decrease in G2/M fractions of cell cycle, fortifying the interpretation of G1 blockage of DNA synthesis and inhibition of proliferation of the mouse splenocytes (Podder et al., 2010). The decrease in white pulp content likewise signifies perturbation of haemopoiesis and accumulation of G0/G1 population; in agreement with the report by Zhou et al. (2009) that excessive fluoride intake seriously damages the specific immune function in rabbits. The nature of altered histoarchitecture of splenic pulps observed herein is in concordance with the findings of earlier workers. High fluoride (400, 800 and 1200 mg/kg) treatment in chickens through their diets decreases the number of splenic nodules, the lymphocyte population within splenic nodules and periarterial lymphatic sheaths in white pulp (Chen et al., 2009).

Podder et al. (2010) attributed the histopathological changes in spleen to fluoride induced cytotoxicity and to apoptosis. Alike our findings they too reported, decreased white pulp, increased red pulp and mononuclear infiltration in red pulp. They also recorded an increase in the percent of dead cells in the spleen. The mononuclear cells in white pulp of spleen of Group IV appear very dark which might be due to fluoride induced cytotoxicity as fluoride has been reported to generate toxic free radicals inducing oxidative stress (Xi et al., 2012). The dose dependent, mild to severe depletion in lymphocytes from white pulp of spleen of rats in Groups II, III and IV also explains the decrease in immune response with respect to the rats of control group and justifies the observations of contact skin sensitivity to rat ear against challenge dose of DNFB in the present study. From the observations and the results of this study, it is concluded that fluoride impairs the cell mediated immunity owing to a marked decrease in the number of leucocytes and the associated stress. Moreover, the immunotoxic potential of fluoride is marked and very significant in sufficiently higher doses. There is an urgent need to get rid off the dreadful and spreading tentacles of fluoride problem globally by exploring the possible alternatives against fluoride induced immunotoxicity.

Abbreviations: NaF, Sodium fluoride; **DNFB,** Dinitroflurobenzene.

REFERENCES

Balogh P, Horvath G, Szakal AK (2004). Immunoarchitecture of distinct reticular fibroblastic domains in the white pulp of mouse spleen. J.

Histochem. Cytochem. 52:1287-1298.

Bancroft JD, Stevens A (1990). Theory and practice of histological techniques. Churchill Livingstone, Edinburgh, pp. 113-305.

Cesta MF (2006). Normal structure, function, and histology of the spleen. Toxicol. Pathol. 34(5):455-465.

Chattopadhyay A, Podder S, Agarwal S, Bhattacharya S (2011). Fluoride - induced histopathology and synthesis of stress protein in liver and kidney of mice. Arch. Toxicol. 85:327-335.

Chen T, Cui Y, Gong T, Bai C, Peng X, Cui H (2009). Inhibition of splenocyte proliferation and spleen growth in young chickens fed high fluoride diets. Fluoride. 42(3):203-209.

Coles EH (1986). Veterinary Clinical Patholgy. W. B. Saunders Company, Philadelphia.

Das SS, Maiti R, Ghosh D (2006). Fluoride-induced immunotoxicity in adult male albino rat: a correlative approach to oxidative stress. J. Immunotoxicol. 3(2):49-55.

De Vos G, Jerschow E, Liao Z, Rosenstreich D (2004). Effects of fluoride and mercury on human cytokine response in vitro. J. Allergy Clin. Immunol. 113:66.

Elmets CA, Bowen KD (1986). Immunological suppression in mice treated with hematoporphyrin derivative photoradiation. Cancer Res. 46:1608-1611.

Griem P, Wulferink M, Sachs B, Gonzalez JB, Gleichmann E (1998). Allergic and autoimmune reactions to xenobiotics: how do they arise? Immunol. Today. 19(3):133-141.

Jain NC (1986). Schalm's Veterinary Hematology. Lea and Febiger, Philadelphia, USA.

Liu J, Cui H, Peng X, Fang J, Zuo Z, Wang H, Wu B, Deng Y, Wang K (2012). Changes induced by high dietary fluorine in the cecal tonsil cytokine content of broilers. Fluoride. 45(2):94-99.

Luna LG (1972). Manual of histological staining methods of the Armed Forces Institue of Pathology. W. B. Saunders, Philadelphia.

Machalinska A, Wiszniewska B, Tarasiuk J, Machalinski B (2002). Morphological effects of sodium fluoride on haematopoietic organs in mice. Fluoride. 35(4):231-238.

Matos TJ, Duarte CB, Goncalo M, Loes MC (2005). DNFB activates MAPKs and upregulates CD_{40} in skin-derived dendritic cells. J. Dermato. Sci. 39:113-123.

Moore DM (2000). Haematology of the Rat (Rattus norvegicus) In: Bernard F. Feldman, Joseph G. Zinkl and Nemi C. Jain, Schalm's Veterinary Haematology Lippincott Williams and Wilkins, Canada. pp. 1210-1218.

Nolte MA, Hamann A, Kraal G, Mebius RE (2002). The strict regulation of lymphocyte migration to splenic white pulp does not involve common homing receptors. Immunology 106:299-307.

Ozsvath DL (2009). Fluoride and environmental health: a review. Rev. Environ. Sci. Biotechnol. 8(1):59-79.

Phanuphak P, Moorhead JW, Claman HN (1974). Tolerance and contact sensitivity to DNFB in mice, In vivo detection by ear swelling and correlation with in vitro cell stimulation. J. Immunol. 112:115-123.

Podder S, Chattopadhyay A, Bhattacharya S, Ray MR (2010). Histopathology and cell alteration in the spleen of mice from low and high doses of sodium fluoride. Fluoride 43(4):237-245.

Raghuvansi S, Mishra AK, Arya M (2010). Fluoride in drinking water: a challenge to public health. Asian J. Exp. Chem. 5(1):7-11.

Rao AVB, Vidyunmala S (2010). Cumulative effect of fluoride on hematological indices of mice, Mus norvegicus albinus. Am. Eur. Toxicol. Sci. 2(2):93-95.

Santhakumari D, Subramanian S (2007). Effect of fluoride intoxication on bone tissue of experimental rats. Res. J. Environ. Sci. 1(3):82-92.

Sharma JD, Sharma MK, Agrawal P (2004). Effect of fluoride contaminated drinking water in albino rats (Rattus norvegicus). Asian J. Exp. Sci. 18:37-46.

Sharma JD, Solanki M, Solanki D (2007). Sodium fluoride toxicity on reproductive organs of female albino rats. Asian J. Exp. Sci. 21(2):359-364.

Snedecor GW, Cochran WG (1968). Statistical methods. Iowa State University Press, Ames, Iowa.

Tamang RK, Jha GJ, Gupta MK, Chouhan HVS, Tiwary BK (1988). In vivo immunosuppression by synthetic pyrethroid (Cypermethrin) pesticide in mice and goats. Vet. Immunol. Immunopathol. 19:299-305.

Wang B, Feliciani C, Freed I, Cai Q, Saunder DN (2001). Insights into molecular mechanisms of contact hypersensitivity gained from gene knockout studies. J. Leukoc. Biol. 70:185-191.

WHO (1984). Guidelines for drinking water quality, Vol. 1. World Health Organization, Geneva, Switzerland.

Xi S, Liu Z, Yan L, Wang F, Guifan S (2012). A role of fluoride on free aadical generation and oxidative stress in BV-2 microglia cells. Mediators of Inflammation. doi:10.1155/2012/102954.

Zhou B, Wang H, Wang J, Zhang J, Yan X, Wang J (2009). Effects of malnutrition and supplemented nutrition on specific immune parameter changes induced by fluoride in rabbits. Fluoride 42(3):216-223.

23

Assessment of udder characteristics of West African Dwarf (WAD) goats reared under different management systems in Makurdi, Benue State, Nigeria

A. H. Abu[1], L. I. Mhomga[2] and E. I. Akogwu[3]

[1]Department of Veterinary Physiology, Pharmacology and Biochemistry, College of Veterinary Medicine, University of Agriculture, P.M.B. 2373, Makurdi, Benue State, Nigeria.
[2]Department of Animal Health and Production, College of Veterinary Medicine, University of Agriculture, Makurdi, Benue State, Nigeria.
[3]College of Veterinary Medicine, University of Agriculture, Makurdi, Benue State, Nigeria.

A study was carried out to assess udder characteristics of 114 West African Dwarf (WAD) goats reared under extensive and semi-intensive systems of management. Udder width (UW), udder circumference (UC), udder length (UL), distance between teats (DBT), right teat diameter (RTD), left teat diameter (LTD), right teat length (RTL) and left teat length (LTL) were measured. The animals were classified into two groups comprising animals below the age of two years and above two years of age respectively. Statistical analysis of the data collected showed a significant difference ($P < 0.05$) in UC of goats reared under the two systems. The means of 22.60 ± 0.78 and 22.69 ± 0.82, 18.82 ± 1.45 and 18.73 ± 1.48 for right and left halves of the udder were recorded for goats reared under the semi intensive and extensive management systems respectively. Goats that were aged two years and above had mean values of 25.44 ± 0.58 and 25.28 ± 0.54 under the semi intensive system for the right and left halves respectively, while extensive system had 23.09 ± 0.68 and 23.20 ± 0.61 for the right and left halves respectively. No significant difference ($P > 0.05$) was observed between the two management systems for the rest of the udder characteristics that were studied in the two groups of goats.

Key words: Udder characteristics, management system, reproduction, goat.

INTRODUCTION

Goat production plays a very vital role in the livelihood of rural populations in Nigeria as sales of the animals and their products help to stabilize household income. The tropical environment, with its characteristic harsh weather conditions, adversely affect meat and dairy performance of animals (El-Hassan et al., 2009). It has been reported that goat milk is a preferred substitute for persons who suffer from allergies to cow milk or other food sources (Van der Horst, 1976), hence more people drink goat milk in preference to milk from other species (Park and

Chukwu, 1989). In many developing countries, there is a growing awareness of using goats as the most efficient animals for milk production.

The udder is a very important gland in reproducing animals and for milk production. Several studies have confirmed that udder and teat characteristics are important determinants of milk yield and ease of milking or milking ability in dairy animals (Akpa et al., 2002; De la Fuente et al., 1999; Rogers and Spencer, 1991). Udder and teat characteristics have been shown to be influenced by several factors such as genotype, breeding and management systems (Milerski et al., 2006). In a study conducted in South Western Nigeria, Amao et al. (2003) showed that age, lactation status and live weight are the major factors that influence udder traits in West African Dwarf goats. The study of udder morphology and physiology should be of special interest for rearing ability, dairy potentials and diversification of some breeds considered for meat purpose. There is lack of wealth of information on udder and teat characteristics of meat breeds of goats. The present study was designed to evaluate the influence of management system on the udder and teat characteristics of the West African Dwarf (WAD) goats.

MATERIALS AND METHODS

Experimental site

The study covered all the districts within Makurdi municipal, the Headquarters of Benue state, Nigeria. Makurdi is located in the savanna zone, north central Nigeria with monthly minimum and maximum temperatures of 22.1 and 32.2°C. It is located within latitude 7°43'50"N and longitude 8°32'10" E. The altitude is about 104 m above sea level.

Experimental animals

Animals used for this study were female WAD goats reared under extensive and semi-intensive systems of management. A total of 114 goats were sampled between December 2011 and January 2012 in various districts of Makurdi metropolis.

Measurement of udder characteristics

The animals were adequately restrained in a standing or recumbent position depending on the parameter to be taken. The external udder measurements were measured as described by Papachristoforou and Mavrogenis (1981). These measurements included udder length (UL) from attachment to middle of udder, and udder circumference (UC) above teats measured with a flexible tape. Right and left teat lengths (TL) were measured from attachment of teat with udder to the end of teats. By using a calliper, teat diameter (TD) was taken in the middle of teats, udder width (UW) above teats at rear of udder and distance between teats (DBT) was between the two teats attachment with udder. Cistern height (CH) was taken as the distance from the base of the teat to a point midway before the attachment of the udder to the abdominal wall, also the teat angle (TA) which was measured with a protractor by placing it on the udder with the midline of the protractor in line with the linea alba which divides the mammary gland into two. An imaginary line was then observed and traced with a broom stick from the teat downwards to the linea alba and this was recorded as the teat angle, and teat floor distance (TFD) was taken using a ruler as the distance between the tip of the teat and the ground.

Statistical analysis

The data were analyzed using Graph Pad In stat where the student T-test was used to compare means of different parameters between the two age groups.

RESULTS AND DISCUSSION

The udder traits (mean ± SEM) measured for female WAD goats under the age of two and two or more years reared under both extensive and semi intensive management systems are presented in Tables 1 and 2 respectively. The results showed that, both the extensive and semi intensive systems had no significant influence on the TD, TL, UW, DBT, TA, and CH. Also there was no significant difference between the right and left udder and teat in terms of DBT, TA, CH, UW, TFD. However, the UC was found to significantly (P < 0.05) vary between extensive and semi intensive systems (Tables 1 and 2).

Animals that store a large proportion of milk in the gland cistern produce more milk and are more able to tolerate extended milking intervals. Udder size has a strong and significant effect on milk yield (Mavrogenis et al., 1988) which makes it an important factor in the machine milkability of dairy breeds (Fernandez et al., 1995).

The udder circumference, width and height have been identified as traits which could replace the udder volume measurement because they are easy to measure and have high repeatabilities (Martinez et al., 2011). The no significant difference observed between the right and left udder and teat in respect of DBT, TA, CH, UW, TFD indicate that these parameters were not affected by the management factor.

The values for the right and left udder circumferences (UC) of WAD goats aged less than two years were 22.60 ± 0.78 and 22.69 ± 0.82 for the semi intensive system (Table 1). On the other hand, under the extensive system the values recorded were 18.82 ± 1.45 and 18.73 ± 1.48 for right and left udders respectively. Similarly, the values for WAD goats aged two years and above were 25.44 ± 0.58 and 25.28 ± 0.54 under the semi-intensive and 23.09 ± 0.68 and 23.20 ± 0.61 under the extensive management systems (Table 2). These results agree with earlier studies by Fernandez et al. (1995) and with the findings of Amao et al. (2003) for non-lactating West African dwarf do. Amao et al. (2003) showed that lactation status significantly influenced the udder circumference. However, in the present study, the differences observed might be as a result of feed supplementation associated with goats reared under the semi intensive system of management (Anurudu et al., 2004). These results may confirm that the udder circumference is a trait of interest in lactation studies. It will be worthwhile to carry out more baseline studies on the udder and teat traits of indigenous breeds of goats with the aim of establishing selection markers for milk improvement.

ACKNOWLEDGMENT

The authors wish to acknowledge the Administration of the College of Veterinary Medicine, University of

Table 1. Udder characteristics of WAD goats below two years of age (Mean ± SEM).

Trait		Management	
		Semi intensive	Extensive
Teat length (cm)	R	1.95 ± 0.09	2.00±0.10
	L	2.00 ± 0.10	2.04 ± 0.10
Teat diameter (cm)	R	0.78 ± 0.03	0.74 ± 0.05
	L	0.78 ± 0.03	0.76 ± 0.05
Teat angle (°)	R	39.29 ± 1.59	40.85 ± 1.20
	L	40.86 ± 1.99	40.77 ± 1.93
Distance between teats (cm)		7.33 ± 0.26	6.72 ± 0.38
Teat floor distance (cm)	R	21.24 ± 0.94	22.36 ± 1.08
	L	21.48 ± 0.92	22.79 ± 1.14
Cistern height (cm)	R	2.90 ± 1.08	2.55 ± 0.30
	L	2.89 ± 0.19	2.71 ± 0.28
Udder depth (cm)	R	5.89 ± 0.36	5.74 ± 0.48
	L	5.78 ± 0.26	5.70 ± 0.37
Udder width (cm)	R	10.32 ± 0.61	8.61 ± 0.63
	L	10.26 ± 0.45	8.86 ± 0.67
Udder circumference (cm)	R	22.60 ± 0.78[b]	18.82 ± 1.45[c]
	L	22.69 ± 0.82[b]	18.73 ± 1.48[c]

Means with different superscripts in a row are significantly different ($P < 0.05$).

Table 2. Udder characteristics of WAD goats two or more years of age (Mean ± SEM).

Traits		Management Systems	
		Semi intensive	Extensive
Teat length (cm)	R	2.40 ± 0.11	2.40 ± 0.07
	L	2.34 ± 0.06	2.43 ± 0.06
Teat diameter (cm)	R	0.97 ± 0.03	0.94 ± 0.03
	L	0.97 ± 0.03	0.95 ± 0.03
Teat angle (º)	R	40.31 ± 1.97	41.22 ± 1.46
	L	41.95 ± 1.39	40.36 ± 1.51
Distance between teat (cm)		8.24 ± 0.26	7.83 ± 0.23
Teat floor distance (cm)	R	18.95 ± 0.64	20.43 ± 0.75
	L	18.80 ± 0.65	20.42 ± 0.70
Cistern height (cm)	R	3.76 ± 0.16	3.42 ± 0.16
	L	3.95 ± 0.17	3.49 ± 0.15
Udder depth (cm)	R	7.44 ± 0.26	7.35 ± 0.28
	L	7.44 ± 0.27	7.25 ± 0.29
Udder width (cm)	R	12.00 ± 0.34	11.48 ± 0.33
	L	11.96 ± 0.37	11.38 ± 0.31
Udder circumference (cm)	R	25.44 ± 0.58[b]	23.09 ± 0.68[c]
	L	25.28 ± 0.54[b]	23.20 ± 0.61[c]

Means with different superscripts in a row are significantly different ($P < 0.05$).

Agriculture, Makurdi, for assigning a final year student for this study.

REFERENCES

Akpa GN, Asiribo OE, Oni OO, Alawa JP, Dim NI, Osinowo OA (2002). Milk production by agro pastoral Red Sokoto goats in Nigeria. Trop. Anim. Health Prod. 34:525-533.

Amao AO, Osinowo OA, Onwuka CFI, Abiola SS, Dipeolu MA (2003). Evaluation of udder traits in West African Dwarf goats. Nig. J. Anim. Prod. 30(2):246-252.

Anurudu NF, Babayemi OJ, Adewumi M (2004). Reproductive performance of West African Dwarf ewes fed Siam weed-based diets. Trop. J. Anim. Sci. 7(1):41-49.

De la Fuente LF, Fernandez G, San Primitivo F (1999). A linear evaluation system for udder traits of dairy ewes. Livestock Prod. Sci. 45:171-178.

El-Hassan K, El-Abid A, Abu -Nikhalia MA (2009). A study on some non-genetic factors and their impact on some reproductive traits of Sudanese Nubian goats. Int. J. Dairy Sci. 4:152-158.

Fernandez G, Alvarez P, San Primitivo F, de la Fuente LF (1995). Factors affecting variation of udder traits of dairy ewes. J. Dairy Sci. 78:842-849.

Martínez ME, Calderón C, de la Barra R, de la Fuente FL, Gonzalo C (2011). Udder morphological traits and milk yield of Chilota and Suffolk down sheep breeds. Chilean J. Agric. Res. 71(1):90-95.

Mavrogenis A, Papachristoforou C, Lysandrides P, Roushias A (1988). Environmental and genetic factors affecting udder characters and milk production in Chios sheep. Génétique Sélection Évol. 20(4):477-488.

Milerski M, Margetin M, Čapistrak A, Apolen D, Španik J, Oravcova M (2006). Relationships between external and internal udder measurements and the linear scores for udder morphology traits in dairy sheep. Czech J. Anim. Sci. 51:383-390.

Papachristoforou C, Mavrogenis A (1981). Udder characteristics of Chios sheep and their relation to milk production and machine milking. Agric. Res. Ins., Cyprus, Technical Paper, 20, ISSN, 0379-0932.

Park YW, Chukwu HI (1989). Trace minerals and Anglo-Nubian breeds during the first five 28 months of lactation. J. Food Comp. Anal. 2:161-167.

Rogers OW, Spencer SB (1991). Relationship among udder and teat morphology and milking characteristics. J. Dairy Sci. 74(12):74418-74431.

Van der Horst RL (1976). Foods of infants allergic to cow milk. South Afr. Med. J. 5:927-932.

24

Reproductive parameters of *Bracon hebetor* Say on seven different hosts

Dabhi, Manishkumar R.[1], Korat, Dhirubhai M.[2] and Vaishnav, Piyushbhai R.[3]

[1]SRA, I.F.T.C., Directorate of Extension Education, Anand Agricultural University, Anand-388 110 (Gujarat), India.
[2]Anand Agricultural University, Anand-388 110 (Gujarat), India.
[3]Department of Agricultural Statistics, BACA, AAU, Anand-388 110 (Gujarat), India.

A study on reproductive parameters of *Bracon hebetor* Say on seven different hosts (Rice moth (*Corcyra cephalonica*), Stainton Angoumois grain moth (*Sitotroga cerealella*), Oliver greater wax moth (*Galleria mellonella*), Linnaeus spotted pod borer *(Maruca testulalis)*, Geyer gram pod borer (*Helicoverpa armigera*) (Hubner), Hardwick tobacco leaf eating caterpillar (*Spodoptera litura* Fabricius) and okra fruit borer (*Earias vittella* Fabricius), studied at ordinary room temperature under laboratory conditions and revealed that *C. cephalonica* was the most suitable host for the development of *B. hebetor* among the host species tested regarding the biological parameters studied (duration of different life stages, fecundity, egg hatching percentage and sex ratio) followed by *S. cerealella, G. mellonella, M. testulalis, E. vittella, H. armigera* and *S. litura.*

Key words: *Bracon hebetor,* reproductive parameters, life stages, different seven hosts.

INTRODUCTION

Bracon hebetor Say is a highly polyphagous gregarious ecto-parasitoid of several species of lepidopteran larvae (Magro and Parra, 2001; Jhansi and Babu, 2002; Fagundes et al., 2005; Yasodha and Natarajan, 2006; Shojaei et al., 2006; Desai et al., 2007; Kyoung et al., 2008; Mohapatra et al., 2008). It attacks a variety of important lepidopterous pests of stored product and pests of field crops (Richards and Thomson, 1932; Athanassiou and Eliopoulos, 2003; Darwish et al., 2003; Gupta and Sharma, 2004; Shojaei et al., 2006). This is a well known parasitoid of several pyralid species, especially those infesting stored grains in various parts of the world. Since the biology of parasitoid differs in different hosts (Landge et al., 2009), it becomes imperative to determine the most suitable host for its mass rearing program. Comparative biology of *B. hebetor* on different

lepidopteran hosts has been studied in former papers (Margo and Parra, 2001; Jhansi and Babu, 2003; Landge et al., 2009). No such attempt has been made by any researchers in Gujarat to determine the best lepidopteran host for its rearing. With this intention we compared the reproductive parameters of *B. hebetor* on different hosts as a basis for improving mass rearing and release programs of the parasitoid in various field crops.

MATERIALS AND METHODS

A study was made on reproductive parameters of *B. hebetor* on seven different hosts (Rice moth (*Corcyra cephalonica*), Stainton Angoumois grain moth (*Sitotroga cerealella*), Oliver greater wax moth (*Galleria mellonella*), Linnaeus spotted pod borer *(Maruca testulalis)*, Geyer gram pod borer (*Helicoverpa armigera*) (Hubner),

Table 1. Development time (days) of different life-stages of *B. hebetor* reared on seven different hosts.

S/N	Host species	Egg	Larva	Pre-pupa	Pupa	Adult		Pre-oviposition	Female		Total life cycle	
						Male	Female		Oviposition	Post oviposition	Male	Female
1	*Corcyra cephalonica* Stainton	0.91	2.66	0.84	3.71	9.33	31.76	0.64	28.41	4.57	15.28	44.30
2	*Galleria mellonella* Linn.	1.12	3.33	0.93	4.64	8.27	24.12	0.74	24.89	4.81	13.36	36.94
3	*Sitotroga cerealella* Oliver	0.97	2.84	0.91	3.86	8.62	26.72	0.66	26.16	4.21	14.22	41.12
4	*Spodoptera litura* Fab.	1.68	3.09	1.22	5.92	5.09	16.02	0.86	15.76	3.07	8.72	20.56
5	*Maruca testulalis* Geyer	1.28	3.42	0.98	5.69	7.39	22.61	0.69	23.98	5.38	10.72	31.32
6	*Earias vittella* Fab.	1.63	3.16	1.02	4.04	6.15	23.80	0.71	19.55	4.14	13.12	29.76
7	*Helicoverpa armigera* (Hubner) Hardwick	1.43	3.84	1.08	5.06	6.73	20.92	0.73	19.87	3.79	10.50	27.64
	SEM ±	0.05	0.17	0.05	0.21	0.31	0.68	-	0.48	0.15	0.40	1.01
	lsd 0.05	0.15	0.48	0.15	0.62	0.89	1.98	NS	1.40	0.45	1.15	2.94
	C. V. (%)	9.29	11.64	11.99	10.20	9.36	6.44	-	4.77	8.04	7.26	6.85

NS = Not significant.

Hardwick tobacco leaf eating caterpillar (*Spodoptera litura* Fabricius) and okra fruit borer (*Earias vittella* Fabricius) studied at ordinary room temperature under laboratory conditions at Biological Control Research Laboratory, Anand Agricultural University, Anand, during 2008. Initial cultures of host larvae were collected from fields and were reared on their respective natural food to obtain healthy and uniform aged (fourth instar) larvae. The mouth of glass jar (15 cm height x 9 cm diameter) containing newly emerged 10 males and 10 females of *B. hebetor* was covered with a piece of white muslin cloth over which 10 full grown larvae of respective hosts were placed. After placing the larvae on the mouth of glass jar again another piece of white muslin cloth of same size was placed over the host larvae and kept in position with the help of rubber bands. Five replicates were used for each host species. After 24 h the parasitized larvae of each host species were removed gently without damage and were kept individually in plastic bowls (4.50 cm height x 3.50 cm diameter) for further study on various biological parameters of *B. hebetor*. The life expectancy of various life stages of *B. hebetor* reared on different hosts are presented in Table 1, whereas the reproductive data on egg hatching (%), fecundity and sex ratio are presented in Table 2. Data on egg hatching and fecundity were analyzed after arc sine and square root transformed values, respectively.

RESULTS

The lowest life expectancy for the egg period of *B. hebetor* was registered with *C. cephalonica* followed by *S. cerealella* larvae as compared to rest of the larvae used for rearing (Table 1). *Maruca testulalis* and *H. armigera* were not significantly different in their duration of egg-period. Among the seven species of host larvae evaluated for comparative biology of *B. hebetor*, significantly longest egg period was recorded in *S. litura* followed by *E. vittella*. Highest number of *B. hebetor* eggs were hatched (Table 2) when it was reared on the larvae of *C. cephalonica* followed by *S. cerealella* and *G. mellonella*.

Lowest hatching percentage was registered in case of *S. litura* over rest of the host larvae evaluated, except *E. vittella*. Significantly, more time was required to complete its duration in case of *H. armigera* in comparison to rest of the host larvae used, except *M. testulalis*. Lowest duration was registered in case of *C. cephalonica* followed by *S. cerealella*. *S. litura*, *E. vittella* and *G. mellonella* exhibited same duration of larvae. Pre-pupal period of *B. hebetor* was found to be highest in case of *S. litura* followed by *H. armigera* and *E. vittella*. Significantly less duration of pre-pupa was registered in *C. cephalonica*, *S. cerealella*, *G. mellonella* and *M. testulalis* in comparison to *S. litura*. Pupal period was lowest in case of *C. cephalonica* followed by *S. cerealella* and *E. vittella*. Shorter pupal period was exhibited in these three larval hosts over rest of the host larvae used, except *G. mellonella*.

Among the different hosts, significantly less duration of male longevity was registered in case of *S. litura*. Not significantly different longevity was recorded in *E. vittella* and *H. armigera*. Highest duration was recorded in *C. cephalonica* followed by *S. cerealella* and differed significantly from rest of the larvae, except *G. mellonella*. As like male longevity, female longevity was also found to be highest in case of *C. cephalonica*. *S. cerealella* stood second in rank by exhibiting 26.72 days as

Table 2. Influence of seven different hosts on egg hatching, fecundity and sex ratio of *B. hebetor*.

S/N	Host species	Egg hatching (%)	Fecundity	Sex ratio (F : M)
1	*Corcyra cephalonica* Stainton	*66.92 (75.34)	12.44 (154.25)**	1 : 1.32
2	*Galleria mellonella* Linn.	63.41 (70.45)	11.55 (132.90)	1 : 1.64
3	*Sitotroga cerealella* Oliver	66.66 (74.99)	12.12 (146.39)	1 : 1.56
4	*Spodoptera litura* Fab.	55.69 (58.89)	8.61 (73.63)	1 : 1.77
5	*Maruca testulalis* Geyer	62.08 (68.52)	10.47 (109.12)	1 : 1.67
6	*Earias vittella* Fab.	59.42 (64.68)	10.85 (117.22)	1 : 1.60
7	*Helicoverpa armigera* (Hubner) Hardwick	60.75 (66.57)	10.10 (101.51)	1 : 1.70
	SEM ±	1.66	0.08	0.05
	lsd $_{0.05}$	4.79	0.23	0.13
	C. V. (%)	5.92	1.66	6.23

* Arc sin transformed values; **, $\sqrt{x + 0.5}$ transformed values; Figures in parenthesis are retransformed values.

female longevity. Both these larval hosts differed from rest of larvae by registering significantly higher longevity over remaining host larvae. On the other hand significantly less duration was recorded on *S. litura*. *Maruca testulalis*, *E. vittella* and *G. mellonella* exhibited not significantly different in longevity and were at par. Pre-oviposition period was found to be 0.64 to 0.86 day on different host larvae and there was no significant difference among the hosts used. Relatively less pre-oviposition period was found in case of *C. cephalonica* and *S. cerealella*. Significantly highest and lowest oviposition period was recorded in *C. cephalonica* and *S. litura*, respectively. With respect to oviposition period, *E. vittella* and *H. armigera* performed equally. Similarly, *G. mellonella* and *S. cerealella* exhibited not significantly different in duration of oviposition. Significantly, highest post-oviposition period was registered in *M. testulalis* than other host larvae. *G. mellonella* stood second in position by registering 4.81 days as oviposition period. On the other hand, *S. litura* exhibited significantly lowest duration of post-oviposition. With respect to post-oviposition period, *C. cephalonica*, *S. cerealella* and *E. vittella* were at par. *B. hebetor* larvae reared on different hosts influenced significantly on egg-laying potential (Table 2). Significantly highest fecundity was registered in case of *C. cephalonica* followed by *S. cerealella*. On the other hand, significantly least number of eggs per female were recorded in *S. litura* followed by *H. armigera*. *M. testulalis*, *E. vittella* and *G. mellonella* exhibited fecundity ranging from 109.12 to 132.90 eggs/ female.

Highest duration to complete one life-cycle (egg to adult) of *B. hebetor* female was recorded in *C. cephalonica* followed by *S. cerealella*. Both these host larvae differed from remaining hosts by exhibiting significantly higher duration. Significantly lowest period for one life-cycle of female was found in *S. litura*. In terms of life-period, *H. armigera* and *E. vittella* were found to be not significantly different and were at par. Similar trend to

complete whole life-cycle on different host larvae was noticed in male insects. Highest duration was registered in case of *C. cephalonica* followed by *S. cerealla*. Among the different hosts, lowest duration of male life-cycle was recorded in *S. litura*. Sex ratio (Female : Male) of *B. hebetor* reared on different host larvae revealed that minimum male population of the parasitoid was registered in case of *C. cephalonica* followed by *S. cerealella*. On the other hand, highest male ratio was predominated when the *B. hebetor* was reared on *S. litura* followed by *H. armigera*. Ratio of female to male was not significantly different in case of *G. mellonella*, *M. testulalis* and *E. vittella* (Table 2).

DISCUSSION

Lowest egg period of *B. hebetor* was registered in *C. cephalonica* followed by *S. cerealella* and *G. mellonella*. This finding agrees with Forouzan et al. (2003) who reported that the average egg period was 1.77 ± 0.03 days in *G. mellonella*. Highest numbers of *B. hebetor* eggs were hatched when it was reared on the larvae of *C. cephalonica* followed by *S. cerealella* and *G. mellonella*. Results of higher (6.82%) percentage of egg hatch on *C. cephalonica* over *M. testulalis* noticed in present study agree with the manuscript of Jhansi and Babu (2003). Larval duration of *G. mellonella* was found as 3.33 days which was not significantly different (3.43 ± 0.04 days) as reported earlier by Forouzan et al. (2003). Pupal period was lowest in case of *C. cephalonica* followed by *S. cerealella*, *E. vittella* and *G. mellonella*. This finding agrees with Forouzan et al. (2003) who reported that the average pupal period was 6.89 ± 0.05 days in *G. mellonella*.

Highest male longevity was recorded in *C. cephalonica* followed by *S. cerealella* and it was differed significantly from rest of the larvae, except *G. mellonella*. This is similar to the values reported by Nikam and Pawar (1993). As like male longevity, female longevity was also found to be highest in case of *C. cephalonica*. *S. cerealella* stood second in rank by exhibiting 26.72 days as female longevity. This finding agrees with Youm and Gilstrap (1993) who reported that female of *B. hebetor* lived an average of 24.7 days when reared on *H. albipunctella* Joannis. Significantly highest fecundity was registered in case of *C. cephalonica* followed by *S. cerealella*. This was not significantly different values (173.7 eggs/female) were found on *H. albipunctella* de Joannis by Youm and Gilstrap (1993). On the other hand, significantly least numbers of eggs per female were recorded in *S. litura* followed by *H. armigera*. *M. testulalis*, *E. vittella* and *G. mellonella* exhibited fecundity ranging from 109.12 to 132.90 eggs/ female. Xie et al. (1989) recorded more number of *B. hebetor* eggs on larger larvae than smaller host larvae which corroborates with the present finding in which significantly higher fecundity of the parasitoid was revealed on larger sized larvae of *C. cephalonica* over *S. cerealella*. In the present study, significantly more number of eggs were laid by females of *B. hebetor* as compared to *M. testulalis* which agree with the results of Jhansi and Babu (2003). Sex ratio (Female : Male) of *B. hebetor* reared on different host larvae revealed that minimum male population of the parasitoid was registered in case of *C. cephalonica* followed by *S. cerealella*. On the other hand, highest male ratio was predominated when the *B. hebetor* was reared on *S. litura* followed by *H. armigera*. Comparatively higher number of females than males was recorded by Jhansi and Babu (2003) on *C. cephalonica* in comparison to *M. testulalis* is contradictory with the present finding. This discrepancy might be due to the variation in food and climatic condition in which the study has been made. None of earlier worker has studied the reproductive parameters of *B. hebetor* on different species of lepidopterous as hosts evaluated in present study except *C. cephalonica*, *S. cerealella*, *G. mellonella* and *M. testulalis*.

From the above results it can be concluded that *C. cephalonica* found to be the best host for laboratory mass rearing of *B. hebetor* followed by *S. cerealella*. This finding is supported by few earlier studies. According to Zohdy (1979), *C. cephalonica* found to be the most suitable host for the development of *B. hebetor*. Jhansi and Babu (2003) reported that with respect to number of eggs laid, percentage of egg hatch, growth index and percentage of adult emergence, *C. cephalonica* found better host than *M. testulalis*. Landge et al. (2009) also revealed superiority of *C. cephalonica* over *Opisina arenosella* Walker.

REFERENCES

Athanassiou CG, Eliopoulos PA (2003). Seasonal abundance of insect pests and their parasitoids in stored currants. Bulletin-OILB/SROP 26:283-291.

Darwish E, El-Shazly M, El-Sherif H (2003). The choice of probing sites by *Bracon hebetor* Say foraging for *Ephestia kuehniella* Zeller. J. Stored Prod. Res. 39:265-276.

Desai VS, Nagwekar DD, Patil PD, Narangalkar AL (2007). Field evaluation of a larval parasite *Bracon hebetor* Say against coconut black headed caterpillar. J. Plantation Crops. 35:188-189.

Fagundes GG, Mohamed H, Solis DR (2005). Biological responses of *Anagasta kuehniella* and its parasitoid, *Bracon hebetor*, to microwaves radiation (2450 MHz). Revista de Agricultura Piracicaba. 80:12-34.

Forouzan M, Sahragard A, Amirmaafi M (2003). Study on the biology of *Habrobracon hebetor* Say under laboratory conditions. J. Entomol. Soc. Iran 22:63-76.

Gupta S, Sharma HB (2004). *Bracon hebetor* Say is the natural enemy of *Ephestia calidella* (Guen.) a pest of stored dry fruits. Uttar Pradesh J. Zool. 24:223-226.

Jhansi K, Babu PCS (2002). Life table studies of *Bracon hebetor* (Say) on *Corcyra cephalonica* (Stainton) and *Maruca testulalis* (Geyer) under laboratory conditions. J. Appl. Zool. Res. 13:22-24.

Jhansi K, Babu PCS (2003). Comparative biology of *Bracon hebetor* Say in two host insects. J. Appl. Zool. Res. 14:165-168.

Kyoung DJ, Ha DH, Nho SK, Song KS, Lee KY (2008). Up regulation of heat stock protein genes by envenomation of ectoparasitoid *Bracon hebetor* in larval host of Indian meal moth, *Plodia interpunctella*. J. Invertebrate Pathol. 97:306-309.

Landge SA, Wakhede SM, Gangurde SM (2009). Comparative biology of *Bracon hebetor* Say on *Corcyra cephalonica* Stainton and *Opisina arenosella* Walker. Int. J. Plant Protect. 2:278-280.

Margo SR, Parra JRP (2001). Biology of the ectoparasitoid *Bracon hebetor* Say on seven lepidopteran species. Scientia Agricola. 58:693-698.

Mohapatra SD, Duraimurugan P, Saxena H (2008). Natural parasitization of *Maruca vitrata* (Geyer) by *Bracon hebetor* Say. Pulses Newsl. 19:11.

Nikam PK, Pawar CV (1993). Life tables and intrinsic rate of natural increase of *Bracon hebetor* Say population on *Corcyra cephalonica* Staint., J. Appl. Entomol. 115:210-213.

Richards OW, Thompson WS (1932). A contribution to the study of genera *Ephestia* (including Strymax, Dyar) and *Plodia interpunctella*, with notes on parasite of the larvae. Trans. Entomol. Soc. London 80:169-247.

Shojaei S, Safaralizadeh M, Shayesteh N (2006). Effect of temperature on the functional response of *Habrobracon hebetor* Say to various densities of the host, *Plodia interpunctella* (Hubner). Pak. Entomol. 28:51-55.

Xie ZN, Li L, Xie YQ (1989). In vitro culture of *Habrobracon hebetor* (Say). Chinese J. Biol. Control. 5:49-51.

Yasodha, P, Natarajan N (2006). Diversity of natural enemies of *Leucinodes orbonalis* Guenee. Entomon, 31:323-326.

Youm O, Gilstrap FE (1993). Population dynamics and parasitism of *Coniesta (Haimachia) ignefusalis*, *Sesamia calamistis* and *Heliocheilus albipunctella* in millet monoculture. Insect Sci. Appl. 14:419-426.

Zohdy NZM (1979). Host selection and host suitability of *Bracon hebetor* Say. Bull. Fac. Sci. Cairo Uni. 48:301-314.

25

Biophysical and the socio-economics of chicken production

Mammo Mengesha

Ethiopian Institute of Agricultural Research, Debre Zeit Agricultural Research Center; P. O. Box 32. Debre Zeit, Ethiopia.

This paper reviews the socio-economics of poultry production with the aim of delivering summarized and synthesized information for the beneficiaries. Poultry production and product consumption are progressively growing in the world. Poultry accounts for about 33% of the global meat consumption and is expected to grow at 2 to 3% per year in the world. Literature stated that large size poultry farms were more efficient, but these farms are few in the developing countries. Although there is prediction that technology favors the intensification of poultry production in these countries, village poultry still is profitable that contributes to poverty alleviation and has no market problems. Women and children are responsible for caretaking of poultry and they are also beneficiaries. The largest off-take rates of birds occur particularly during holidays and festivals in Ethiopia. Chicken population and *per capita* consumption of egg and poultry meat has been declining to the face of population growth in Ethiopia. Livestock production is likely to be increasingly affected by climate change; however poultry industry has a natural advantage over others livestock because of its low global warming potential. Thus, poultry meat and egg production is the most environmentally efficient animal protein production system. But, poultry production has been facing with a problem of food-feed competition and other critical gaps that need to be filled by the institutions of research and development. The study showed that such poultry technologies should be compatible with local socio-economical interests. Further improvement would be possible by lowering the prices at the consumer level and by improving the profitability of producers. It is concluded that poultry production has so many socio-economical advantages in satisfying the demands of animal source foods.

Key words: Health and climate change, poultry production, protein foods, socio-economics.

INTRODUCTION

Shortages of protein availability are a well-known problem in Africa. Poultry is by far the largest group of livestock species (FAO, 2000a) contributing about 30% (Permin and Pedersen, 2000) of all animal protein consumed in the world. From the total poultry population, chicken constituted around 98% in Africa (Gueye, 2003) and almost 100% (Alemu, 1995) in Ethiopia. Moreover, local chickens are widely distributed in the rural areas of the tropics. Indigenous chickens in Africa are hardy and they can adjust themselves to the fluctuations (Kitalyi, 1998). Importance of indigenous poultry breeds for subsistence farmers in many developing countries (David, 2010; FAO, 2006) combined with many consumers' preference for their eggs and meat suggests that these genetic resources are not under immediate threats.

The poultry sector is characterized by its industrialization, faster growth in consumption and trade than any other major agricultural sectors in the world. According to Bos and De Wit (1996), poultry provides an acceptable form of animal protein to most people throughout the world. Moreover, FAO (2010) reported that poultry meat represents about 33% of the total global meat production. Scanes (2007) reported also that to reduce child mortality and to improve maternal health, in sub-Saharan Africa, poultry products provide an excellent source of nutrient (e.g., B_{12}). Wiebe (2007) reported also that poultry meat and egg production is the most environmentally efficient animal protein production system. Poultry and egg sales are decided by women that they provide immediate income and source of self-reliance for them. Moreover, poultry are used for strengthening marriage partnerships and social relationships of the society. Chicken population in Ethiopia is estimated to be around 34.2 million (CSA, 2007) that constitutes 60% of the total population of East Africa (Mekonnen et al., 1991).

From which 97.82% of them are local birds kept in extensive production systems (FAO, 2008). Dawit et al. (2009) reported also that the role of poultry in Ethiopia has been becoming more important over time. Therefore, the need of reviewing chicken production and the socio-economic in Ethiopia is a prioritized issue that for improved production of chicken a country. Moreover, reviewing the successful experiences of chicken production and its socioeconomics, and thereby delivering synthesized form of information for beneficiaries is also another milestone to improving the production of poultry in the country. Based on this outlined background, the objectives of this paper were:

1. Review chicken production scenarios and the socio-economics of chicken production
2. Avail the socio-economic essentials of chicken productions for the beneficiaries

Most of the research findings, which focused on socio-economics of poultry productions, were reviewed. Various research findings that focused on the social and economical impacts of poultry production were reviewed. Effects of food-feed competitions on chicken production types of chickens were also reviewed, depicted and sourced. Trends of production, trade and consumption of poultry were reviewed. Comparative studies of family member participations in poultry production and their share of benefits were reviewed. Moreover, the impacts of climate change on poultry production were also reviewed and synthesized.

OUTLINED DESCRIPTIONS OF POULTRY PRODUCTION

Poultry production can be described not only as the production of meat and eggs, but also distribution and retailing of these products. Based on its level of biosecurity and birds/products marketed, poultry production sector is classified into 4 viz; industrial, commercial, medium-commercial and village chicken productions systems (Jonathan et al., 2004) in the world. Most poultry production, in developing countries is possibly described as a scavenging/ village systems (Kitalyi, 1998) sector. Kryger et al. (2010) reported also that around 80 percent of rural households in developing countries engage in smallholder poultry production. According to Sonaiya and Swan (2004), in sub-Saharan Africa, 85% of poultry sector is managed in village production systems. Moreover, around 97.82% of chicken production in Ethiopia is traditionally managed (FAO, 2008).

Although traditional practices continue to dominate domestic poultry production in Ethiopia; there has been a shift to industrial production (FAO, 2008). According to ILCA (1993) reports, chicken population in Ethiopia was around 56.5 million; however, this number was declined to 34.2 (CSA, 2007) millions. About 99% of the annual poultry meat and egg production comes from the indigenous chickens kept under the traditional systems (FAO, 2008) in Ethiopia. Exotic breeds of chicken are only 2.18% of the total poultry population in a country (CSA, 2005). All of these, except some managed by small scale intensive, are used by large scale (private and government) commercial poultry farms (FAO, 2008) in Ethiopia.

Indigenous chickens are characterized by low performances: Average annual egg production is estimated to be around 60 small eggs with thick shells and a deep yellow yolk colour (Alemu and Tadelle, 1997). Average body weight gained by male birds was also around 1.5 kg at 24 weeks of age (Teketel, 1986). Estimated egg and poultry meat per capital consumption (in the mid 1990s) was 57 eggs and about 2.85 kg (ILCA, 1993), respectively in Ethiopia. However, *per capita* consumption of egg and poultry meat has been declining (2003) by 0.12 and 0.14, respectively (USAID, 2006) to the face of population growth in Ethiopia. Inputs and technologies are always required to improve poultry production and thereby to satisfy the socio-economical needs of the producers.

INPUTS AND TECHNOLOGIES TO IMPROVE POULTRY PRODUCTION

Inputs and appropriate poultry technologies can be local practices or be adopted from other countries. According to Muchenje et al. (2001), poultry technologies should be compatible with local interest. Aichi (1996) observed that many developed production technologies are not compatible with the socio-economic circumstances in the village chicken production system. Furthermore, Adebayo and Adeola (2005) in Nigeria reported that a national support is required in the area of finance and input to strengthen the poultry production. In Sudan, Abda and Amin (2010) reported that feed cost was the main production cost (90.87%) in different poultry farm types and sizes. Similarly, Achoja et al. (2006) in Nigeria reported that the price per bag and poor road network (market access condition) were the major problems affecting efficient marketing of poultry feeds in the study area.

The high cost of commercial poultry feed discourages farmers from supplementing local chicken; therefore farm feed formulations using locally available materials should be encouraged (Njue et al., 2004). Alemu and Tadelle (1997) reported that the quality of mixed feed for commercial poultry production is generally poor in Ethiopia.

According to Vincent et al. (2010), the resources used in poultry production were not properly utilized. The authors recommended that farmers should use inputs more efficiently by reducing their levels of employment. Danielle et al. (2009) reported from Tanzania that vaccinations did increase chicken production and egg and meat consumption. Various scholars are always advising poultry producers to adopt appropriate technologies and practices in order to boost production.

ADOPTION OF IMPROVED POULTRY PRODUCTION PRACTICES

Generally, adoption of improved poultry production practices may involve the transfer of appropriate new technologies and local experiences to be used in improving productivity of the stocks. The pace of adopting new technologies by farmers can vary due to controversial reasons. Teklewold et al. (2006) reported in Ethiopia that farmers' decision on the extent of adoption of exotic poultry breed was positively influenced by age of the household head, experience in adoption of poultry technology, expected benefit from poultry and market problem. Moreover, Truong and Yamada (2002) reported from Vietnam that Farmers' changes of technology use are influenced by technical training, meeting, oral transmission, trust on technician and belief level on technology. Moreover, Nnadi and Akwiwu (2005) reported that there is a varied rate of adoption of the improved poultry production practices in Nigeria. The same authors recommended that extension education campaign should be intensified to avail rural women of new practices in poultry.

According to Augustine (2010), the socioeconomic characteristics of the poultry farmers collectively have positive but low relationship with the cost of inputs adopted by the farmers. However, Eze and Okudu (2008) reported from Nigeria that /ncome; stock and educational levels were the most valuable variables determining the

poultry farmer technology adoptions potential. Moreover, Olaniyi et al. (2008) identified the major constraints to utilization of poultry production technology was access to get capital and inadequate extension contact.

According to Khandait et al. (2011) from many technologies, marketing were highly adopted followed by feeding and watering in the backyard poultry practices of India. However, the same author stated that the least adoption was for health care practices. Truong and Yamada (2002) reported that factors that trigger adoption of new technologies comprise of young and educated male farmers. On the other hand, the author stated the factors, which could limit adoption of poultry technologies included conservative old men, and weak belief on ensure high yield of new technology. Olaniyi et al. (2003) reported that age, awareness and education shows negative relationship with constraints to utilization of poultry production technology. According to Truong and Yamada (2002) usually men use technologies for rice, fruit and fish production; whereas, women use technology for pig and chicken production. In Nigeria, Musa et al. (2008) reported the wide use of ethno veterinary remedies by rural chicken farmers in Plateau State.

Augustine (2010) reported that the socioeconomic characteristics of the farmers collectively have positive but low relationship with medication practices adopted by the farmers. In Ethiopia, Dana et al. (2006) emphasis the need to transform the traditional piece meal approach of poultry technology transfer into promotion of carefully selected and packaged technologies in a multi-institutional framework with due consideration to the input-output market if the potential role of poultry research in development is to be realized. Adoption of improved poultry production practices and transfer of such appropriate new technologies and local experiences may vary according to poultry production systems and socio-economic status.

POULTRY PRODUCTION SYSTEMS AND THE SOCIOECONOMICS

Socio-economical status of farmers can vary according to poultry production systems. Kryger et al. (2010) reported that there are emerging signs of restructuring—with a shift away from small-scale commercial production towards larger-scale production. Likewise to this, Mekonnen (2007) from Ethiopia reported that efforts have to be made to shift the chicken production paradigm to semi intensive in Ethiopia with a holistic support of services such as health, housing and feed to make it productive and sustainable. According to Abda and Amin (2010) the large size farms are more efficient than other sizes and type of the poultry farms. Jugessur et al. (2004) reported from Mauritius that about 65% of the households wished to expand family poultry production as they found this system of production more profitable than rearing improved commercial broilers and layers.

Kryger et al. (2010) reported that due to its provision of utility among all wealth groups, village chicken production is likely to persist within their livelihoods of country's rural population. Permin et al. (2004) by using a holistic approach, it is possible to improve village poultry development, which may help the poorer farmers in developing their skills and creating a sustainable income with very few inputs.

According to Scanes (2007), village poultry contribute to poverty alleviation and improvement of food security. Njue et al. (2004) reported that supplementation and vaccination of local birds was economically profitable. In Bangladesh, Sumy et al. (2010) reported that backyard chicken rearing is profitable for farmers. Vaccinations and balanced diets have a decisive effect on chicken rearing, providing quality products for human consumption and reducing nutritional deficiencies and poverty of the country.

Adrian and Michael (2009) in South Africa observed that small-scale poultry production has the ability to initiate economic growth. Small scale poultry producers face higher transaction costs. Also, FAO (2011) reported that advances in technology favours the intensification of poultry production in developing countries. In Kenya, Vincent et al. (2010) reported that poultry production is one of the most important economic activities to the smallholder farmers. Technical and physical constraints are evident which have resulted in low production of poultry and poultry products to meet population demand and for socio-economic sustainability of the livelihoods.

PROTEIN FOOD DEMANDS AND TRENDS OF CHICKEN PRODUCTIONS

The demand of protein food is progressively growing with the improvements of society's income and population growth that affects trends of chicken production. The enormous surges and the driving force behind livestock and poultry sectors growing at an unprecedented rate is a combination of population growth, rising incomes and urbanization. Similarly, FAO (2009) reported that there is a strong positive relationship between the level of income and the consumption of animal proteins. According to Daghir (2009) the current growth of poultry production and consumption makes a good case for the need and desire for future growth of the poultry industry. Dave (2007) also reported that poultry consumption is expected to grow at 2 to 3% per year.

According to David (2010), chicken meat and eggs are the best source of quality protein for those who are under-nutrition in sub-Saharan Africa (SSA) and South Asia. Muchenje et al. (2001) reported that poultry provide major opportunities for increased protein production and incomes for smallholder farmers. Abedullah and Bakhsh (2007) noted that the major contribution of poultry

consumption in improving *per capita* nutrients level is well documented.

Village chicken in Ethiopia provides 12.5 kg of poultry meat per capita per year, whereas cattle provide only 5.34 kg" (Kitalyi, 1997). According to Windhorst (2008), an increase in egg and poultry meat consumption for least developing countries is 26 and 2.4%, compared with only 2.4 and 1.6% in the most developed countries. FAO (2010) reported also that chicken meat is relatively healthier than others; containing low total fat and it has high desirable monounsaturated fats. Costa (2009) described the attributes of chicken meat to its intensively-based and vertically integrated operation. Furthermore, Paweł (2005) reported that consumption of poultry and fish has not been found to be associated with increased risk of cancers. Arrey (2009) reported also that the possibility of village poultry as a viable sector to boost protein deficiencies in Cameroon is documented. To improve chicken production and to satisfy the demands of protein foods, participation of family members in the household is highly required in the phenomena of poultry productions.

POULTRY PRODUCTION AND PARTICIPATIONS OF FAMILY MEMBERS IN THE HOUSEHOLDS

All family members are participating in poultry production. However, Aichi (1996) noted that subsistence village chicken production in Africa is the domain of women and children. Similarly, Permin et al. (2004) reported that women are the caretakers of poultry in most of the poor countries. The contribution of smallholder poultry production to the income and internal household position of women is widely recognized (Kryger et al., 2010). Scanes (2007) observed that poultry may be one of the few, or only, sources of cash income for women and children. Okoh et al. (2010) reported that women enjoyed about 53.6% level of involvement in commercial poultry farm decision-making and only about 28.7% rate of accessibility to farm resources. According to Dawit et al. (2009), at the micro-economic level poultry is very important especially for women in Ethiopia. Mammo et al. (2008) reported that in Ethiopia shared and individual ownerships are the main mode of ownership of village chickens in the family. Also, Hunduma et al. (2010) stated that around 92.4% of the households' village chicken production is accomplished by women and children. Similarly, about 70% of overall care-taking of chickens, feeding of chickens, cleaning of birds-quarter, treating of sick birds, decision for off take of poultry products in Ethiopia were the responsibility of women (*mammo* et al. 2008).

Kryger et al. (2010) reported that poultry development interventions offering women beneficiaries much more than economic profits; along with economic empowerment come social empowerment. On the other hand, Eze and Okudu (2008) argued that women are more affected by the avian influenza crisis since they are the ones directly involved in the care and handling of poultry particularly in small-scale backyard production. It is sensible that the active participation of family members, in the household of poultry production should always be with their purpose of home consumption, social purpose or incomes generation.

PURPOSE OF POULTRY PRODUCTION AND THE ECONOMICS

Every type of poultry production system has its own purpose, while rearing birds such as: Home consumption, social purpose or to generate incomes. According to Kryger et al. (2010) reports, income and consumption have been considered the main rationale for keeping village poultry. Hunduma et al. (2010) reported from Ethiopia that most of village chicken keepers used chickens and their by products for home expenditure followed by home consumption. Tadelle et al. (2003) reported from the same country that maximum amounts of the eggs produced were used for hatching followed by for sale and home consumption, while more birds were used for sale followed by replacement and consumption. Similarly, Jugessur et al. (2004) reported from Mauritius that almost every household keeps of semi-scavenging poultry for food and to generate additional incomes. However, Adebayo and Adeola (2005) reported that finance and input are essential for substantial improvements in the contribution of the poultry enterprise to household food production and economic well being of poor farmers.

According to Jugessur et al. (2004) reports, family poultry had a guaranteed market. All the poultry merchants who marketed the family poultry found their business profitable, and wished to expand it (Jugessur et al., 2004). That is why Maqbool and Bukhsh (2007) reported from Kenya that commission agents were earning 47% of the total profit in poultry industry, followed by retailers (28%) and producers (25%). Based on this result, these authors conclude that it would be impossible to improve the contribution of poultry in total nutrients uptake of human beings in the country without reversing the trends in profit share. Thus, Maqbool and Bukhsh (2007) reported that inequitable distribution of profit share was assumed to be one of the major obstacles in the expansion of poultry industry

Mammo et al. (2008) in Ethiopia reported that fluctuation in the prices of the village chickens and chicken-products were mainly due to purchasing power of the consumers, fasting and availability of products. However, Mekonnen (2007) reported that more than half of the respondents did not have any information about the price of the chickens. According to Tadelle et al. (2003), the overall gross return as percent of initial values

and gross return per breeding female per year were 67.5% and 12.48 Birr, respectively. However, Solomon (2008) reported that the export market for poultry products in Ethiopia is very limited, but it may be worthwhile studying consumer preferences in neighboring countries to determine if niche markets exist for extensively raised indigenous birds and their eggs. The major criteria used to determine the price of local chickens were body weight, plumage colour and comb-type (Mammo et al., 2008).

According to Ajala and Otchere (2007), incomes from sales of birds and eggs serve as reserve for important household expenditures. In addition, Jugessur et al. (2004) reported that the profit obtained from the sale of chickens and eggs, and the monetary value of sale and home consumption of these commodities represented 9 and 18% of the total income of the family, respectively. Olasimbo (2006) observed that the variation in returns from table egg production was comprised from numbers of birds kept and type of production system used. Generally, Adrian and Michael (2009) reported that government policies should focus on absorbing transaction costs of small-scale poultry producers and interventions like provision of mentoring and training services to stakeholders.

POULTRY PRODUCTS AND PHENOMENA OF CONSUMPTIONS

Livestock and poultry sector globally is highly dynamic, particularly in developing countries that are evolving in response to rapidly increasing demand for animal products. Thus, with this fatly growing poultry production, trade and consumptions, human health should also be considered. Poultry production and consumption has increased in the world (Philip, 2011). Poultry meat accounts for about 33% (87% chicken and 6.7% Turkey) of the global meat consumption (FAO, 2010). Similarly, Costa (2009)) reported that the consumer demands for chicken meat has been growing steadily over the last decade. FAO (2010) reported that the human population benefits greatly from poultry meat and eggs, which provide food containing high-quality protein, and a low level of fat with a desirable fatty acid profiles.

According to Maqbool and Bukhsh (2007), the major contribution of poultry consumption in improving *per capita* nutrients level is well documented, however, further improvement would be possible by lowering the prices at the consumer level and by improving the profitability of producers through up-taking of poultry technologies. Moreover, David (2010) reported that semi-scavenging backyard indigenous poultry are extremely important in providing income and high-quality protein in the diets of rural people whose traditional foods are typically rich in carbohydrate but low in protein.

In Serbia, Rodić et al. (2010) reported that poultry accounts for about 12% of the total value of country's livestock production. The author added that poultry provides relatively cheap food of high quality to the people. Solomon (2008) in Ethiopia reported that the largest off-take rates flocks occur particularly during holidays and festivals and during the onset of disease outbreaks. The periods of low bird sales and consumption in Ethiopia coincide with the pre-Easter fasting period and pre-Christmas fasting period. On the other hand, Danielle et al. (2009) in Tanzania reported that shortages in protein availability are a well. Nutritional importance of animal-source protein is increasingly being recognized in village economics to solving the neurological problem, especially in Africa. Protein food demands increments and progressively growing of the phenomena of poultry product consumption needs to be given emphasis for reduction of climate change.

HARVESTING, CONSERVATION AND VALUE ADDITION OF POULTRY PRODUCTS

According to Deogade et al. (2008), market for value added chicken meat products are popular in local market, but now a day's big player of market are launching their products mainly ready to eat and ready to cook type. Froning (1998) reported that although shell egg consumption has declined, the growth of value added egg products has been very encouraging. The author added that approximately 30% of our egg consumption is in the form of value egg products. For world's poultry (2009) egg processing serves a rapidly increasing market worldwide. Meanwhile, there is an increase in demand for product differentiation to meet separate market requirements. On the other hand, Hafez (1999) reported that post-processing food handling is also a very important factor in reducing foodborne infections.

CLIMATE CHANGE POULTRY PRODUCTION

Both climate change and animal production have always negative impacts on each other. Climate change could affect animal production due to the impacts of increasing air temperature, feed-grain availability and favouring the diseases (Adams et al., 1990; Bowes and Crosson, 1993). To adopt the improved poultry production practices, socio-economic characteristics of the target population as well as bio-physical environmental considerations are needed (Nnadi and Akwiwu, 2005). David (2010) reported that poultry production has a less detrimental impact on the environment than other livestock and uses less water. Moreover, Wiebe (2007) reported that poultry meat and egg production are the most environmentally efficient animal protein production system. Intensive poultry production has much less impact on global warming than organic or free-range production (Wiebe, 2007). However, Guèye (2009)

reported that despite efforts to develop intensive poultry production, family poultry remains important in the developing countries of Africa. Similar to this argument, Sungno and Robert (2006) reported that small farms of livestock (poultry) are better able to adapt to warming. However, organic egg production needs more energy than non-organic that increases environmental burdens. The disadvantage, from environmental point of view, is litter-free breeding of birds, which causes great amounts of liquid manure. Obayelu and Adeniyi (2006) reported also that climate change has an effect on poultry feed intake, encourages outbreak of poultry diseases, which invariably reduce egg production. However, Gilbert et al. (2009) reported that little is known about the direct effect of climate change factors on highly pathogenic avian influenza transmission of domestic birds and persistence to allow inference about the possible effects.

Jan and Henning (2008) reported that the global poultry industries have faced competition for feed ingredients including the prospect of future ethanol production.

Philip (2011) reported that livestock production is likely to be increasingly affected by carbon constraints, environmental and animal welfare legislations. However, Costa (2009) observed that the poultry industry has a natural advantage over other livestock industries because of its low global warming potential. For sustainable poultry production, a collaborative effort of research and development institutions is required.

COLLABORATION OF RESEARCH AND DEVELOPMENTAL INSTITUTIONS FOR POULTRY IMPROVEMENT

With collaborative efforts of research and developmental institutions, improvements of poultry production would have to be achieved. Smallholder farming systems are bio-economically complex involving several kinds of resources and input/output flows such that it would not be advisable to study poultry production only without considering other crop-livestock components of the farming system (Muchenje et al., 2001). According to Adrian and Michael (2009) in South Africa reported that alleviating constraints for a large number of small enterprises of poultry productions is expected to impact more positively on the rural economy than if a few larger enterprises were encouraged to grow bigger. Critical gap of poultry production still needs to be filled by the research extension outfit in combining the technical and socio-economic aspects of poultry production in order to boost Nigeria's egg and meat production (Adebayo and Adeola, 2005). Njue et al. (2004) reported that sustainable cost effective interventions are necessary if full potential is to be realized in local chicken production of Kenya. Sonaiya and Swan (2004) reported also that research and development in the field of family poultry (FP) must first examine the social, cultural and technical constraints faced by this sector. Muchenje et al. (2001) reported that the economic importance of poultry is not adequately appreciated by researchers and decision-makers because poultry products in the smallholder farming sector only passes through non-formal marketing channels. Moreover, Kryger et al. (2010) reported that the smallholder poultry production is practiced by most rural households throughout the developing world; despite the fact that its contribution to livelihoods appears to be of little nominal value when observed by researchers and other outsiders. Rural poultry sector in Serbia is important; however, it has actually no institutional support for many years (Rodić et al., 2010). Past research has not explicitly included user differentiation and participation in the technology development process (Aichi, 1996). Social and capital aspects of smallholder poultry production have been given little attention in research and or in development projects (Kryger et al., 2010).

However, Scanes (2007) reported that there is tremendous scope for chicken improvements. Okeke (2001) reported that the special features of the innovative poultry brooding technology of the project are that it harnesses solar energy and uses locally available materials; it can be adapted to both rural and urban poultry production.

CONCLUSION AND RECOMMENDATIONS

Poultry production and consumption is highly growing agricultural sector in the world. Moreover, the socio-economics of poultry production is also acceptable and feasible. This all makes poultry production the preferred and recommend production types to supplying the future protein demand of the society.

In most cases, poultry production in developing countries is mainly traditional; however, technologies favour the intensification of poultry production in such countries, but if not managed, environment and health issues will be the concerns of the future. A grain yield is adversely affected by warming that leads to food-feed competitions. This competition gives rise to search for alternative feeds and other ingredient utilization techniques for birds. It is therefore concluded that intensifying poultry production will reduce negative impacts of climate changes and it will also satisfy the demands of proteinaceous foods.

REFERENCES

Abda AE, Amin MH (2010). Economics of egg poultry production in Khartoum State with emphasis on the open-system in Sudan. Afr. J. Agric. Res. 5(18):2491-2496.

Abedullah MA, Bukhsh K (2007). Issues and Economics of Poultry Production: A Case Study of Faisalabad, Pakistan. P. Vet. J. 27(1):25-28.

Achoja FO, Ofuoku AU, Okoh RN (2006). Linkages between Socio-Economic Variables and the Efficient Marketing of Poultry Feeds in

Adams RM, Rosenweig C, Peart RM, Ritchie JT, McCarl BA, Glyer JD, Curry RB, Jones JW, Boote KJ, Allen LH JR (1990). Global climate change and US Agriculture. Nature 345:219-224.

Adebayo OO, Adeola RG (2005). Socio-Economic Factors Affecting Poultry Farmers in Ejigbo Local Government Area of Osun State. J. Hum. Ecol. 18(1):39-41.

Adrian WT, Michael LC (2009). Rural Economic Growth Linkages and Small Scale Poultry Production: A Survey of Poultry Producers in Kwazulu Natal. Contributed Paper Presented at the 41st Ann. Conf. Agri. Ec. Asso. S. Afr. October 2-3, 2003, Pretoria, South Africa. P. 16.

Aichi JK (1996). Socio-Economic Aspects of Village Chicken Production in Africa: The Role of Women, Children and Non-Governmental Organizations. XX World's P.Cong.s. FAO. Rome, Italy P. 70.

Ajala Nwag MKBI, Otchere EO (2007). Socio-Economics of Free-Range Poultry Production Among Agropastoral Women In Giwa Local Government Area Of Kaduna State, Nigeria. Niger. V. J. 28(3):11-18.

Alemu Y (1995). Poultry production in Ethiopia. World's p. sc. J. 51:197-201.

Alemu Y, Tadelle D (1997). The Status of Poultry Research and Development in Ethiopia, pp. 40-60. In: Fifth National Conference of Ethiopian Soc. Anim. Prod., 15-17 May 1997, Addis Ababa Ethiopia

Arrey MI (2009). Why Village Poultry is a necessary route for Cameroon? African Centre for Community and development. P. 9.

Augustine UJ (2010). Adoption of Improved Poultry Technologies by Poor Resource Farmers in Nigeria: Implications to Meat Protein Availability in the 21st Century Agric. Conspec. Sci. 75(3):133-139.

Bos JFFP, Wit De J (1996). Environmental Impact Assessment of Landless Monogastric Livestock Production Systems. Livestock and the Environment Finding a Balance. The Netherlands. Int. Agri. C. P. 92.

Bowes M, Crosson P (1993).Consequences of climate change for the MINK economy: Impacts and responses. Climatic Change 24:131-158.

Costa ND (2009). Climate Change: Implications for Water Utilization in Animal Agriculture and Poultry, in Particular. Aus. P. Sci. Symp. 20th Ann. Australian P. Sci. Sympos. Sydney, New South Wales. 9 - 11th February 2009. Australia.

CSA (Central Statistical Authority) (2005). Agricultural Sample Survey 2004/05. Volume II. Report on Livestock and livestock characteristics. CSA. Addis Ababa, Ethiopia. Stat. Bull. P. 331.

CSA (Central Statistical Authority) (2007). Agricultural Sample Survey 2006/07. Vol. I. Report on Livestock and livestock characteristics. Stat. Bull. A. Ababa. Ethiopia.

Daghir NJ (2009). Poultry Production in Hot Climates. Book Reviews. J. Appl. P. Res. 18:131-134.

David F (2010). The role of poultry in human nutrition. Poultry Development review. pp. 90–104.

Dana N, Duguma R, Teklewold H, Aliye S (2006). Transforming village poultry systems into small agro-business ventures: a partnership model for the transfer of livestock technologies in Ethiopia. Liv. Res. Ru. Dev. 18(12).

Danielle K, Peter C, Ayubu OM, Peter M, David M, Carol C (2009). Translinks-Case Study. Improving Poultry Production for Sustainability in the Ruaha Landscape, Tanzania. Report Prepared For WCS Translinks Program. P. 24.

Dave H (2007). Perspectives on the global markets for poultry products. Poultry in the 21 Century. Inter. conf. Bangkok 5-7 Nov. 2007. pp 37-38,

Dawit A, Tamirat D, Setotaw F, Serge Nz, Devesh R (2009). Overview and Background Paper on Ethiopia's Poultry Sector: Relevance for HPAI Research in Ethiopia. A Collaborative Research Project. Africa/Indonesia Team Working Paper. 1:60.

Delta State, Nigeria: Implication for Extension Services. World's P. S. J. 62:709-715.

Deogade AH, Zanjad PN, Raziuddin M (2008). Value Added Meat Products: REVIEW. Ve. World.1(3):88-89.

Eze CI,Okudu PO (2008).Discriminant Analysis Of Poultry Farmers Technology Adoption Potentials In Abia State Nigeria . Global Approaches to Extension Practice. J. Agr. Ext. 4:2.

FAO (2000a). FAOSTAT. Statistical database of Food and Agriculture

Organization of the United Nations, Rome. Italy.

FAO (2006). Livestock a Major Threat to the Environment: Remedies Urgently Needed. FAO. Rome Italy.

FAO (2008). An Analysis of the Poultry Sector in Ethiopia. Poultry Sector Country Review. FAO. Rom. Italy. P. 48.

FAO (2009). The State of Food and Agriculture. Livestock in Balance.

FAO (2010). Poultry meat and Eggs. Agribusiness handbook. Director of Investment Centre Division. FAO. Rome. Italy. P. 77.

FAO (2011). Agriculture and Consumers Protection Departments. Animal Roduction and health division. FAO. Rome Italy. P. 180.

Froning GW (1998). Recent Advances in Egg Products Research and Development. Presented at the University of California. Egg Processing Workshop. Riverside and Modesto on June 2-3.1998.

Gilbert MJ, Slingenbergh, Xiao X (2009). Climate change and avian influenza. Rev. Sci. Tech. 27(2):459-466.

Gueye EF (2003) Production and consumption trends in Africa. World's P. Sc. 19(11):12-14.

Guèye EF (2009). The role of networks in information dissemination to f.Poultry farmers. Small-Scale Family Poultry Production. World. P. Sci. J. 65:115-124.

Hafez HM (1999). Poultry meat and food safety: pre- and post-harvest Approaches to reduce food borne pathogens. World's P. Sc. J. 55:269-280.

Hunduma D, Regassa Ch, Fufa D, Samson L, Endale B (2010). Socio-economic importance and management of village chicken production in rift valley of Oromia, Ethiopia. Liv. Res. Ru. Dev. 22(11).

ILCA (International Livestock Center for Africa) (1993) (1-35) Annual report Program ILCA. Addis Ababa. Ethiopia. P. 98.

Jan H, Henning St (2008). Global feed issues affecting the Asian poultry industry. Proc. Int. P. Conf. Poultry in the 21st Century Avian Influenza and Beyond. FAO Bangkok, November 2007. Italy. pp. 33-35.

Jonathan R, Rommy V, Emmanuelle GB, Anni Mc (2004). Impact of avian influenza outbreaks in the poultry sectors of five S-E Asian countries. FAO. Rome, Italy. P. 25.

Jugessur VS, Pillay MM, Ramnauth R, Allas MJ (2004). the Socio-Economic Importance of Family Poultry Production in the Republic of Mauritius: Improving Farmyard Poultry Production in Africa: Pages 164-178 in Proc. of a Final Res. Coo. Meeting Organized by Joint FAO/IAEA Held in Vienna, 24-28 May, 2004.

Khandait VN, Gawande SH, Lohakare AC, Dhenge SA (2011). Adoption evel and Constraints in Backyard Poultry Rearing Practices at Bhandara District of Maharashtra (India). Res. J. Agr. Sc. 2(1):110-113. India.

Kitalyi AJ (1997). Village chicken production systems in developing countries: What does the future hold? World Anim. Rev. P. 89.

Kitalyi AJ (1998) .Village chicken production systems in rural Africa. House-holds Food and gender issues. Animal Production and Health paper 142, FAO: Italy. P. 81.

Kryger KN, Thomsen KA, Whyte MA, Dissing M (2010). Smallholder Poultry Production – Livelihoods, Food Security and Socio-cultural Significance, Smallholder Poultry Production. FAO. Rome Italy. P. 76.

Mammo M, Berhan T, Tadelle D (2008). Socio-economical contribution and labor allocation of village chicken production in Jamma district, South Wollo, Ethiopia. 20:60.

Maqbool AA, Bukhsh K (2007). Issues and Economics of Poultry: A Case Study of Faisalabad, Pakistan. Pak. Vet. J. 27(1):25-28.

Mekonnen G, Teketel F, Alemu G, Dagnatchew Z, Anteneh A (1991). The Ethiopian Livestock Industry: Retrospect and prospects. Proc. of 3rd Nat. Live. Improve. Confe./1991, Inst. Agri. Res. Addis-Ababa, Ethiopia.

Mekonnen GM (2007). Characterization of Smallholder Poultry Production and Marketing System of Dale, Wonsho and Loka Abaya Weredas of S. Ethiopia. Msc Thesis, Hawassa University, Ethiopia. P. 111.

Muchenje V, Manzini MM, Sibanda S, Makuza SM (2001). Sustainable Animal agriculture and crisis mitigation in livestock-dependent systems in S. Africa: Socio-economic and biological issues to consider in smallholder poultry development and research in the new millennium. Proc. of the regional conf. held at Malawi Inst. of Mana, Lilongwe, 30 October to 1 November, 2000. Zimbabwe.

Musa U, Abdu PA, Dafwang II, A.Katsayal U, Edache JA, Karsin PD

(2008). Ethnoveterinary Remedies Used for the Management of Newcastle Disease In Some Selected Local Government Areas Of Plateau State Nigeria, March, 2008. Niger. J. Pharm. Sci. 7:1.

Njue SW, Kasiiti JL, Gacheru SG (2004). Assessing the Economic Impact of Commercial Poultry Feeds supplementation and Vaccination against Newcastle Disease in Local Chickens, In Kenya: Improving Farmyard Poultry Production in Africa: in Proc. of a Final Res. Coo. Meet. Organ. by the JointFAO/IAEA Held in Vienna, 24-28 May, 2004. pp. 116-124.

Nnadi FN, Akwiwu CD (2005). Adoption of Improved Poultry Production Practices by Rural Women in Imo State. Ani. Prod. Res. Adva. 1(1):32-38.

Obayelu AE, Adeniyi A (2006). The Effect of Climate on Poultry Productivity in Ilorin Kwara State, Nigeria. Int. J. P. Sci. 5(11):1061-1068.

Okeke CE (2001). Raising Healthier Poultry: Examples of Successful Uses of Renewable Energy Sources in the South, Nigeria. 8:70-76.

Okoh SO, Rahman SA, Ibrahim HI (2010). Gender participation in Commercial poultry production in Karu and Lafia Areas, Nasarawa State, Nigeria. Live. Res. Rur. Dev. 22:160.

Olaniyi OA, Adesiyan IO, Ayoade RA (2008). Constraints to Utilization of Poultry Production Technology among Farmers in Oyo State, Nigeria. J. Hum. Ecol. 24(4):305-309

Olasimbo MA (2006). Analysis of Table Egg Production as a Livelihood Activity in Ekiti State. J. Agri. Soc. Res. 6:2.

Paweł PM (2005). Nutritional potential for improving meat quality in poultry. Animal Science Papers and Reports, Presented at the Conf. held at the ANIMBIOGEN" Centre Excellence Genom. Biotech. 23(4):303-315.

Permin A, Pedersen G (2000). Problems related to poultry production at Village level. Possibilities. Proc. of smallholder poultry projects in Eastern and Southern Africa, 22-25; May 2000, Morogoro, Tanzania.

Permin A, Riise JC, Kryger KN, Assoumane I, Schou TW (2004). Experiences In Using Poultry as a Tool for Poverty Alleviation at Village Level: Improving Farmyard Poultry Production in Africa: pp 42-48 in Proc. of a Final Res. Coo. Meeting Organized By the joint FAO/IAEA Held in Vienna, 24-28 May 2004.

Philip TK (2011). Livestock production: recent trends, future prospects. Online ISSN: 1471-2970.

Rodić V, Perić L, Pavlovski Milošević Z (2010). Improving the poultry sector in Serbia: major economic constraints and opportunities, Serbia. World's P. Sc. J. 66:241-250.

Scanes CG (2007). Contribution of Poultry to Quality of Life and Economic development in the Developing World. Poult. Sci. 86(11):2289-2290.

Solomon D (2008). An analysis of the poultry sector in Ethiopia. A review Report on HPAI prevention and control strategies in Eastern Africa. FAO. Italy. P. 48.

Sonaiya EB, Swan ESJ (2004). Small scale poultry production technical guide. FAO, Animal Production and Health Manual. 1:114. Rome, Italy.

Sumy MC, Khokon MSI, Islam MM, Talukder S (2010). Study on the socio Economic condition and productive performances of backyard chicken in some Selected areas of Pabna district. Bangladesh. J. Bangla. Agri. Univ. 8(1):45-50.

Sungno NS, Robert M (2006). The Impact of Climate Change on Livestock Management in Africa: A Structural Ricardian Analysis, -23XCentre for Environ. Econ. Policy in Africa. ISBN: 1-920160, USA.

Vincent N, Lagat BK, Korir MK, Ngeno EK, Kipsat MJ (2010).Resource Use Efficiency in Poultry Production in Bureti District, Kenya. Paper Presented at the Joint 3rd African Association of Agric. Economists (AAAE) and 48th Agricultural Economists Association of South Africa (AEASA) Conference, Cape Town, South Africa.

Tadelle D, Million T, Alemu Y, KJ Peters (2003). Village chicken Production systems in Ethiopia: Use patterns and performance valuation and chicken products and socio-economic functions of chicken, Ethiopia. Liv. Res. Rur. Dev. 15(1).

Teketel F (1986). Studies on meat production potential of some local strains of chicken in Ethiopia. PhD Thesis, J L Giessen University. Germany.

Teklewold H, Dadi L, Yami A, Dana N (2006). Determinants of adoption of poultry technology: a double-hurdle approach. Livestock Research for Rural Development. 18:40. Retrieved January 22, 2007, from http://www.cipav.org.co/lrrd/lrrd18/3/tekl18040.htm.

Truong TN, Ryuichi Y (2002). Factors affecting farmers' Adoption of technologies in farming system: A case study in OMon district, Can Tho province, Mekong Delta. Omonrice. Vietnam. 10:94-100.

USAID (2006). Partnership for Safe Poultry in Kenya (Pspk) Program Value Chain: Analysis of Poultry in Ethiopia. P. 42.

USAID (2009). Improving Poultry Production for Sustainability in the Ruaha Landscape, Tanzania.

Wiebe VDS (2007). Intensive Poultry Production. World poultry. Farming futures Climate change series: fact sheet-9. Focus on poultry. 23:12.

Windhorst HW (2008). A projection of the regional development of egg production until 2015. World's P. Sc. J. 64(3):356-376.

26

Prevalence of camel (*Camelus dromedaries*) mastitis in Jijiga Town, Ethiopia

Abdi Husein[1,2], Berihu Haftu[1], Addisalem Hunde[1] and Asamenew Tesfaye[1]

[1]Wollega University, School of Veterinary Medicine, P. O. Box 395, Nekemte, Ethiopia.
[2]Jijiga Regional Veterinary Laboratory, Jijiga, Ethiopia.

A cross sectional study of camel mastitis was conducted on 384 lactating camels from Jijiga between November 2011 to April 2012 to estimate the prevalence and causes of mastitis, as well the risk factors involved on disease. Prevalence of mastitis was assessed by using California mastitis test (CMT). An overall prevalence of camel mastitis was found to be 30.2% (116/384) out of which, 4.9% (19/384), 25.3% (97) were clinical and sub-clinical mastitis, respectively. The overall quarter level prevalence was 25.8% (397/1536). There was significant (P<0.05) in prevalence between camels with teat lesion, tick infestation, parity or age to mastitis than those without these factors. Microbiological examination of 174 randomly selected CMT positive milk samples from clinical quarters, revealed that the majority of the isolates were coagulase negative *Staphylococci* (39.6%), followed by *Streptococcus dysagalactiae* (22.2%), *Corynebacteria* spp. (9%), *Bacillus* spp. (7.6%), *Streptococcus uberis* (7.6%), *Escherichia coli* (6.3%), *Staphylococcus aureus* (4.2%) and *Streptococcus agalactiae* (3.5%). The prevalence of camel mastitis in the study area was found to be significantly high. Therefore, implementation of integrated approaches has great importance in the study sites for the prevention and control of mastitis hence minimizing economic loss and prevents significant public health risks.

Key words: Camel, prevalence, lactating, mastitis, Jijiga.

INTRODUCTION

The camel (*Camelus dromedaries*/one humped camel) is the most dominant and widely distributed animal in the tropical and subtropical continents of Africa and Asia. It makes an important contribution to human survival and utilization of these in dry and arid land (Abdurahman and Younan, 2004).

In Ethiopia, camels are kept in arid and semi-arid low lands of Borana, Somalia and Afar regions, which cover 50% of the pastoralist area in country. The major ethnic groups owing camels in Ethiopia are the Somali, Borana and Afar (Teka, 1991). The annual camel milk production in Ethiopia was estimated to be 75,000 tones (Felleke, 2003). In most pastoralists, camel milk is always consumed either fresh or in varying degrees of sourness in the raw state without heat treatment and, can pose a health hazard to the consumer. In their natural desert habitat, where camels are usually raised particularly during the long dry season, camels are subjected to severe stress conditions which render them susceptible to many diseases and ailments (Abbas et al., 1993; Agab, 1993). Although, camels were considered in the past, and for a fairly long time, as resistant to many disease causing factors (Dalling et al., 1988), it has been proved that camels are susceptible, to similar diseases that affecting the livestock or other animal species (Wilson et al., 1982; Abbas and Tilley, 1990; Saint-Martin et al., 1992; Abbas and Agab, 2002).

Mastitis is a complex disease occurring worldwide

among dairy animals with heavy economic losses. Mammary infections results in milk compositional changes such as increase in leukocyte counts, leakage of plasma proteins into the milk and cell damage, resulting in leakage of intracellular constituents into milk, change in ion composition and decrease in milk production (Korhonen and Kaartinen, 1995). Bacterial infections are considered the primary cause of mastitis in domestic animals.

The causative agents of bovine mastitis are well defined There is extensive literature on bovine mastitis and to a lesser extent on ovine and caprine mastitis. In contrast, there is paucity of information about the etiological agents associated with camel mastitis. Few available studies indicate that *Staphylococcus aureus, streptococcus spp.* (Barbour et al., 1985; Abdurahman et al., 1995; Al-Ani and Al-Shareefi, 1997; Younan et al., 2001), *Micrococcus spp.* (Barbour et al., 1985; Al-Ani and Al-Shareefi, 1997), *Streptococcus agalactiae* (Abdurahman et al., 1995; Younan et al., 2001), coagulase negative *staphylococci* (Abdurahman et al., 1995), *Staphylococcus epidermides, Pasteurella haemolytica* (Al-Ani and Al-Shareefi, 1997), *Escherichia coli* (Abdurahman et al., 1995; Al-Ani and Al-Shareefi, 1997) and *Corynebacterium spp* (Barbour et al., 1985) have been implicated as causes of mastitis in camels.

There is extensive literature on bovine mastitis and to a lesser extent on ovine and caprinemastitis; however, little is known about mastitis in camels.

Likewise there is limited information on the prevalence and causative agents of camel mastitis in Ethiopia. The prevalence and causes of mastitis differ markedly due to geographical area and individual herd management (Guidry, 1985). To establish an efficient mastitis control program in a dairy herd, baseline information on the nature of mastitis and economic impact of the problem need to be known (Honkanen-Buzalski and Pyörälä, 1995). Therefore, the objectives of the study were to determine the prevalence of mastitis in the study area, isolate the possible causes of the diseases and to identify the possible risk factors of the diseases. These can generate baseline information on status of the disease that could serve as an input for possible interventions programs on the problem by the regional government or at national level.

MATERIALS AND METHODS

Study area

The study was conducted from November 2011 to April 2012 around Jijiga in Eastern part of Ethiopia. Jijiga is located approximately 80 km East of Harar and 60 km West of the border with Somalia and located at distance of 628 km Eastern of Addis Ababa. The areas are geographically found at a latitude and longitude of 9°21'N and 42°48' E, respectively and characterized by unreliable and erratic rainfall with a precipitation ranging from 300 to 600 mm per annum, high ambient temperature 30°C, sparsely

distributed vegetation dominated by *Acacia* species, cactus and bushy woodlands (Tafesse, 2001). These are arid and semi-arid lowlands lying at an elevation of 500 to 1500 m above sea level and are not suitable for crop production.

In these areas, camels are herded by nomadic pastoralists who rely mainly on livestock husbandry for their livelihood. A single-visit, multiple-subject diagnostic survey (ILCA, 1990) was used to assess the occurrence of mastitis and traditional management practices used to control mastitis in camels. A total of 53 households who own camels and who are familiar with camel husbandry were selected from Jijiga region using purposive sampling technique. Households at each location were selected based on accessibility of the village and willingness of the camel owners to take part in the interview.

The camels were at different stages and numbers of lactation, and they were of various age groups. Information about traditional management, herd size, milking frequency, milking procedure, occurrence of mastitis, and traditional mastitis control methods was obtained from camel owners by means of a semi-structured questionnaire. The camels were fed exclusively on natural browse, watered on the average every 3 to 4 days, herded during the daytime on communal grazing lands and kept at night in traditional enclosures (Corral) made of thorny bushes and tree branches as protection from predators. The camels were milked on the average three times a day.

Study animals

The study animals are lactating camels that were kept under traditional management from different areas of in and around Jijiga region. A total of 384 lactating cow-camels destined for inspection of prevalence of clinical and subclinical mastitis accordingly.

Sampling and study type

A cross-sectional study was conducted on 384 lactating camels in which case study animals were visited once for data collection and sample taking. Regarding the sampling procedure area around Jijiga region were selected based on present camels population and accessibility of information thereby, accordingly collecting the sample were achieved.

Sample size

The desired sample size for the study was calculated using the formula given by Thrusfield (1995) with 95% confidence interval (CI) and 5% desired absolute precision. Accordingly, the estimated sample size was 384 camels.

Physical examination of the udder

Mastitis was detected using California mastitis test (CMT) result of clinical inspection of the udder (Table 1). In this study, the clinical cases were defined based on Radostits et al. (2000) which is characterized by swollen, reddened, hardened udder, painful upon palpation and alteration in the color and consistency of milk depending on the degree of inflammation. Thus, the General udder abnormalities, the size of rear and forequarters and fibrinosis were examined by deep palpation. Tick infestations, presence of lesion were also noted. The milk was examined for its consistency, color and other visible abnormalities. The clinical mastitis was recognized by abnormal milk, sign of udder infection and detection of mastitis by positive culture result. In contrast, sub-clinical mastitis was recognized by apparently normal milk and increased in leukocyte

Table 1. Interpretation for California mastitis test.

CMT score	Interpretation	Visible reaction	Total cell count
0	Negative	Milk fluid is normal	0-200,000 (0-25% neutrophils)
T	Trace	Slight precipitation	$(1.5-5) \times 10^5$ (30-40% neutrophils)
1	Weak positive	Distinct precipitation but not gel formation	$(4-15) \times 10^5$ (40-60 neutrophils)
2	Distinct positive	Mixture thickens with gel formation	$(8-50) \times 10^5$ (60-70% neutrophils)
3	Strong positive	Strong gel that is cohesive with a convex surface	$\geq 5,000,000$ (70-80% neutrophils)

Source: Quinn et al. (1999).

number as evident by CMT and positive culture result.

Milk sample collection

The camel calves were allowed to suckle in order to stimulate milking and milk samples were collected from all CMT positive quarters during screening for sub-clinical mastitis. The teat of affected quarter was carefully washed with clean water and soap, dried and teat ends were disinfected with cotton swabs soaked in 70% alcohols and allowed to dry. Approximately 10 ml of milk was collected aseptically after discarding the first stream of milk. Samples were placed immediately into an ice box (4-8°C) and brought to the Regional Veterinary Diagnostic and Research laboratory for processing and storage.

California mastitis test (CMT)

Sub-clinical mastitis was diagnosed based on CMT results and the nature of coagulation and viscosity of the mixture (milk and CMT), which show the presence and severity of the infection, respectively. Before sample collection for bacteriological examination, milk samples were examined for visible abnormalities, they were screened by CMT according to Quinn et al. (1999) from each quarter of the udder, a squirt of milk sample was placed in each of the cups on CMT paddle and an equal amount of 3% CMT reagent was added to each cup and mixed well. The interpretation was in such a way that CMT score: 0 was taken as negative, while CMT scores trace, 1+, 2+ and 3+, were considered positive, thus forming five categorical classes. All milk samples considered positive irrespective of CMT results were bacteriological examined.

Bacteriological examinations

Among the CMT positive milk samples (369) and milk samples collected from clinical quarters (48), 174 samples were randomly selected and used for bacteriological analysis. A loopful of each milk sample was streaked on defibrinated sheep (5%) blood agar. Plates were incubated at 37°C for 48 h. Among the 260 colonies grown, 174 colonies selected randomly and subjected to the following tests as recommended by the National Mastitis Council

(NMC) (1987): morphology, haemolysis pattern and Gram stain. Gram-positive cocci were tested for catalase, and catalase-positive isolates further tested with coagulase test. Streptococci were identified by performing CAMP, esculin, reffinose, salicin, mannitol, and inulin tests. Gram-negative rods were further differentiated by testing for motility, lactose fermentation (growth on MacConkey agar) and by using oxidase test.

Questionnaire survey

A general questionnaire survey was carried out in which of age of camel, parity number, housing, feeding, source of water, economic importance of mastitis, milking order of lactating, camels traditional husbandry system used by camel owners, stage of lactation, pre milking udder preparation and hygiene were included in questionnaire.

Data analysis

The data were recorded in Microsoft excel spread sheet for statistical analysis. Descriptive statistics was used to summarize the data and calculate same of sample statistics and various proportions. Additionally, Chi-square test was used to see the presence and strength of association of the potential risk factors with occurrence of mastitis using SPSS.

RESULTS

Animal level prevalence of mastitis

A total of 384 clinical as well as sub-clinical cases of lactating camels were examined during study period by using CMT. The overall mastitis prevalence was 30.2% (116/384) out of which, 19 (4.9%), 97 (25.3%) camels showed clinical and sub-clinical mastitis, respectively (Table 2). From the 384 camels, 130 (33.9%) camels had varying degree of tick infestation and 10 (2.6%) had lesion on the teat and udder. Out of these 130 ticks

Table 2. Prevalence of mastitis both at animal level and quarter level based on CMT and grown culture.

Sample	CMT		
	Number tested	Number positive	Prevalence (95%)
Camel level	384	97	25.3
Quarter level	1508	369	24.2

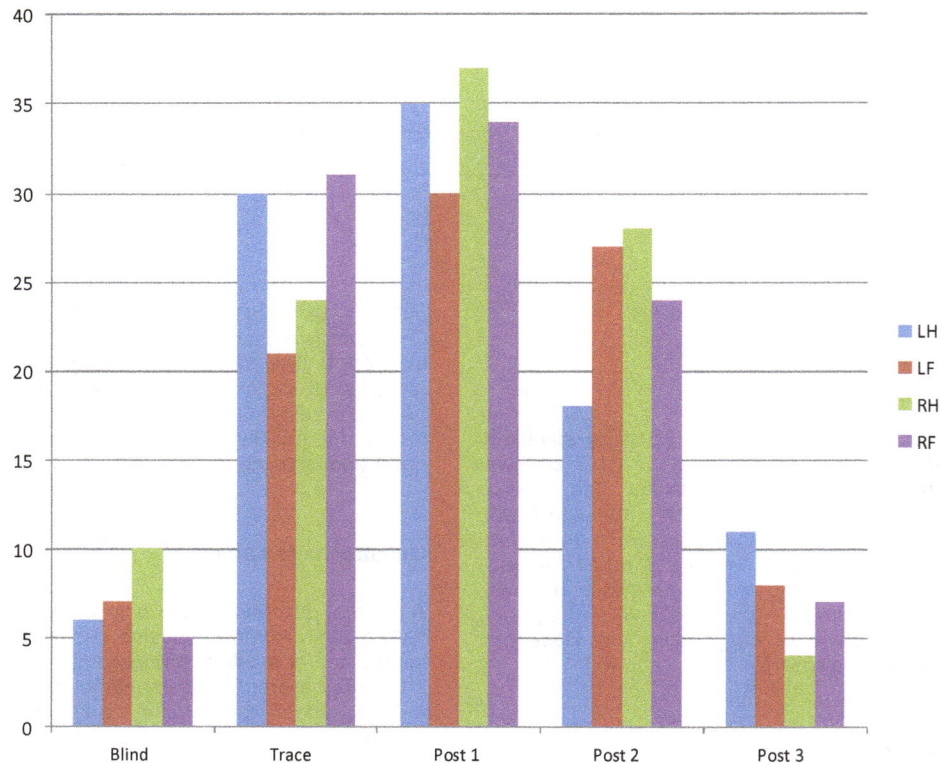

Figure 1. The relative proportion (number) of positives (either trace, +, ++, +++) in relation with quarters.

infested udder of lactating camels, and 68 where positive for mastitis. This indicates high percentage (52.3%) of mastitis was found among the tick infested group.

Quarter level prevalence of mastitis and microbiological culture

Out of the total 1536 examined teats, 369 (24.5%) teats were CMT-positive. In Table 2, the quarter level calculated prevalence of 24.5% (369/1508) was only based on the CMT result excluding the blind teats from which the milk samples could not be taken. Otherwise, the overall quarter level prevalence of mastitis was 25.8% (397/1536). The relative proportion (number) of positives (either trace, +, ++, +++) in relation with quarters is presented in (Figure 1 and Table 3). In addition to this,

among the milk samples subjected to bacteriological examination, 144 (82.8%) yielded mastitis pathogens (Table 4). From these, 144 growths, the most prevalent mastitis causing agent were Coagulase negative staphylococci (39.6%; n = 57/144) and least one was *Streptococcus agalactiae* (3.5%; n=5/144).

Analysis of risk factors

Mastitis is prevalent in the area and its incidence is influenced by age, parity number, hygiene of milking process, and presence of lesion on udder or teats were found significantly associated (p < 0.05) with the prevalence of mastitis in lactating camel. There was the lowest prevalence (5.2%) of mastitis in she-camels of 5 to 7 years of age, while the highest (51.7%) in the animal

Table 3. CMT result with regard to each teats.

Quarters	No of blind teats	CMT result						
		No of positive teats in each category				Total positive	Total negative	Total
		Trace	+	++	+++			
Left behind	6	30	35	18	11	100	284	384
Left front	7	21	30	27	8	93	291	384
Right behind	10	24	37	28	4	103	281	384
Right front	5	31	34	24	7	101	283	384
Total	28	106	136	97	30	397	1139	1536

Table 4. Bacterial species isolated from quarter milk samples (n = 174) obtained from traditionally managed camels in and around Jijiga.

Bacterial species	Number of isolates	% of total isolates
Coagulase negative staphylococci	57	39.6
Streptococcus dysagalactiae	32	22.2
Corynebacteria spp.	13	9.0
Bacillus spp.	11	7.6
Streptococcus uberis	11	7.6
Escherichia coli	9	6.3
Staphylococcus aureus	6	4.2
Streptococcus agalactiae	5	3.5
Total	144	10

aged between 14 to 16 years in Table 5.

Questionnaire result

Camel owners were interviewed from different area and 70% of them responded that mastitis was a disease they are aware of and this disease is known by different names in the study areas (Table 5). "Gofla" is the predominant type of camel mastitis in the study areas and it causes a significant decline in milk yield as reported by the respondents. It is a clinical type characterized by swelling of the udder. 'Arar' (Carcar) is a mild type and the second prevalent type of camel mastitis in the areas. It causes swelling of the udder and release of pus from the teats. Jid was the third abundant type of mastitis in camels. It is a chronic form and causes blind teats. However, they were not aware of sub-clinical mastitis. They thought that they can control the spread of the disease by milking cow-camels at the end of milking, but most respondent (57%) did not have the awareness of the way of transmission of the disease. They milk the entire herd in the same container (that is made of wood).

In the area, milking procedure is usually carried out by one person. Almost more than half of the respondents indicated that while preparing utensils for milking they wash and smoke milking utensils with wood called *Oliva*

africana which is locally called "Ugay" before milking camels and they explained that this keeps milk for longer period of time. Of the twenty camel owners interviewed, majority of respondents (98%) reported that they do not practice washing their hand prior to milking and 96.7% of camel owners explained that the use of anti-suckling material to prevent calf from suckling. They do this by tying two pair of teats together with fiber which definitely causes trauma to the udder and predispose to mastitis. Furthermore, most of them explained the effect of ecto-parasite (ticks) as a causative agent of udder and teat lesion (Table 6).

DISCUSSION

The overall prevalence (30.2%) of mastitis in camel herds as determined by the CMT and clinical examinations of the udder and the milk samples is lower than that reported by Obeid et al. (1996) who found an overall mastitis prevalence of 66.8% in Sudanese camel herds and 59.8% report of Afar Region, North Eastern Ethiopia by Bekele and Molla (2001). However, the present finding is consistent with the findings of Osman et al. (1991) who found an overall mastitis prevalence of 29% in Jijiga zone, Somali Regional State.

Table 5. Risk factor associated with occurrence of mastitis.

Factor	Mastitis	Non-mastitis	Total	P-value
Tick free	48	206	254	
Tick infested	68	62	130	
Total	116	268	384	
Parity				
1st	49	108	148	
2nd	40	102	151	
3rd and more	27	58	85	
Total	116	268	384	
Age (Yr)				
5 to 7	6	98	104	
8 to 10	20	67	87	
11 to 13	30	57	87	
14 to 16	60	46	104	
Total	116	268	384	
Lactation stage in month (m)				
1 to 2	50	110	P160	
3 to 9	38	98	136	
10 to 18	28	60	88	
Total	116	268	384	

Table 6. Indicates respondent answers and tick infestation.

Milking procedure	% of total respondents in and around Jijiga
Wash udder/teats before milking	None
Wash hands before milking	96.7% not practiced
Wash/smoke milk utensils before milking	98% practiced
Let the calf to suckle before milking	100%

Tick infestation (% of total herd)	Mastitis	Non-mastitis	Total
Tick infested	68	62	130
Tick free	48	206	254
Total	116	268	384

On the other hand, the reported clinical (4.9%) and sub-clinical mastitis (25.3%) reported in the current study is consistent with the finding of Magarsa (2010) who reported prevalence of sub-clinical mastitis ranged from 28.6 to 37.6% and clinical mastitis ranged from 10 to 17%, respectively during minor wet, major wet and dry season in dromedary camels in Borana area of Southern Ethiopia. Furthermore, the finding of this study regarding the clinical and sub-clinical mastitis also agree with finding of Abdurahman et al. (1995) who reported prevalence of (5.9%) in Sudan also reported 8.3% prevalence of clinical mastitis in Jijiga. Higher result of

clinical mastitis were also reported by Barbour et al. (1985) 15%, Magarsa (2010) 17% in minor wet season and Obied et al. (1996) 19.5% in Saudi Arabia, Borana and Sudan, respectively.

From the present study, the prevalence of sub-clinical mastitis at quarter level was 24.5%, which is agreeable with that of 20.5% reported by Almaw and Molla (2000). Comparable result (15.8%) is also reported by Abdurahman and Bornstein (1991) in Jijiga and higher rate of CMT result were reported by Taketelew and Bayleyegn (2001) 47.3% in Afar.

Tick burden, together with thorny plant of desert and

ant suckling material, seems to be risk factor to the occurrence of mastitis in camels in the study area. The udder is predilection site for tick infestation which causes skin and teat lesions. This is one of the factors that predispose camels to mastitis, since lesions caused by ticks facilitate bacterial entry and cause permanent tissue damage and influenced by poor udder hygiene Megersa, (2010). Similar to this fact, the current study also revealed that the presence of tick infestation on udder is one of the potential risk factors for the occurrence of mastitis. As mentioned earlier by many of the researchers this could be due to the fact that tick infestation can predispose the udder area by creating a conducive situation for the entrance of majority of mastitis causing microorganisms.

Concerning to udder lesion, penetrating and non-penetrating superficial ski lesion of the teat and udder were observed and out of 10 camels having udder lesion, all of them (100%) were mastitis positive compared to the prevalence of those camels without udder lesion. High prevalence of mastitis 72.2% in camels with udder lesion was reported by Teketelew and Bayleyegn (2001) in afar region. Woubit et al. (2001) also recorded that the udder or teat skin scratches can be caused by thorny plant of the desert. Generally, trauma may be responsible directly to mastitis which can result injury and predispose to bacteria invasion of the udder.

A positive relation was observed between mastitis and lactation stage. Prevalence of mastitis in early stage of lactation was significantly higher. This was sometimes due to the fact that most new infection occurs during the early part of dry period and in the first two month of lactation, especially with environmental pathogens (Radostits et al., 2000).

The high percentage of mastitis pathogens isolated from camel milk samples examined in the present study is consistent with the findings of Woubit et al. (2001) who reported that 74% of the CMT positive quarter milk samples of camels in Borena area of southern Ethiopia yielded pathogenic bacteria. Gram-positive *cocci* were the main cause of mastitis in the camels and constituted 93.8% of the total isolates. This finding is in line with that reported by Obied et al. (1996) and Woubit et al. (2001). Among the bacterial isolates, coagulase negative staphylococci (CNS) were identified as the predominant mastitis causing organisms in the camels studied. This agrees with the report of Abdurahman (2006) who found that CNS and *S. aureus* represented 61.1 and 38.9%, respectively of the total isolates and considered as the main organisms that cause mastitis in the Bactrian camel. *Streptococcus dysagalactiae* was the second most common cause of mastitis in the camel herds examined in this study.

This finding agrees with that reported previously by other researchers (Woubit et al., 2001; Abdurahman, 2006; Guliye et al., 2002). *Streptococcus agalactiae* and *S. aureus* were reported to be the most common causes

of camel mastitis in Eastern Sudan (Obeid et al., 1996) and Kenya (Younan, 2004).

The bacteria isolated from camel milk samples in the present study are types that cause both contagious and environmental mastitis. Correct and good milking techniques are essential in the prevention of both environmental and contagious mastitis. The teats must be cleaned with individual clothes dipped in hot water. The fact that the pathogens isolated from camel milk samples in the present study are bacteria that causes both environmental and contagious mastitis suggest that proper management and adequate hygienic condition of the environment (enclosures) are required in order to minimize occurrence of mastitis in the study area.

Conclusion

This study revealed high prevalence of mastitis in camel herds in the sampled area. The high prevalence of mastitis was attributed to inadequate hygienic condition of the dairy environment and tick infestation. Additionally, it was observed that the occurrence of camel mastitis significantly vary with stage of lactation indicating a higher prevalence during early stage of lactation. Finally, among the important mastitis causing bacteria, coagulase negative staphylococci, Streptococcus dysgalactiae, Corynebacteria spp were found the most common. Therefore, good management practices with proper sanitation and tick control measures are required to prevent the incidence of mammary infection in camels in the study areas. The isolation of genera of pathogenic bacteria from the camel milk samples suggests the need for strict hygienic measures during the production and handling of camel milk to reduce public health hazards. Furthermore, public education should be given to improve their awareness about the importance of proper herd health management and hygienic milking practices in order to minimize the adverse effect of mastitis on the yield, quality of milk and zoonotic impact of the pathogen.

ACKNOWLEDGEMENTS

The authors thank Wollega University, for allowing its financial support during the study period. Managers and technical staff of the Jijiga Veterinary Laboratory and veterinary clinic are also acknowledged for their help and collaboration in realization of the research.

REFERENCES

Abbas B, Agab H (2002). Review of camel brucellosis. Prev. Vet. Med. 55:47-56.
Abbas B, Saint-Martin G, Planchenauct D (1993). Constraints to camel production in eastern Sudan: a survey of pastoralistsconception. Sud. J. Vet. Sci. Anim. Husb. 32(1):31-41.
Abbas B, Tilley P (1990). Pastoral management for protecting

ecological balance in HalaibDistrict, Red Sea Province, Sudan. Nomadic People. 29:77-86.

Abdurahman OA Sh (2006). Udder health and milk quality among camels in the Errer valley of Eastern Ethiopia. LRRD. 18:110.

Abdurahman OASH, Agab H, Abbas B, Åström G (1995). Relations between udder infection and somatic cells in camel (*Camelus dromedarius*) milk. Acta Vet. Scand. 36:423-431.

Abdurahman OASH, Bornstein S (1991). Diseases of camels(*Camelus dromedarius*) in Somalia and prospects for better health. Nomadic People. 29:104-112.

Abdurahman OA-Sh, Younan M (2004). The udder health, In: Farah and Fisher (editor). Milk and meat from the camel handbook on product And processing *vdfhochschulerleg* AG and der ETH Zurich. pp. 73-76.

Agab H (1993). Epidemiology of Camel Diseases in Eastern Sudan withEmphasis on Brucellosis. M. V. Sc. Thesis, University of Khartoum. P. 172.

Al-Ani FK, Al-Shareefi MR (1997). Studies on mastitis in lactating one-humped camels (*Camelus dromedarius*) in Iraq. J. Camel Pract. Res. 4:47-49.

Almaw G, Molla B (2000). Prevalence and etiology of mastitis in camels (*Camelus dromedarius*) in Eastern Ethiopia. J. Camel Pract. Res. 71: 97-100.

Barbour EK, Nabbut NH, Frerichs WM, Al-Nakhli HM, Al-Mukayel, AA (1985). Mastitis in *Camelus dromedarius* in Saudi Arabia. Trop. Anim. Health Prod. 17:173-179.

Bekele T, Molla B (2001). Mastitis in lactating Camels (*Camelus dromedarius*) in Afar Region, north-eastern Ethiopia. Berl.Munch. Tierz. Woch. 114:169-172.

Dalling T, Robertson A, Boddie G, Spruell J (1988). Diseases ofcamels. In: The International Encyclopedia of Veterinary Medicine.Edinburgh, U.K.; W. Green and Son. P. 585.

Felleke G (2003). A Review of the Small Scale Dairy Sector - Ethiopia. FAO Prevention of Food Losses Program: Milk and Dairy Products, Post-harvest Losses and Food Safety in Sub-Saharan Africa and the Near East. Retrieved on April 10, 2007 from http://www.fao.org/ag/againfo/projects/en/pfl/documents.html.

Guidry AJ (1985). Mastitis and the immune system of the mammary gland. In: Larson, B. L. (Ed.), Lactation. The Iowa State University Press, Iowa. pp. 229-262.

Guliye AY, Van Creveld C, Yagil R (2002). Detection of subclinical mastitis indromedary camels (*Camelus dromedarius*) using somatic cell counts and the N-acetylbeta- D glucosaminidase test. Trop. Anim. Heatlh Prod. 34:95-104.

Honkanen-Buzalski T, Pyörälä S (1995). Monitoring and management of udder health at the farm. In: Sandholm, M., Honkanen-Buzalski, T., Kaartinen, L. and Pyörälä, S. (Eds.), Thebovine udder and mastitis. University of Helsinki, Finland. pp. 252-260.

ILCA (1990). Livestock systems research manual. ILCA Working Paper 1.International Livestock Centre for Africa, Addis Ababa, Ethiopia.

Korhonen H, Kaartinen L (1995). Changes in the composition of milk induced by mastits. In: Sandholm, M., Honkanen-Buzalski, T., Kaartinen, L. and Pyörälä, S. (Eds.), The bovine udder and mastitis. University of Helsinki, Finland. pp. 76-82.

Megersa B (2010). An epidemiological study of major camel diseases in the Borana Lowland, Southern Ethiopia. Dry lands Coordination Group Report No. 58, 09, 2010. pp. 32-33.

NMC (1987). Laboratory and field handbook on bovine mastitis. National Mastitis Council (NMC) Inc., Madison.

Obeid AI, Bagadi HO, Mukhtar MM (1996). Mastitis in *Camelus dromedarius* and the somatic cell content of camels' milk. Res. Vet. Sci. 61:55-58.

Osman Kh Sh, Abdi AM, Abdurahman O, Bornstein S, Zakrisson G (1991). Prevalence of mastitis among camels in Southern Somalia A pilot study. Camel forum, working paper 37:1-9.

Quinn PJ, Carter ME, Markey B, Carter GR (1999). Clinical Veterinary Microbiology, (Wolf publishing, London, England). P. 327.

Radostits OM, Gay CC, Blood DC, Hinchliff KW (2000). Veterinary Medicine: A Textbook of the Diseases of Cattle, Sheep, Pigs, Goats and Horses, 9th edition Philadelphia: WB Saunders. pp. 563-613.

Saint-Martin G, Delmet C, Zubeir ARY, Peyre de Fabriques B, Harbi MSMA, Bagadi HO (1992). Camel Project ofButana : Final Report. MaisonAlfort, France, IEMVT. P. 128.

Tafesse B (2001). Studies on Cephalopinatitillator, the case of 'Senecal' in camels (*Camelus dromedarius*) in semi-arid areas of Somali state, Ethiopia. Trop. Anim. Health Prod. 33:489-500.

Teka T (1991). The dromedary in Eastern Africa countries. Nomad. People. 29:31-41.

Teketelew B, Bayeleyeg M (2001). Mastitis in lactating camels (*Camelus dromedaries*) in Afar Region, North Eastern Ethiopia. BerlmunchTierztlwochenschr. 1145:169-72.

Thrusfield M (1995). Sampling in: veterinary epidemiology 2nded.London: Blackwell Science Ltd. pp. 179-283.

Wilson AJ, Schwartz HJ, Dolan R, Field CR, Roettcher D (1982). EpidemiologischeAspektebedeutenderKamelkrankheiteninausgewaehltengebietenKenias. Der praktische Tierarzt. 11:974-985.

Woubit S, Bayleyegn M, Bonnet PET, Jean-Baptiste S (2001). Camel (*Camelus dromedarius*) mastitis in Borena lowland pastoral area, South Western Ethiopia, Revue d'Elevage et de Médecine Vétérinaire des Pays Tropicaux 54:207-212.

Younan M (2004). Milk Hygiene. In: Milk hygiene and udder health. pp. 67-72.

Younan M, Ali Z, Bornstein S, Müller W (2001). Application of the California mastitis test in the intramammary *Streptococcus agalactiae* and *Staphylococcus aureus* infections of camels (*Camelus dromedarius*) in: Kenya. Prev. Vet. Med. 51:307-316.

27

Threats, attempts and opportunities of conserving indigenous animal genetic resources in Ethiopia

Kefyalew Alemayehu

Department of Animal Production and Technology, Faculty of Agriculture and Environmental Sciences, Bahir Dar, University P. O. Box 21 45, Bahir Dar, Ethiopia.

Loss of genetic diversity among animal populations occurs due to genetic introgression, crossbreeding, inbreeding, climate change and its related factors. Therefore, the objective of the study was to quantify threats, previous conservation attempts and opportunities of conserving indigenous animal genetic resources in Ethiopia. The consequences of genetic introgressions include reduced survival and fitness of the first and second generations, accelerated growth rate, decreased predator avoidance behaviors and increased agonistic behaviors. Inbreeding allows rare, harmful recessive alleles to become expressed in the homozygous form, with resulting harmful effects and reduces genetic variability and performances on the offspring. The cause threatening the survival of the adapted indigenous animal breeds in Ethiopia is indiscriminate crossbreeding with exotic germplasm. On the other hand, due to climate changes, indigenous animal breeds are changing their distribution patterns. Animals specially the wild are shifting their ranges, altering their phenology, changes in population dynamics, and some are facing extinction, or have become extinct. Different organizations have made attempts to conserve 3 indigenous cattle breeds and one sheep breed in Ethiopia. Their attempts are failing due to the gaps of information for sustainable utilization and conservation of animal genetic resources. Setting priorities to conserve, develop and utilize the available genetic resources as well as *ex situ* conservation can be taken as an opportunity for the maintenance of nucleus flocks as a repository of the pure breed.

Key words: Climate change, crossbreeding, genetic introgression, inbreeding, Ethiopia.

INTRODUCTION

Conservation of animal resources should ideally be undertaken at global level, because of the existence of non- and trans-boundary breeds. However, national conservation programs better serve specific local interests, such as conservation with the objectives of improving indigenous breeds. There are obviously many reasons why genetic conservation should be considered, e.g. cultural, historic, and scientific interests and on the one hand for more practical and economic considerations.

It is necessary to characterize and conserve animal genetic resources, population size trends and their distributions are components of characterization and should precede any conservation efforts, of valuable breeding stocks. In developed countries, most conservation programs are based with strong collaboration between gene banks and the animal breeding industry. In these countries, few breeds of the major species are covered by conservation programs, and the programs are of variable quality; the focus is typically on *in situ* conservation (FAO, 2007a). Strategies for setting conservation priorities for livestock populations depend on the nature and severity of the threats, the species and available technologies. The argument is conservation for sustainable utilization of the resources currently and for posterity (Solomon et al., 2008). If risk strategy were to be adopted for setting conservation priorities for Ethiopian animal genetic resources, then most of the genetic diversity would have been conserved.

Loss of genetic diversity among populations occurs due to high rate of gene flow from other populations (Staines, 1991; Abbott, 1992; Tesfaye, 2004), genetic introgression (McGinnity et al., 2003; Hutchings and Fraser, 2008; Kidd et al., 2009), inbreeding (Keller and Waller, 2002; Fredrickson et al., 2007) and climate change (McCarthy et al., 2001, 2003; Thomas et al., 2006; Boko et al., 2007).

According to the Institute of Biodiversity Conservation, IBC (2004) Farm animals are an integral part of the Ethiopian agricultural fabric. They are sources of food, traction, manure, raw materials, investment, cash, security, foreign exchange earnings, social and cultural identity. The Ethiopian agriculture is based on availability and/or contribution of farm animals and no agricultural practices other than mowing are conducted without the involvement of farm animals. Land cultivation is mainly done by oxen, horses, donkeys or their combinations. Farm animals serve to level the ploughed field, shortly before and after sowing. In some areas, farm animals are involved in weeding, especially maize and sorghum fields. Transportation of the harvested crops to and from threshing sites, threshing itself, transportation to and from the market is conducted by the farm animals. Similarly, transportation of water, firewood, mobile houses, construction materials and other goods is conducted by farm animals and they are the main means of human transport. Therefore, the objective of this review was to quantify the threats, previous conservation attempts and opportunities which exist for conserving indigenous animal genetic resources in Ethiopia.

THREATS TO INDIGENOUS ANIMAL GENETIC RESOURCES OF ETHIOPIA

Threads caused by breeding processes

Genetic introgression

In both domestic and wild animals, genetic introgression between invasive organisms with exotic germplasm and local populations would enhance genetic homogenization, leading to the disintegration of the components of genetic diversity generated by divergent adaptation to heterogeneous habitats (Randi, 2008). Introgressive hybridization among local, wild and invasive populations together with habitat degradation and loss of ecological structure as well as unsustainable selective pressures for adaptation to global climate changes, over exploitation and loss of community structure, are major threats to conservation of animal genetic resources (Allendorf et al., 2001; Randi, 2008). Though it is important for farm ani-

mals up to some generation, the consequences of genetic introgressions of domestic and exotic germplasm are extensive and includes reduced survival and fitness of the F_1 and F_2 generations, accelerated growth rate, decreased predator avoidance behaviors and increased agonistic behaviors (McGinnity et al., 2003; Wessel et al., 2006; Hutchings and Fraser, 2008). In Ethiopia, genetic introgression has occurred between Simien fox and Ethiopian wolfs, the trepidation of genetic introgression between Walia ibex (Capra walie) and domestic goats (Capra hircus) (Kefyalew et al., 2011a) at and near the national parks are becoming some of the problems which threat both wild and domestic indigenous animal genetic resources.

Inbreeding

Loss of genetic diversity and an increase in inbreeding rates are also critical genetic issues to be considered in animal genetic resources (Ballou and Lacy, 1995). Inbreeding and loss of genetic variation are inevitable consequences of small population sizes (Frankham et al., 2002). The extent to which a population becomes inbred or loses genetic diversity over time depends on a number of factors, including immigration, effective population size, generation length, and selection intensity (Jamieson et al., 2006). Small populations face inbreeding, genetic drift (Pronounced effects of random genetic drift that lead to erratic fluctuations in allele frequencies) and high susceptibility to catastrophes, diseases and environmental stochasticity (Frankham et al., 2004; Von et al., 2007; Kefyalew et al., 2011c). Saccheri et al. (1998) and Fredrickson et al. (2007) disclosed that inbreeding reduces fitness and increases the risk of population extinction. Significant impacts of inbreeding depression on extinction risk in populations with carrying capacities of up to two thousand individuals have been noted (O'Grady et al., 2006). Increased inbreeding rates also have a strong and significant effect on overall population growth rate of small and isolated populations and are associated with inbreeding depression (Smith et al., 1998). Moreover, an increased rate of inbreeding also means an increased risk of loss of genetic diversity (Meuwissen, 1991) and a reduced additive genetic variance is expected (Falconer and Mackay, 1996). Small population size animals and the primary factors contributing to extinction are habitat loss, introduced species, overexploitation and pollution. These factors are caused by humans, and related to human population growth (Frankham et al., 2002).

The consequence of the inbreeding process is the reduction in the genetic variability within a population and in performance mainly in traits that are associated with the fitness of an individual (Maiwashe et al., 2006). High inbreeding rates have higher contribution to over all inbreeding of all the populations and hence inbreeding depression (Marshall et al., 2002). Inbreeding allows the rare, harmful recessive alleles to become expressed in the homozygous form, with resulting harmful effects on the offspring such as reduction in fertility, fecundity, offspring size, growth and survival, and physical deformities (Lomker and Simon, 1994). Inbreeding can potentially reduce population growth rates and increase extinction (Newman and Pilson, 1997). In Ethiopia, due to lack of grazing lands, the indigenous animals are kept together and interbreed themselves with harmful effects on the offspring. All these lead the farmers to abandon the local animal genetic resources and shifting to exotic germplasm, which again reduce their performances especially after the third generations.

Crossbreeding

Crossbreeding may increase the overall genetic diversity as it introduces new genes in the population and new genotypes (e.g. synthetic breeds). However, the major threat to the adapted indigenous breeds in Africa is indiscriminate or irrational crossbreeding. Crossbreeding can be considered as "a necessary evil" as it delivers the much desired fast growth in livestock productivity and at the same time threatens the indigenous breeds through breed replacement (Solomon et al., 2008).

Rege and Gibson (2003) suggest that, the use of exotic germplasm, changes in production systems, producer preference because of socio-economic factors, and a range of disasters (drought, famine, disease epidemics, civil strife/war) as the major causes of genetic erosion. Tisdell (2003) suggested major causes for threats of animal populations which are development interventions, specialization (emphasis on a single productive trait), genetic introgression, the development of technology and biotechnology, political instability and natural disasters. For at-risk animal breeds in Africa, (Rege and Gibson, 2003) lists as major cause's replacement by other breeds, crossbreeding with exotic breeds or with other indigenous breeds, conflict, loss of habitat, disease, neglect and lack of sustained breeding programs among the threats. The increased demand for livestock products in many parts of the developing world drives efforts to increase the output of meat, eggs, and milk for the market (Delgado et al., 1999).

Effects of crossbreeding on animal genetic resources in Ethiopia

Indigenous livestock are well adapted to tropical conditions and have high degree of heat tolerance, which are partly resistant to many of the diseases prevailing in Ethiopia and have the ability to survive long periods of feed and water shortage. These attributes have been acquired through natural selection over hundreds of

generations. They are all essential for successful animal production (Rege and Lipner, 1992). Indigenous stocks represent a genetic resource which should not only be conserved for future use, but should also be fully exploited for short-term benefits (Rege and Lipner, 1992). Due to the low genetic potential of indigenous cattle, milk-meat production and productivity remain low in Ethiopia (Shiferaw et al., 2003). Improvement of the genetic potential of indigenous cattle was achieved by cross breeding with high producing cattle of temperate origin. Of course, crossbreeding was and still is perceived as "the way forward" to improve productivity of indigenous livestock under smallholder conditions and development policies has largely ignored the adapted farm animal genetic resources (ILRI, 1999). In Ethiopia, crossbred cattle mainly cross of zebu with Holstein-Friesian cattle have been used for milk production for decades (Negussie et al., 1998). However, crossbreeding with exotic breeds clearly is a major factor contributing to the erosion of locally adapted animal genetic resources (Köhler-Rollefson, 2004). Crossbreeding also results inconsistent and rapid loss of genetic diversity by dilution of the autochthonous genetic makeup. Therefore, designing of a crossbreeding program in Ethiopia needs to take into consideration a mechanism that ensures conservation of animal genetic resources (Aynalem et al., 2011).

Major causes threatening diversity of genetic resources in Ethiopia include poorly designed and managed introduction of exotic genetic materials, droughts and consequences of drought associated indiscriminate restocking schemes, political instability and associated civil unrest, and weak development interventions (Nigatu et al., 2004). The effects of the misguided and uncontrolled introduction of exotic genes and that of interbreeding among indigenous breeds might require application of molecular genetics for purposes of precision. In extreme scenarios, however, it could have a drastic effect leading to extinction of a breed within few generations (ESAP, 2004).

Even indigenous genotypes may well be adequate and able to respond sufficiently to reasonable economic improvements in their production system (Workneh et al., 2003), trends of improving indigenous cattle already exist in Africa and the population of pure indigenous cattle breeds is likely to diminish, because of crossbreeding or neglect of local animal genetic resource (Rege, 1992). As a result, some of the animal genetic resources are endangered, and unless urgent concerted efforts are made to characterize and conserve these breeds, they may be lost even before they are described and documented (Zewdu et al., 2008). Loss of genetic diversity increases the risk of the subsistence for the millions of livestock keepers who depend on these resources to secure their livelihoods (Fedlu et al., 2007). Unique germplasm is threatened by replacement of breeds with more productive or popular stocks, dilution of breeds through crossbreeding programs, and decreased diversity within highly selected breeds or lines that have a small number of breeding individuals (FAO, 2007). The application of artificial insemination (AI) in indigenous cattle using semen from exotic cattle breeds is, for instance, resulting in unforeseen substitution of indigenous genes by exotic genes (ESAP, 2004; IBC, 2004). The application of these technologies for germplasm propagation and dissemination may contribute to the erosion of diversity. Besides, exotic sheep breeds, such as Awassi, Hampshire, Blue-de-main, Merino, Romney, Corriedale, and Dorper were introduced to Ethiopia for their wool and mutton production. Crossbreeding of the Menz breed with the 5 exotic breeds, namely Awassi, Hampshire, Bleu-de-Main, Romney and Dorper are being used for development and research activities (Solomon, 2008). However, the mating of indigenous sheep and cattle exotic genetic resources was indiscriminate. According to FAO (1999), about 30% of the world's farm animal breeds are subjected to the risk of extinction due to this.

Threads caused by climate change

Challenges such as climate change underline the importance of retaining a diverse portfolio of livestock breeds (FAO, 2007b). Livestock production both contributes to and is affected by climate change (Hoffmann, 2010). There are evidences that livestock and environmental trade-offs are currently substantial and that these will increase significantly in the future as a result of the increased demand for livestock products from the growing population (Herrero et al., 2009a). Some of the most important impacts are those associated with land use change for feed production both for ruminants and monogastrics, which have significant simultaneous impacts on a range of environmental dimensions (land use, emission of gases , water cycles, nutrient balances, biodiversity) (FAO, 2007b; Herrero et al., 2009a). Moreover, livestock mitigation measures could include technical and management options to reduce emissions from livestock as well as the integration of livestock into broader environmental service approaches (Thornton et al., 2007).

Adapting to global climate change is likely to present a serious challenge to many livestock producers over the coming decades. The pastoral systems of the world's dry lands are among the most vulnerable, with climate change taking place against the background of natural environments that are already experiencing resource degradation. In general, climate change is likely to present significant problems for production systems where resource endowments are poorest and where the ability of livestock keepers to respond and adapt is most limited (FAO, 2007b).

Climate change is adding to the considerable develop-ment challenges (Thornton et al., 2007) and this will require more efficient animal production systems, careful husbandry of natural resources and measures to reduce waste and environmental pollution (FAO, 2010). Climate change impacts such as rising temperatures and declining rainfall in combination with other stresses could result in the shifting of ecological zones, loss of flora and fauna and an overall reduction in ecological productivity (Thomas et al., 2006; Boko et al., 2007). There is compelling evidence that in response to on-going changes in regional climates, species are already shifting their ranges (Walther et al., 2005; Lavergne et al., 2006; Thomas et al., 2006), altering their phenology (White et al., 2003), changes in population dynamics (Balmford et al., 2003; Forsman and Monkkonen, 2003), fluctuation in population growth rates (Lande et al., 2003; Grosbois et al., 2008) and alters ecosystem functioning (Doran et al., 2002; Hays et al., 2005) and that some species are facing extinction, or have become extinct (Foden et al., 2007).

In addition to the physiological effects of higher temperatures on individual animals, the consequences of climate change are likely to include increased risk that geographically restricted rare animal breed populations will be badly affected by disturbances (Hoffmann, 2010). The IPCC, (2007a) predicted that by 2100 the increase in global average surface temperature may be between 1.8 and 4.0°C. With global average temperature increases of only 1.5 to 2.5°C, approximately 20 to 30% of plant and animal species are expected to be at risk of extinction (FAO, 2007).

Climate change in Ethiopia

Both instrumental and proxy records have shown that, there is significant variations in the spatial and temporal patterns of climate in Ethiopia. According to NMA (2006), the country experienced 10 wet years and 11 dry years over 55 years analyzed, demonstrating the strong inter-annual variability. Between 1951 and 2006, annual minimum temperature in Ethiopia increased by about 0.37°C every decade. The climate change profile for Ethiopia also shows that the mean annual temperature increased by 1.3°C between 1960 and 2006, at an average rate of 0.28°C per decade (McSweeney et al., 2008).

The temperature increase has been most rapid from July to September (0.32°C per decade). It is reported that the average number of hot days per year has increased by 73 days (an additional 20% of days) and the number of hot nights has increased by 137 days (an additional 37.5% of nights) between 1960 and 2006. IPCC"s mid-range emission scenario show that compared to the 1961 to 1990 average mean annual temperature across Ethiopia will increase by between 0.9 and 1.1°C by the year 2030 and from 1.7 to 2.1°C by the year 2050. The temperature across the country could rise by between 0.5 and 3.6°C by 2080, whereas precipitation is expected to show some increase (NMA, 2006). Unlike the temperature trends, it is very difficult to detect long-term rainfall trends in Ethiopia, due to the high inter-annual and inter-decadal variability. According to NMA (2006), between 1951 and 2006, no statistically significant trend in mean annual rainfall was observed in any season. As compared to the 1961 to 1990, annual precipitation showed a change of between 0.6 and 4.9% and 1.1 to 18.2% for 2030 and 2050, respectively (NMA, 2006). The percentage change in seasonal rainfall is expected to be up to about 12% over most parts of the country (ICPAC, 2007).

Climate change projections suggest also that, further selection for breeds with effective thermoregulatory control may be needed (Hoffmann, 2010). Animal breeding indices should include traits associated with thermal tolerance, low quality feed and disease resistance, and give more consideration of genotype-by-environment interactions to identify animals most adapted to specific conditions (Hoffmann, 2010). The conservation of existing animal genetic resources and diversity as a global insurance measured against unanticipated change (Thornton et al., 2007).

NMSA (2001) has also stressed that, mean temperature and precipitation in Ethiopia have been changing over time. Over the past 60 years, some of the years have been characterized by dry rainfall conditions resulting in drought and famine where as the others are characterized by wet conditions. Droughts in Ethiopia can shrink household farm production by up to 90% of a normal year output (World Bank, 2003).

All these have effects on animal genetic resources directly on heat stress (Thornton et al., 2009), emergence, spread, and distribution of livestock diseases ((Baylis and Githeko, 2006), feed and water availability (Thornton and Gerber, 2010) and genetic mechanisms which influence fitness and adaptation (Hoffmann, 2010).

PREVIOUS ATTEMPTS TO CONSERVE ANIMAL GENETIC RESOURCES IN ETHIOPIA

Conservation of local breeds of farm animal genetic resources should be a part of animal management and the communities should be informed by pertinent parties for the distribution, structures, and trends, productive and adaptive performances of populations of the existing breeds. Although much information is lacking, conservation of farm Animal Genetic Resources (AnGR) in the Ethiopian perspective should be viewed from the rational utilization and protection of existing genotypes from genetic erosion (IBC, 2004). Unfortunately, except for limited activities that are meant to maintain pure stocks of 3 cattle breeds and 1 sheep breed, no conservation activities of farm AnGR have so far been practiced in the country. To date, Borana cattle bred at Did Tuyura Ranch, Horro cattle breed at Horro Ranch, Fogera cattle breed at Metekel Ranch and Andassa Agricultural Research Centre, and Menz sheep bred at Amed Guya Research Centre (IBC, 2004) are the only conservation attempts made in Ethiopia.

Conservation measures for threatened breeds have already been established in some countries (FAO, 2007a) and are a priority of the global plan of action for animal genetic resources (FAO, 2007b). In Ethiopia, institutions involved in in situ conservation of biodiversity include the IBC, the Ethiopian Agricultural Research Organization (EARO), Regional Agricultural departments, Higher Learning Institutions, etc. However, the impact of their work on conservation of biological resources in practical terms is very limited (FDRE, 2005). Loss of animals as a result of droughts and floods, or disease epidemics related to climate change may thus increase (FAO, 2008). If breeds are geographically isolated (endemic) as is the case for some local and rare breeds, there is a risk of their being lost in localized disasters. Therefore, identifying the status of the livestock genetic resources and designing conservation strategies based on the priorities is crucial.

Constraints and lessons learnt

In Ethiopia, information for sustainable utilization and conservation of the farm animal genetic resources are very limited and, if available, are full of gaps (IBC, 2004). It would no doubt be of interest to future generations of animal breeding specialists, as well as to interested laymen, if it were possible to maintain representative samples of some of the once important animal breeds especially in ex situ conservation, which are now on the verge of disappearing (FAO, 1990). Practical and economic needs ought to be the most important reason for conservation in future; one could perhaps argue that, farmers with economically competitive breeds or genetic types should take care of their own preservation.

Numerous attempts made to introduce 'improved' breeds with poor success in terms of achieving genetic potential. Fertility and longevity of introduced breeds so poor that continual importation of exotic breeds necessary. Rare breeds often crossed with 'improved' breeds due to small population, dilution of breed characteristics and creation of gene pool from which it is then difficult to identify and utilize favorable local breeds genetic characteristics are also the main threats to animal genetic resources in Ethiopia. Unfortunately, the situation is more complex. The economic and environmental conditions are changing and genetic types which are superior under one set of conditions may be inferior under a different set of conditions. As the changes are gradual and different breeds or types are not generally compared under exactly the same conditions, the individual breeder or leader of a breeding program has usually no interest in, or possibility of, conserving for future use animals which, at any given time, considers slightly inferior to those selected for breeding (Rendel, 1975). However, sustainable use of genetic resources should effectively deal with semen and embryos preservation as part of the ongoing utilization and improvement programs.

OPPORTUNITIES

In situ conservation of livestock breeds is primarily the active breeding of animal populations and their continued use as part of an ongoing livelihood strategy (Solomon et al., 2008). Village-based breed improvement programs must be a complementary to *in situ* livestock conservation objectives with the concept conservation through sustainable utilization. In such a context, it can be viewed as part and parcel of a comprehensive conservation plan, and not as a separate genetic improvement activity, that entails significant additional costs. In the mean time, there are more feasible conservation methods at hand under the current circumstances including *in vivo* conservation. *In vivo* conservation includes *in situ* and *ex situ* methods. *Ex situ in vivo* conservation is the maintenance of pure-bred nucleus flocks in an organized government farms or research farms which can form a repository of the pure breed. A conservation-based breeding program should be based on broader breeding objectives that incorporate the needs and perceptions of the community and maintenance of the genetic diversity such as adaptation traits. Involvement of the farmers in the design and implementation of the breeding program in line with the principles of *in situ* conservation of genetic resources is one of the options which must be considered.

CONCLUSION

From the review, it was possible to see that, genetic introgression, crossbreeding, inbreeding and climate changes and related factors are the threats for animal genetic resources in Ethiopia. The genetic introgression between wild and the domestic, interbreeding among closely related animals, reduction in effective population size and lack of grazing lands all lead to reduction in fertility, fecundity, offspring size, growth and survival, and physical deformities. All these pilot the farmers to abandon the local animal genetic resources and shifting to exotic germplasm. Subsequently, replacing local breeds by range of high-yielding breeds is widespread. Nonetheless, due to the existing climate change effects and gene segregation especially after third generation, high yielding animals could not be used sustainably. Adapting to present climate change and related factors is a serious challenge to many animal producers. Admixture

of genes of indigenous animal population with exotic germplasm of conspecifics and increasing temperature will trigger loses of animal genetic resources in Ethiopia. Therefore, to conserve and use sustainably animal genetic resources, controlled and monitored cross-breeding with appropriate records, habitat management of the wild and domestic animals, interbreeding among related but not closely related breeds and tackling climate change related factors are some of the most important ones.

REFERENCES

Abbott RJ (1992). Plant invasions, interspecific hybridization and the evolution of new plant taxa. Tree 7:401-405.

Allendorf FW, Leary RF, Spruell P, Wenburg JK (2001). The problems with hybrids: setting conservation guidelines. Trends Ecol. Evol. 16:613-622.

Aynalem H, Azage T, Workneh A, Noah K, Tadelle D (2011). Breeding strategy to improve Ethiopian Boran cattle for meat and milk production. Working Paper No. 26.

Ballou JD, Lacy RC (1995). Identifying genetically important individuals for management of genetic diversity in pedigreed populations. In: Population Management for Survival and Recovery: Analytical Methods and Strategies in Small Populations Conservation (eds) Ballou, J.D., Gilpin, M., Foose, T.J). Columbia University Press. New York. pp. 76-111.

Balmford A, Green RE, Jenkins M (2003). Measuring the changing state of nature. Trends Ecol. Evol. 18:326-330.

Baylis M, Githeko AK (2006). The effects of climate change on infectious diseases of animals. Report for the Foresight Project on Detection of Infectious Diseases, Department of Trade and Industry, UK Government. P. 35.

Boko M, Niang I, Nyong A, Vogel C, Githeko A, Medany M, Osman-Elasha B, Tabo R, Yanda P (2007). Africa. In, M. L., Parry, O. F., Canziani, J. P., Palutikof, P. J., van der Linden and Hanson, C. E. (2007) Climate Change (Eds.): Impacts, Adaptation and Vulnerability. Contribution of Working Group II to the Fourth Assessment Report of the Intergovernmental Panel on Climate Change, Cambridge University Press, Cambridge, United Kingdom.

Delgado C, Rosegrant M, Steinfeld H, Ehui S, Courbois C (1999). Livestock to 2020 the Next Food Revolution. Food, Agric. Environ. Discuss. P. 28.

Doran PT, Priscu JC, Berry LW, Walsh JE, Fountain AG, Mc-Knight DM, Moorhead DL, Virginia RA, Wall DH, Clow GD, Fritsen CH, Mckay CP, Parsons AN (2002). Antarctic climate cooling and terrestrial ecosystem response. Nature 415:517–520.

ESAP (Ethiopian Society of Animal Production). (2004). Farm Animal Biodiversity in Ethiopia: Status and Prospects. Asfaw Yimegnuhal and Tamrat Degefa (eds.). Proceedings of the 11th Annual conference of the Ethiopian Society of Animal Production (ESAP) held in Addis Ababa, Ethiopia, August 28 –30, 2003. ESAP, Addis Ababa.

Falconer DS, Mackay TFC (1996). Introduction to quantitative genetics. 4th ed. Longman Scientific and Technical, Harlow, UK. pp. 1529-1536.

FAO (1990). Animal genetic resources A global programme for sustainable development Proceedings of an FAO Expert Consultation Rome, Italy.

FAO (1999). The global strategy for the management of farm animal genetic resources. Executive brief. FAO, Rome, Italy. P. 43.

FAO (2007a). The State of the World's Animal Genetic Resources for Food and Agriculture – in brief. Pilling, D. and Rischkowsky B., (eds). Food and Agriculture Organization of the United Nations, Rome.

FAO (2007b). Global Plan of Action for Animal Genetic Resources and the Interlaken Declaration. Food and Agriculture Organization of the United Nations (FAO), Rome.

FAO (2008). Climate change and food security: A framework document http://www.fao.org/docrep/010/k2595e/k2595e00.htm, 5 March 2011.

FAO (2010). Commission on Genetic Resources for Food and Agriculture. Intergovernmental Technical Working Group on Animal Genetic Resources for Food and Agriculture. Draft Guidelines for the Cryoconservation of Animal Genetic Resources, Sixth Session. Rome. 24-26 November 2010.

FDRE (Federal Democratic Republic of Ethiopia) (2005). Institute of Biodiversity Conservation, National Biodiversity Strategy and Action Plan. Addis Ababa, Ethiopia.

Fedlu H, Endashaw WA, Tadelle D (2007). Genetic variability of 5 indigenous Ethiopian cattle breeds using RAPD markers. Afr. J. Biotechnol. 6(19):2274-2279.

Foden W, Midgley GF, Hughes GO, Bond WJ, Thuiller W, Hoffman MT, Kaleme P, Underhill LG, Rebelo AG, Hannah L (2007). A changing climate is eroding the geographical range of the Namib Desert tree Aloe through population declines and dispersal lags. Diversity Distrib. 13:645–653.

Forsman JT, Monkkonen M (2003). The role of climate in limiting European resident bird populations. J. Biogeogr. 30:55-70.

Frankham R, Ballou JD, Briscoe DA (2002). Introduction to conservation genetics. Published in the United States of America by Cambridge University Press, New York. pp. 1-10.

Frankham R, Ballou JD, Briscoe DA (2004). Primers of conservation genetics: A brief introduction to the general principles of conservation genetics, First edition, Cambridge University Press, UK. P. 238.

Fredrickson R, Siminski P, Woolf M, Hedrick P (2007). Genetic rescue and inbreeding depression in Mexican wolves. Proceeding, Royal Soc. London. 274:2365-2371.

Grosbois V, Gimenez O, Gaillard JM, Pradel R, Barbraud C, Clobert J, Møller AP, Weimerskirch H (2008). Assessing the impact of climate variation survival in vertebrate populations. Cambridge philosophical society. Biol. Rev. 83:357-399.

Hays GC, Richardson AJ, Robinson C (2005). Climate change and marine plankton. Trends Ecol. Evol. 20:337-344.

Herrero M, Thornton PK, Gerber P, Reid RS (2009a). Livestock, Livelihoods and the environment: Understanding the trade-offs. Current Opin. Environ. Sustainability. 1:111-120.

Hoffmann I (2010). Climate change and the characterization, breeding and conservation of animal genetic resources. Anim. Genet. 41(1):32-46.

Hutchings JA, Fraser DJ (2008). The nature of fisheries and farming-induced evolution. Mol. Ecol. 17:294-313.

IBC (Institute of Biodiversity Conservation). (2004). The state of Ethiopia's Farm Animal Genetic Resources: A contribution to the first report on the state of the world's animal genetic resources. May 2004, Addis Ababa, Ethiopia.

ILRI (1999). Economic Valuation of Animal Genetic Resources. Proceedings of a FAO/ILRI workshop held at FAO Headquarters, Rome, Italy. P. 80.

IPCC (Intergovernmental Panel on Climate Change) (2007a). Summary for Policymakers. In, S. Solomon, D. Qin, M. Manning, Z. Chen, M. Marquis, K. B. Averyt, M. Tignor and H. L. Miller (Eds.), Climate Change 2007: The Physical Science Basis. Contribution of Working Group I to the Fourth Assessment Report of the Intergovernmental Panel on Climate Change, Cambridge University Press, Cambridge, UK and New York, NY, USA.

IPCC (2007). Forth Assessment Report, Climate Change 2007 (AR4), Working Group II summary for policy makers.

Jamieson IG, Wallis GP, Briskie JV (2006). Inbreeding and Endangered Species Management: Is New Zealand Out of Step with the Rest of the World? Conserv. Biol. 20:38-47.

Kefyalew A, Tadelle D, Solomon G, Aynalem H, Yoseph M (2011a.). The probable genetic introgression between Walia ibex (Capra walie) and domestic goats (Capra hircus) at Simien mountains national park (SMNP) in Ethiopia. Afr. J. Agric. Res. 6:856-865.

Kefyalew A, Tadelle D, Yoseph M (2011c). Effects of Habitat Loss and Limitation on Effective Population Size and Inbreeding Rates of Walia Ibex (Capra walie) in Ethiopia. Afr. J. Ecol. 50:125-130.

Keller LF, Waller DM (2002). Inbreeding effects in wild populations. Trends Ecol. Evol. 17:230-241.

Kidd AG, Bowman J, Lesbarrères D, Schulte-Hostedde AI (2009). Hybridization between escaped domestic and wild American mink (Neovison vison). Mol. Ecol. 18:1175-1186.

Köhler-Rollefson I (2004). Farm Animal Genetic Resources. Safeguarding National Assets for Food Security and Trade. A Summary of Workshops on Farm Animal Genetic Resources Held in the South African Development Community (SADC). GTZ, BMZ, FAO, CTA., SADC. Eschborn, Germany.

Lande R, Engen S, Saether BE (2003). Stochastic population dynamics in ecology and conservation. Acta. Biotheoretica. 52(3):219-220.

Lavergne S, Molina J, Debussche M (2006). Fingerprints of environmental change on the rare Mediterranean flora: a 115-year study. Global Change Biol. 12:1466-1478.

Lomker R, Simon DL (1994). Costs of and inbreeding in conservation strategies for endangered breeds of cattle. In: Proceedings of the 5th World Congress on Genetics Applied to Livestock production. 14:434-442.

Maiwashe A, Nephawe KA, Westhuizen RR, Mostert B, Theron HE (2006). Rate of inbreeding and effective population size in four major South African dairy cattle breeds. South Afr. J. Anim. Sci. 36:1.

Marshall TC, Coltmanr't DW, Pemberton JM, Slate J, Spalton JA, Guinness FE, Smith JA, Pilkington JG, Clutton-Brock TH (2002). Estimating the prevalence of inbreeding from incomplete pedigrees. Proceeding Royal Society of London – Biological Science. 269:1533-1539.

McCarthy JJ, Canziani OF, Leary NA, Dokken DJ, White KS (2001). Climate Change 2001: Impacts, adaptation and vulnerability. contribution of working group ii to the third assessment report of the intergovernmental panel on climate change, Cambridge University Press, Cambridge, UK and New York, USA.

McCarthy MA, Andelman SJ, Possingham HP (2003). Reliability of relative predictions in population viability analysis. Conserv. Biol. 17:982-989.

McGinnity P, Prodoehl P, Ferguson A (2003). Fitness reduction and potential extinction of wild populations of Atlantic salmon, Salmo salar, as a result of interactions with escaped farm salmon. Proceedings of the Royal Society London. Biol. Sci. 270:2443-2450.

Meuwissen THE (1991). Expectation and variance of genetic gain in open and closed nucleus and progeny testing schemes. Anim. Prod. 53:133-141.

National Meteorological Agency (NMA) (2006). National Adaptation Program of Action of Ethiopia (NAPA). National Meteorological Agency, Addis Ababa.

NMSA (National Meteorological Services Agency) (2001). Initial National Communication of Ethiopia to the United Nations Framework Convention on Climate Change (UNFCCC). NMSA, Addis Ababa,Ethiopia.

Negussie E, Brannang E, Banjaw K, Rottmann OU (1998). Reproductive performance of dairy cattle at Asella Livestock Farm, Arsi, Ethiopia. I. Indigenous cows versus their F1 crosses. J. Anim. Breeding Genet. 115:267-280.

Newman D, Pilson D (1997). Increased probability of extinction due to decreased genetic effective population size: experimental populations of Clarkia pulchella. Evolution 51:354-362.

Nigatu A, Getachew G, Drucker AG (2004). Reasons for the loss of animal genetic resources (AnGR) and the importance of indigenous knowledge in AnGR Management. Proceedings of the 11th Annual Conference of the Ethiopian Society of Animal Production (ESAP), August 28-30, 2003, Addis Ababa, Ethiopia. pp. 31-45.

O'Grady JJ, Brook BW, Reed DH, Ballou JD, Tonkyn DW, Frankham R (2006). Realistic levels of inbreeding depression strongly affect extinction risk in wild populations. Biol. Conserv. 133:42-51.

Randi E (2008). Detecting hybridization between wild species and their domesticated relatives. Mol. Ecol. 17:285-293.

Rege JEO (1992). Background to ILCA characterization project. Proceedings of the Research Planning Workshop, 19-21 February 1992, ILCA, Addis Ababa, Ethiopia.

Rege JEO, Gibson JP (2003). Animal genetic resources and economic development: issues in relation to economic valuation. Ecol. Econ. 45:319-330.

Rege JEO, Lipner ME (1992). African Animal Genetic Resources: Their characterization, Conservation and Utilization. Proceedings of the Research Planning Workshop held at ILCA, Addis Ababa, Ethiopia 19–21 February 1992.

Rendel J (1975). The utilization and conservation of the world's animal genetic resources. Agric. Environ. 2:101-119.

Saccheri IJ, Kuussaari M, Kankare M, Vikman P, Fortelius W, Hanski I (1998). Inbreeding and extinction in a butterfly metapopulation. Nature 392:491-494.

Shiferaw Y, Tenhagen BA, Bekana M, Kassa T (2003). Reproductive Performance of Crossbred Dairy Cows in Different Production Systems in the Central Highlands of Ethiopia. Trop. Anim. Health Prod. 35:551-561.

Smith LA, Cassell BG, Pearson RE (1998). The effects of inbreeding on the lifetime performance of dairy cattle. J. Dairy Sci. 81:2729-2737.

Solomon G (2008). Sheep resources of Ethiopia: Genetic diversity and breeding strategy. PhD Thesis, Wageningen University, The Netherlands.

Solomon G, Hans K, Jack J, Wing OH, Johan AM, Van Arendonk (2008). Conservation priorities for Ethiopian sheep breeds combining threat status breed merits and contributions to genetic diversity. Gene. Sel. Evol. 40:433-447.

Staines BW (1991). Red deer, Cervus elaphus, in the handbook of hritish mammals, edited by G. B. Corbett and S. Harris. Blackwell Scientific Publications, Oxford. pp. 492-504.

Tesfaye A (2004). Genetic characterization of indigenous goat populations of Ethiopia using microsatellite DNA markers. PhD dissertation presented to National Dairy Research Institute (Deemed University), Karnal (Haryana), India. P. 279.

Thomas CD, Franco AMA, Hill JK (2006). Range retractions and extinction in the face of climate warming. Trends Ecol. Evol. 21:415-416.

Thornton P, Herrero M, Freeman A, Mwai O, Rege E, Jones P, McDermott J (2007). Vulnerability, Climate change and Livestock – Research Opportunities and Challenges for Poverty Alleviation. International Livestock Research Institute (ILRI), Kenya. SAT eJ. 4:1.

Thornton P, van de Steeg J, Notenbaert MH, Herrero M (2009). The impacts of climate change on livestock and livestock systems in developing countries: A review of what we know and what we need to know. Agricultural Systems 101: 113-127.

Thornton PK, Gerber P (2010). Climate change and the growth of the livestock sector in developing countries. Mitigation Adapt. Strateg. Glob. Change 15, 169–184.

Tisdell C (2003). Socioeconomic causes of loss of animal genetic diversity: analysis and assessment. Ecol.l Econ. 5:365-437.

Von HA, Bassano B, Festa-Bianchet M, Luikart G, Lanfranchi P, Coltman D (2007). Age dependent genetic effects on a secondary sexual trait in male Alpine ibex, Capra ibex. Mol. Ecol. 16:1969-1980.

Walther GR, Berger S, Sykes MT (2005). An ecological "footprint" of climate change. Proceeding Royal Society of London, Biol. Sci. 272:1427-1432.

Wessel ML, Smoker WW, Fagen RM, Joyce J (2006). Variation of agonistic behavior among juvenile Chinook salmon (Oncorhynchus tshawytscha) of hatchery, hybrid, and wild origin. Can. J. Fish. Aquatic Sci. 63:438-447.

White MA, Brunsell N, Schwartz MD (2003). Vegetation phenology in global change studies. In: Schwartz, M.D. (Ed.), Phenology: An Integrative Environmental Science. Kluwer Academic Publishers, Dordrecht. pp. 453-466.

Workneh A, King JM, Bruns E, Rischkowsky B (2003). Economic evaluation of smallholder subsistence livestock production: lessons from an Ethiopian goat development program Ecological Economics. 45(3):473-485.

World Bank (2003). Ethiopia: Risk and Vulnerability Assessment. Draft Report.

Zewdu W, Workneh A, Soelkner J, Hegde BP (2008). The Mahibere Silassie composite: a new cattle breed type in northwestern Ethiopia. Eth. J. Anim. Prod. 8(1):22-38.

28

Influence of electrolytes and ascorbic acid supplementation on serum and erythrocytic indices of broiler chickens reared in a hot environment

Majekodunmi B. C., Sokunbi O. A., Ogunwole O. A.* and Adebiyi O. A.

Physiology Unit, Department of Animal Science, University of Ibadan, Ibadan, Oyo State, Nigeria

Effect of supplementing drinking water with ammonium chloride, sodium bicarbonate, calcium chloride and ascorbic acid on haematology and selected serum indices of broiler chicks reared during high temperature and humified period was evaluated. A total of 200 one-day old Arbor acre broiler chicken strain was randomly allotted to five treatments: T1 (control-water without any supplement), T2 (0.5% ammonium chloride), T3 (0.5% sodium bicarbonate), T4 (0.5% calcium chloride), T5 (300 ppm ascorbic acid) in a completely randomized design. The packed cell volume, haemoglobin and albumin values obtained at day 28 and 49 varied significantly ($p < 0.05$) between treatments with consistently higher ($p < 0.05$) values for birds on T5. The serum total proteins at day 28 and alanine amino transferase at day 49 were also significantly different ($p < 0.05$) among treatments. Glucose values (g/dl) of groups at day 28 were not significantly ($p > 0.05$) different, while glucose levels of control group was significantly ($p < 0.05$) higher than those on T3 at day 49. Ascorbic acid supplementation enhanced the haematological profile of birds while supplemental electrolytes ameliorated heat stress with more profound effects on serum indices of birds on sodium bicarbonate.

Key words: Heat stress, ascorbic acid, haematological indices, serum metabolites, electrolytes, tropic environment.

INTRODUCTION

In the tropics, high environmental temperature is one of the most important stressors affecting production performance of broilers and compromises the ability of birds to maintain homeostasis. Reduction in feed intake, growth rate, egg production and feed efficiency are immediate consequences of heat stress in poultry production (Siegel, 1995).

It has been reported by Cier et al. (1992) that supplemental ascorbic acid alleviates the effect of heat stress on the performance of broiler chicks and increased immunoglobulin G levels (Tras et al., 2001). Though, ascorbic acid was reported to be synthesized by poultry and therefore not required to be supplemented in the diet under normal conditions, there has nevertheless been considerable interest in its possible role in maintaining homeostasis. The requirements of ascorbic acid may be elevated under hot environmental conditions (Pardue and Thaxton, 1986) when endogenous synthesis may not be adequate to meet the physiological needs of the birds. Also, electrolytes of various sources may have different effects on the physiology and blood characteristics of heat stressed birds, depending on whether they were included in the feed or water (Teeter and Smith 1985; Branton et al., 1986; Borges et al., 2007).

The blood system is particularly sensitive to changes in temperature, being an important indicator of physiological responses in birds to stressing agents. Borges et al. (2007) surmised that quantitative and morphological

---

*Corresponding author. E-mail: oluogunwole@yahoo.com.

Table 1. Gross composition of diets fed to birds given supplemental electrolytes and ascorbic acid.

Feed ingredients (%)	Starter diet	Finisher diet
Maize	52.00	52.00
Wheat offals	7.73	7.73
Soyabeans	35.00	30.00
Palm kernel meal	-	4.50
Palm oil	2.50	3.00
Oyster shell	0.50	0.50
Dicalcium phosphate	1.50	1.50
Salt	0.25	0.25
Methionine	0.15	0.15
Lysine	0.06	0.06
Broiler premix	0.25	0.25
Avatec	0.06	0.06
	100	100
Calculated analysis		
Crude protein	22.97	19.06
Metabolizable energy (kcal/kg)	3007.7	3018.9
Calorie:protein ratio	136.4:1	152.8:1

1 kg of premix contains: Vitamin A-10,000,000 IU; Vitamin D3-2,000,000; Vitamin E-20,000 IU; Vitamin K-2,250 mg; Thiamine B1-1,750 mg; Riboflavin B2- 5,000 mg; Pyridoxine B6-2,750 mg; Niacin-27,500 mg; Vitamin B12-15 mg; Pantothenic acid-7,500 mg; Folic acid-7500 mg; Biotin-50 mg; choline chloride-400 g; Antioxidant-125 g; Magnesium-80 g; Zinc-50 mg; Iron-20 g; Copper-5 g; Iodine-1.2 g; Selenium-200 mg; Cobalt-200 mg.

changes in blood cells were associated with heat stress and were translated by variations in haematocrit value, leukocyte counts, erythrocyte and haemoglobin contents. Vecerek et al. (2002) reported decreased haemoglobin levels and increased total blood leukocytes counts in chicken due to gradually increasing temperature. Maxwell et al. (1990) studied the effect of feed restriction on erythrocyte characteristics and reported significant changes in hematocrit, haemoglobin, mean corpuscular volume (MCV), mean corpuscular haemoglobin (MCH) and leukocyte count. Junqueira et al. (1999) evaluated sodium bicarbonate supplementation together with ammonium chloride on the drinking water and found no effect of treatments on the hematological parameters.

The effect of electrolytes and ascorbic acid supplementation on the erythrocytic and serum indices of heat stressed broiler is not adequately documented. This study therefore evaluated the effect of electrolytes and ascorbic acid supplementation in drinking water on erythrocytic and selected serum indices of broiler chickens exposed to heat stress.

MATERIALS AND METHODS

The study was undertaken at the Teaching and Research Farm of the University of Ibadan, Nigeria during the hot dry season of the year for a period of seven weeks. The site is located between latitudes 6°10" and 9°10" North of the equator and longitudes 30 and 60 of the Greenwich.

Experimental birds and management

Two hundred one day-old Arbor acre broiler chicken strain was used for the experiment. After one week of brooding, they were weighed and randomly allotted to five treatments. Each treatment was replicated four times with 10 birds per replicate in a completely randomized design (CRD). The birds were raised on conventional deep litter open sided house. Ambient temperature and humidity of the poultry house were recorded daily at 8.00, 13.00 and 20.00 h using thermo hygrometers.

Formulated broiler starter and finishers' diets contained 3000 Kcal/kg ME and 23% CP; 3000 kcal/kg ME and 19% CP respectively were offered to birds ad libitum in the course of the experiment. Composition of the experimental diets is shown in Table 1. Clean water in which test electrolytes or ascorbic acid has been added was provided ad libitum from day 15 to day 49 of the experiment. Treatment 1 (control) was without any supplement added, Treatment 2 (0.5% ammonium chloride), Treatment 3 (0.5% sodium bicarbonate), Treatment 4 (0.5% calcium chloride), and Treatment 5 (300 ppm ascorbic acid). Routine vaccinations and other medications were administered.

Blood collection

At day 28 and 49 of the experiment, three birds per replicate were randomly selected and bled at the jugular vein from which blood was drained and collected into heparinized bottles for haematological study. Blood for serum indices determination was collected into plain vacutainer tubes without EDTA for serum separation. Blood samples were centrifuged, serum separated out, decanted, and deep frozen for analysis.

Estimation of haematological variables

Packed cell volume (PCV) was determined by micro-hematocrit method (Schalm et al., 1975). Haemoglobin (Hb) concentration was measured spectrophotometrically using cyanomet haemoglobin method (Schalm, 1975). Red blood cell (RBC) count was estimated using haemocytometer (Schalm et al., 1975). Mean corpuscular volume (MCV), mean corpuscular haemoglobin (MCH) and mean corpuscular haemoglobin concentration (MCHC) were calculated from Hb, PCV and RBC (Jain, 1986)

Determination of serum parameters

Serum total protein (STP) was determined using Biuret method as described by Kohn and Allen (1995). Albumin was determined using Bromocresol green (BCG) method (Peter et al., 1982). The globulin concentration was obtained by subtracting albumin from the total protein. Cholesterol determination was as described by Roschlan et al. (1974). Aspartate amino transferase (AST) and alanine amino transferase (ALT) activities were determined by spectrophotometric methods (Rej and Hoder, 1983)

Statistical analysis

Data generated were analysed using analysis of variance (SAS, 1999). Significant treatment means were compared using Duncans option of the same software. P<0.05 level was accepted statistically significant level.

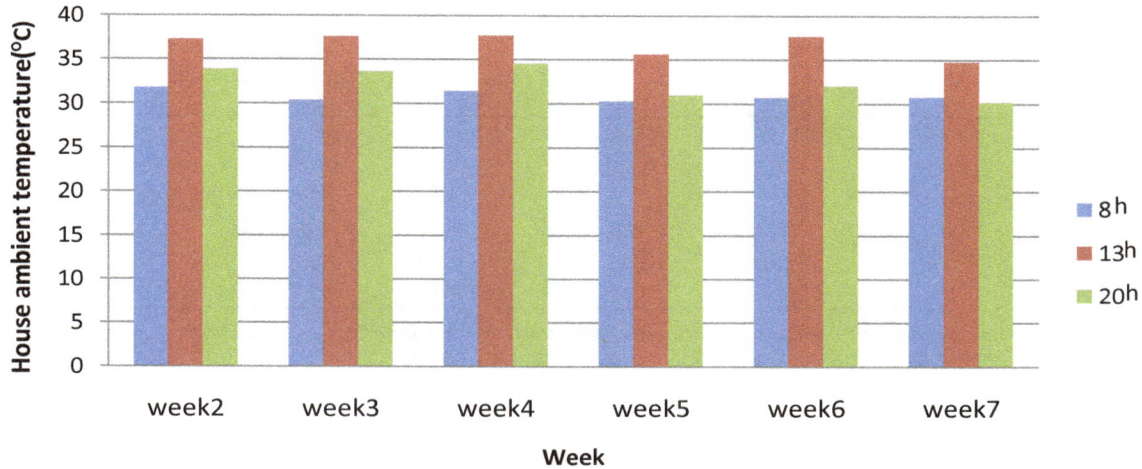

Figure 1. Weekly poultry house ambient temperature.

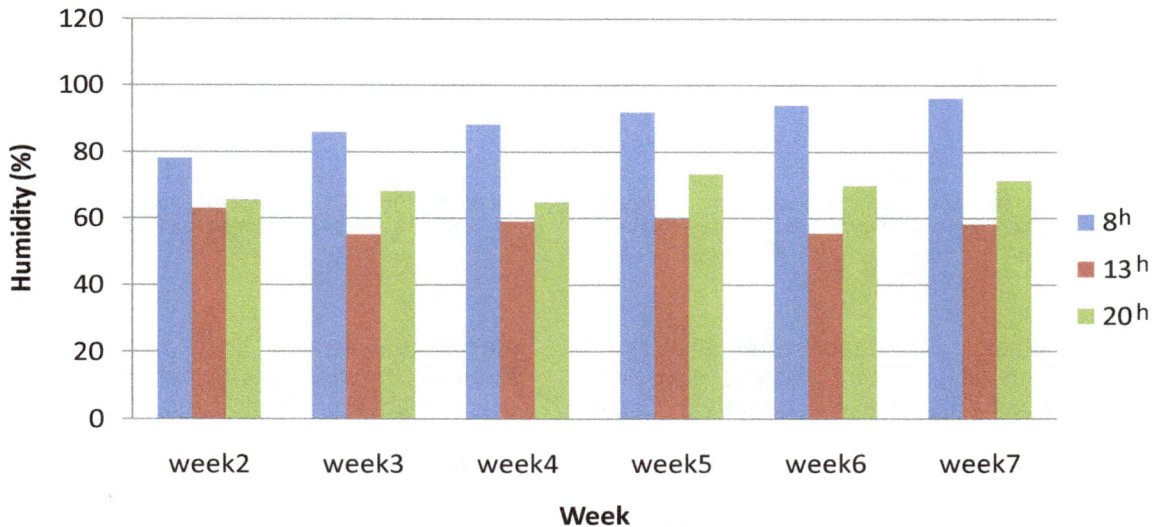

Figure 2. Weekly poultry house humidity.

RESULTS AND DISCUSSION

Environmental condition

Figure 1 shows the average weekly house ambient temperature. The recorded ambient temperature in the experimental period indicated a range (30.90 to 36.73°C) well above the thermo neutral zone (18 to 22°C) for broiler (Charles, 2002) indicating exposure of birds to perpetual heat stress. Behavioural responses such as panting, couching near cool surfaces and wide spreading of the wings were observed during the experimental period. According to Gray et al. (2003) panting would normally be expected to occur when the ambient temperature is near or above 30°C.

Figure 2 shows the average weekly relative humidity of the poultry house. A range of 58.48 to 89.24% relative humidity was observed which was quite high. High humidity above 60% has been reported to impair heat transmission from body core to the peripheral at 35°C but facilitated it at 30°C in broiler chicken of 4-week age (Lin et al., 2006). The prevailing weather conditions were clearly indicative that the birds were perpetually under stress.

Poultry house temperature and relative humidity

Table 2 shows the haematological profile of heat stressed broiler given water supplemented with electrolytes and ascorbic acid at day 28. PCV (%) and haemoglobin levels of treatment 5 were higher (p<0.05) higher than treatments

Table 2. Haematological parameters of heat stressed broiler given water supplemented with electrolytes and ascorbic acid at day 28.

Parameter	T1	T2	T3	T4	T5	SEM
Packed cell volume (%)	27.5[b]	27.3[b]	29.0[ab]	27.7[b]	29.7[a]	0.18
Haemoglobin (g/dl)	9.16[b]	9.11[b]	9.66[ab]	9.24[b]	9.91[a]	0.06
Red blood cell ($\times 10^6$/mm^3)	2.36	2.07	2.14	2.20	2.26	35.6
MCV(μ^3)	117	136	155	128	137	3.54
MCHC (%)	33.3	33.3	33.3	33.3	33.3	0.00
MCH ($\mu\mu$g)	39.2	45.3	51.8	42.7	45.7	1.18

[a,b]Means on the same row with different superscripts are significantly different (p<0.05). MCV, Mean corpuscular volume; MCHC, mean corpuscular haemoglobin concentration; MCH, mean corpuscular volume. SEM, Standard error of mean; T1, control; T2, ammonium chloride; T3, sodium bicarbonate; T4, calcium chloride; T5, ascorbic acid.

Table 3. Haematological parameters of heat stressed broiler given water supplemented with electrolytes and ascorbic acid at day 49.

Parameter	T1	T2	T3	T4	T5	SEM
Packed cell volume (%)	31.1[ab]	30.5[ab]	28.6[b]	28.5[b]	34.6[a]	0.62
Haemoglobin (g/dl)	10.3[ab]	10.16[ab]	9.54[b]	9.49[b]	11.4[a]	0.20
Red blood cell (x10^6/mm^3)	2.16	1.96	2.21	2.24	2.28	0.05
MCV(μ^3)	146	157	132	130	156	4.05
MCHC (%)	33.3[a]	33.3[a]	33.3[a]	33.3[a]	33.1[b]	0.02
MCH ($\mu\mu$g)	48.8	52.5	44.0	43.4	51.6	1.33

[a,b]Means on the same row with different superscripts are significantly different (p<0.05) MCV, Mean corpuscular volume; MCHC, mean corpuscular haemoglobin concentration; MCH, mean corpuscular volume. SEM, Standard error of mean. T1, control; T2, ammonium chloride; T3, sodium bicarbonate; T4, calcium chloride; T5, ascorbic acid.

1, 2 and 4. The haematological response of heat stressed broiler at day 49 is shown in Table 3. Increased (p<0.05) PCV and haemoglobin values of treatment 5differ significantly (p<0.05) from treatments 3 and 4.

The Hb, PCV and MCHC values obtained in the present study were consistent with the reported range for broiler (Mitruka and Rawnsley, 1977). Iheukwumene and Herbert (2003) reported values of 6.0 to 13.0%, 29.0 to 38.0% and 33.0 to 35.0 pg, respectively for these parameters. Islam et al. (2004) noted that commercial and local chickens reared inSylhet region in Bangladesh recorded Hb value of 7.06 to 9.37%, PCV value of 26.56 to 34.60% and MCV value of 84.27 to 163.56 fl which conformed with the report of this study. The PCV range of 26.25 to 34.62% obtained for birds in this study was lower than the average value of 38.52% reported by Awotuyi (1990) for adult domestic chicken. The observed difference in the value of PCV may be due to breed and age difference. As earlier documented (Kubena et al., 1972; Oyewale, 1987) that birds in the tropics tend to have lower haemoglobin values compared to those reared in the temperate.

The noted increase in the values of PCV and hemoglobin of birds on Treatment 5 could be attributed to the effect of vitamin C in protecting the membrane integrity of the erythrocytes as earlier reported (Candan et al., 2002; Adenkola et al., 2010). This function directly affected the haemoglobin concentration of birds on Treatment 5 which was higher than the values obtained for other groups. The increase in haemoglobin concentration could also be attributed to the role of vitamin C in increasing the absorption of iron from digestive system (Harper et al., 1979)

Table 4 shows the serum metabolites and enzymes of heat stressed broiler given water supplemented with electrolytes and ascorbic acid at day 28. The values obtained for serum enzymes did not vary significantly (p>0.05). However, significant variations (p<0.05) were observed in the serum total protein and albumin values. Birds on treatment 3 had the higher total protein value which was not significantly different (p>0.05) from values obtained for birds on treatments 1, 2 and 5. Significant variations (p<0.05) were also observed in the values obtained for albumin. Birds on treatment 5 had higher value of albumin but not significantly different (p>0.05) from the values obtained for birds on treatments 1, 2 and 3.

Serum metabolites and enzymes of heat stressed broiler treated with electrolytes and ascorbic acid at day 49 is shown in Table 5. Cholesterol, triglyceride, total protein, globulin and alanine aminotransferase values of birds on different treatments were not significantly affected by supplemental electrolytes and ascorbic acid. Significant difference (p<0.05) was observed in the

Table 4. Serum metabolites of heat stressed broiler treated with electrolytes and ascorbic acid at week four.

Parameter	T1	T2	T3	T4	T5	SEM
Glucose (g/dl)	196	181	182	187	184	3.18
Cholesterol (mg/dl)	96.4	96.3	106	103	98.0	3.02
Triglycerides (mg/dl)	299	295	306	282	309	0.08
Total protein (g/dl)	3.29[ab]	3.42[ab]	3.78[a]	2.91[b]	3.47[ab]	0.04
Albumin (g/dl)	2.42[ab]	2.43[ab]	2.53[a]	2.22[b]	2.65[a]	0.08
Globulin (g/dl)	0.87	0.99	1.25	0.69	0.82	1.85
Aspartate Aminotransferase (AST) (i.u/l)	102	104	106	111	110	0.19
Alanine aminotransferase (ALT) (i.u/l)	7.04	7.34	7.84	7.58	7.50	5.23

[a,b]Means on the same row with different superscripts are significantly different(p<0.05). SEM, Standard error of mean. T1, control; T2, ammonium chloride; T3, sodium bicarbonate; T4, calcium chloride; T5, ascorbic acid.

Table 5. Serum metabolites of heat stressed broiler treated with electrolytes and ascorbic acid at week 7.

Parameter	T1	T2	T3	T4	T5	SEM
Glucose(g/dl)	180[a]	174[ab]	143[b]	172[ab]	157[ab]	3.80
Cholesterol(mg/dl)	128	135	134	124	113	4.73
Triglyceride(mg/dl)	243	250	271	260	263	11.7
Total protein(g/dl)	4.19	3.83	3.89	3.58	3.79	0.13
Albumin(g/dl)	1.89[a]	1.78[ab]	1.58[ab]	1.25[b]	1.59[ab]	0.07
Globulin(g/dl)	2.29	2.05	2.31	2.32	2.20	0.13
Aspartate aminotransferase AST (i.u/l)	132	125	129	136	130	2.11
Alanine aminotransferase ALT (i.u/l)	9.78[ab]	7.68[b]	9.46[ab]	10.3[ab]	11.8[a]	0.35

[a,b]Means on the same row with different superscripts are significantly different (p<0.05), SEM, Standard error of mean. T1, control; T2, ammonium chloride; T3, sodium bicarbonate; T4, calcium chloride; T5, ascorbic acid.

glucose values of birds on treatments 1 and 3. Birds on treatment 1 had higher (p<0.05) glucose value while birds on treatment 3 had the lower glucose value which was similar (p>0.05) to those obtained for treatments 2, 4 and 5. According to Borges et al. (2007) increase in glucose concentration was one of the direct responses of birds to greater adrenaline, noradrenalin and glucocorticoid secretion in stressful conditions which was needed to prepare birds for a "fight and flight" response. The lower values of serum glucose observed for birds on treatments 2, 3, 4 and 5 compared with control group indirectly connoted reduced level of stress among the birds brought about by the effect of supplemental electrolytes and ascorbic acid. The lowest value of glucose obtained for birds on treatment 3 indicated that the said treatment had greater mitigating effect on heat stress.

Significant variation (p<0.05) was also observed in the albumin values, with birds on Treatment 1 having higher value which was similar (p>0.05) to the values obtained for birds on treatments 2, 3 and 5. Significant difference (p<0.05) was observed in the values of alanine amino transferase (ALT) with treatment 5 having higher value which was significantly different (p<0.05) from the value obtained for birds on treatment 2. ALT catalyzes the transfer of an amino group from alanine to α-ketoglutarate, the products of this reversible transamination

reaction being pyruvate and glutamate. This index has been an indication of liver function test and elevated levels monitored liver malfunction (Murray et al., 1990).

Conclusion

Erythrocytic indices of birds reflected the positive enhancing effects of supplemental ascorbic acid. Also, the obtained pattern of serum glucose concentration was indicative of the mitigating effect of both supplemental electrolytes and ascorbic acid on stressed birds which was more pronounced in birds on sodium bicarbonate.

REFERENCES

Adenkola AY, Kaankuka FG, Ikyume TT, Ichaver IF. Yaakugh IDI (2010). Ascorbic acid effect on erythrocyte osmotic fragility, haematologicalparameters and performance of weaned rabbits at the end of rainy season in Makurdi Nigeria. J. Anim. plant Sci. 9(1):1077-1085

Awotuyi EK (1990). A study of baseline haematological values of domestic and commercial chickens in Ghana. Bull. Anim. Health. Prod. Afr. 38:453-458.

Borges SA, Fischer da Silva, Maiorka A (2007).Acid- base balance in broilers. World's Poultry Sci. J. 63:73-81

Branton SL, Reece FN, Deaton JW (1986). Use of ammonium chloride and sodium bicarbonate in acute heat exposure of broilers. Poult. Sci. 65(9):1659-1663.

Candan F, Gultekin F, Candan F (2002). Effect of Vitamin C and Zinc on fragility and lipid peroxidation in Zinc-deficient haemodialysispatent. Cell Biochem. Function 20:95-98.

Charles DR (2002) Responses to the thermal environment.In: Poultry Environment Problems, A guide to solutions (Charles, D.A. and Walker, A.W. Eds.), Nottingham University Press, Nottingham, United Kingdom. pp. 1-16.

Cier D, Rimsky Y, Rand N, Polishuk O, Gur N, Shoshan AB, Frish Y, Moshe AB (1992). The effect of supplementing ascorbic acid on broiler performance under summer conditions. Proc. 19th World Poult. Cong. Amsterdam, Netherlands, September 19-24(1):586-589.

Gray D, Butcher DV, Richard M (2003). Heat stress management in broilers.Veterinary Medicine large Animal clinical science department, Florida cooperative Extension Service.Institute of food and Agricultural Science, University of Florida. http:/edis. IfasUfl. Edu.

Harper HA, Rodwell, Mayer VW, Peter A (1979). Review of physiological chemistry. 17th (ed.), London. pp. 159-160.

Iheukwumene FC, Herbert U (2003). Physiological responses of broiler chickens to quantitative water restrictions: Haematology and Serum biochemistry. Int. J. Poul. Sci. 2(2):117-119.

Islam MS, Lucky NS, Islam MR, Ahad A, Das BR, Rahman MM Siddiui MSI (2004). Haematology parameters of Fayumi, Assil and local chikens reared in Sylhet Region in Bangladesh. Int. J. Poult. Sci. 3:144-147.

Jain NC (1986). Scanning electron micrograph of blood cells. In: Schalm's Veterinary haematology. 4th ed. P Lea &Febiger. Philadelphia 4:63-70.

Junqueira OM, Fonseca LEC, Araujo LF, Sakomura NK, Faria DE.(1999) Desempenho e parâmetroshematológicos de frangos de cortesubmetidos   à restriçãoalimentarrecebendosoluçõeshidroeletrolíticas.RevistaBrasileira de CiênciaAvícola 1(1):55-59. In: Junqueira OM, Fonseca LEC,Araújo LF, Duarte KF, Araújo CS da S, Rodrigues EAP (2003). Feed restriction on performance and blood parameters of broilers fed diets with different sodium levels. Rev. Bras. Cienc. Avic. 5:2 (Campinas May/Aug. 2003)

Kohn RA, Allen MS (1995). Enrichment of proteolysis activity relative to nitrogen in preparations from the rumen for in vitro studies. Anim. Feed Sci. Technol. 52:1-4

Kubena, LF, May JD, Reece FN, Deaton JW (1972). Hematocrit and hemoglobin of broilers as influenced by environmental temperature and dietary iron level. Poult. Sci. 51:759.

Lin H, Jiao HC, Buyse J, Decuypere E (2006). Strategies for preventing heat stress in poultry. World's Poult. Sci. J. 62:71-85.

Maxwell MH, Robertson GW, Spence S, McCorquodale CC (1990) Comparison of hematological values in restricted and ad libitum-fed domestic fowls: red blood cell characteristics. British Poult. Sci. 31:407-413.

Mitruka BM, Rawnsley HN (1977). Clinical Biochemical and Haematological Reference values in normal experimental Animals. Masson, New York.

Murray RK, Mayas AM, Granner DK, Rodwel VW (1990). Harper Biochemistry.BarisKitabevi/Appleton and Lange, USA.

Oyewale JO (1987). Haematological studies on apparently healthy Nigerian domestic chickens (Gallus domesticus). Bull. Anim. Health. Prod. Afr. 35: 108. In: Awotuyi EK (1990). A study of baseline haematological values of domestic and commercial chickens in Ghana. Bull. Anim. Hlth. Prod. Afr. 38:453-458.

Pardue SL, Thaxton JP (1986). Ascorbic acid in poultry.A review.World's Poult. Sci. 42:107-123.

Peter T, Biamonte GT, Doumas BT (1982). Protein (Total protein) in serum, urine and cerebrospinal fluids; Albumin in serum. In: Selected method of clinical chemistry p.9 (Paulkner WR, Meites S Eds) American Association for clinicalchemistry,Washington, D.C.

Rej R, Holder H (1983). Aspartate transaminase. In: Methods of Enzymatic Analysis. 3rd ed (H.U.Bergmeyer, J. Bergmeyer and M.Grassl, Eds). WeinheinVerlag- Chemie 3:416-433.

Roschlan P, Bernet E, Guber W (1974). Enzymatischebestimmung des gesant cholesterium in serum. J. Clin. Biochem. 12:403-407.

SAS (1999).SAS/STAT User's Guide.Version 8 for windows. SAS Institute Inc., SAS Campus Drive, Cary, North Carolina, USA.

Schalm OW, Jain NC, Carrol EG (1975). Veterinary haematology. Leo and Febiger,Philadelphia. pp.140-152.

Siegel HS (1995). Stress, strains and resistance.Br. Poult. Sci. 36:3-22.

Teeter RG, Smith MO (1985). High chronic ambient temperature stress effects on broiler acid base balance and their response to supplemental ammonium chloride, potassium chloride and potassium carbonate. Poult. Sci. 65(9):1777-1781.

Tras B, Inal F, Bas AL, Altunok V, Elmas M, Yazar, E (2001). Effects of continuous supplementations of ascorbic acid, aspirin, vitamin E, and selenium on performance, immune response, and some biochemical parameters under normal environmental and management conditions in broilers. Arch Geflugelk. 65: 187-192.

Vecerek V, StrakovaE, Suchy P, Voslarova E (2002). Influence of high environmental temperature on production and haematological and biochemical indexes in broiler chickens. Czech J. Anim. Sci. 47:176-182.

29

Effect of feeding graded levels of camel blood-rumen content mixture on nutrient digestibility and carcass measurement of growing rabbits

G. Mohammed, S. B. Adamu, J. U. Igwebuike, N. K. Alade and L. G. Asheikh

Department of Animal Science, University of Maiduguri, Maiduguri, Nigeria.

A ten-week feeding trial was conducted to investigate the effect of camel blood-rumen content mixture (CBRCM) on the nutrient digestibility and carcass measurement of cross-bred (Dutch x New Zealand) growing rabbits aged between five and seven weeks. Forty five rabbits were randomly allocated to five dietary treatments of nine per treatment in a complete randomized block design experiment. The CBRCM was included at 0, 10, 20, 30 and 40% levels in diets 1, 2, 3, 4 and 5 respectively. The response showed that the digestibility of dry matter, crude protein, crude fibre, ether extract and Total ash were significantly affected (P<0.05) by the test material in the diets but only nitrogen-free extract (NFE) digestibility was not significantly affected (P>0.05) by the test material in the diets. The slaughter weight, dressed weight, dressing percentage, rack, thighs, head, tail, skin, feet, heart, lungs, kidneys, caecum, large intestine, small intestine, stomach and body length were not significantly (P>0.05) different among the treatment groups. The shoulder, loin and liver were significantly (P<0.05) different amongst the treatment groups. The study indicated that CBRCM can be included in rabbit diet up to 40% inclusion level.

Key words: Growing rabbits, camel blood-rumen content mixture, nutrient digestibility, carcass characteristics.

INTRODUCTION

In Nigeria as in most developing countries the daily intake of animal protein (3.24 g) falls grossly short of the recommended 27 g animal protein per day (FAO, 1993). This observed low animal protein consumption may be attributed to the declining animal protein production as a result of high cost of livestock feed which usually accounts for up to 70% of total cost of production (Ijaiya and Awonusi, 2002). Therefore, effort targeted at reducing the cost of livestock feed should be possible remedies. The production of highly prolific and fast maturing animal such as rabbit can provide remedy for the shortage of animal protein for human consumption, because livestock like cattle, pigs, goat and sheep take

longer period to mature (Ajayi et al., 2007).

Maize account for about 45% of the diet of rabbit, hence such price changes will induce a classic increase in the price of finished feed as reported by Adeniji (2008). There is need to research into the use of cheap and unconventional feed ingredients for compounding rabbits diets so as to boost rabbit production. One of such unconventional ingredient is blood-rumen content mixture which is by-products from abattoir and slaughter houses. Investigation had revealed the composition and potential of blood-rumen content mixture as a good source of protein in monogastric (Adeniji and Balogun, 2001). The study was aimed at determining the effect of feeding

Table 1. Ingredients compositions of the diets.

Ingredients (%)	Level of CBRCM in the diets (%)				
	0	10	20	30	40
Maize	40.98	39.12	37.41	35.24	24.35
Wheat offal	17.00	17.00	17.00	17.00	17.00
CBRCM	0.00	10.00	20.00	30.00	40.00
Groundnut cake	23.37	15.23	6.94	0.00	0.00
Fish meal	3.00	3.00	3.00	2.11	3.00
Groundnut haulms	13.00	13.00	13.00	13.00	13.00
Bone meal	2.00	2.00	2.00	2.00	2.00
Common Salt (NaCl)	0.50	0.50	0.50	0.50	0.50
Premix*	0.15	0.15	0.15	0.15	0.15
Total	**100.00**	**100.00**	**100.00**	**100.00**	**100.00**

* Premix (grow fast) manufactured by Animal Care Service Consult (Nig) Ltd. Lagos, Supplied the following per kg of premix: Vitamin A, 5000,00 IU; Vitamin D₃ 800,000 IU; Vitamin E, 12,000 mg; Vitamin K, 1,5000 mg; Vitamin B₁, 1,000 mg; Vitamin B₂, 2,000 mg, Vitamin B₆, 1,500 mg; Niacin, 12,000 mg; pantothenic acid, 20.00 mg; Biotin,10.00 mg; Vitamin B₁₂, 300.00 mg; folic acid, 150,000 mg; choline, 60,000 mg; manganese, 10,000 mg; iron;15,000 mg, zinc 800.00 mg; Copper 400.00 mg; Iodine 80.00 mg; cobalt 40 mg; selenium 8,00 mg. CBRCM= Camel blood-rumen content mixture.

camel blood-rumen content mixture on nutrient digestibility and carcass measurement of growing rabbits.

MATERIALS AND METHODS

Experimental animals and management

A total of forty five Dutch x New Zealand white rabbits, aged between five and seven weeks were used for the feeding trial which lasted for 10 weeks. Before commencement of the experiment, a one-week adjustment period was observed. During this period, the rabbits were treated against internal and external parasites by subcutaneous injection of Ivomec (0.2 ml/rabbit). The rabbits were individually weighed and divided into five groups. Each group was replicated thrice with three rabbits per replicate in such way to ensure uniformity of average weight and sex of each group (six males and three females per treatment). The groups were randomly assigned to five dietary treatments. Each rabbit was individually housed in a wire cage measuring 38 × 33 × 45 cm. The cages, in rows, were raised 45 cm above the ground to facilitate cleaning. Each cage cell was equipped with plastic drinkers and metal feeding troughs. The experimental diets (in mash form) and clean drinking water were provided *ad libitum* throughout the experimental period. A total of 100 g feed was supplied to each rabbit per day at rate 50 g in the morning (8.00 am) and 50% in the evening (3.30 pm).

Source of camel blood-rumen content mixture

The camel rumen content was collected in Maiduguri abattoir. Blood was collected from camel in a clean container during slaughter and the blood and camel rumen content weighed in a ratio of 1:3 (that is, 1 kg of blood and 3 kg of rumen content) into a drum. The blood and the rumen content were mixed in the drum and boiled for 30 min with constant stirring to ensure a uniform mixture. The boiled camel blood–rumen content mixture was sun-dried for 5 days on a clean dry slab. The dried sample was ground with a hammer mill and analysed for proximate composition before inclusion into diets.

Experimental diets

The ingredient composition and the calculated analysis of the experimental diets are shown in Table 1. The camel blood–rumen content mixture was incorporated at levels of 0, 10, 20, 30 and 40% in diets 1 (control) 2, 3, 4 and 5 respectively. The diets were formulated to supply 19% crude protein (CP) and 2800-3000 K/cal of ME on dry matter basis.

Digestibility study

The nutrient digestibility study was conducted at the end of the week of the experiment. Faecal samples were collected from three rabbits per treatment (that is, one from each replicate) for a period of seven days using fine wire mesh trays placed under the cage cells. The amount of faeces voided daily was weighed and allowed to dry for 24 h at 80°C in an oven. The dried faecal samples were stored in air -tight bottles for chemical analysis. The proximate composition of the diets and faecal samples were determined according to AOAC (2000).

Carcass measurements

At the end of the experiment, three rabbits (one rabbit from each replicate based on average weight) from each treatment, were selected for slaughter. They were deprived of feed for 12 h as recommended by Joseph et al. (1994) but drinking water was provided. Withholding feed for 12 h before slaughter reduced the volume of gut contents and hence bacteria, and therefore reduced the risk of contamination of the carcass during dressing without adversely affecting meat yield and quality (FAO, 1991; Joseph et al., 1994). The rabbits were weighed in the morning and slaughtered by cutting transversely across the trachea, oesophagus, large carotid arteries and jugular veins to ensure maximum bleeding (Mann, 1960). They were later opened and dressed as described by Blasco et al. (1993). The dressed carcass is the portion of the rabbit remaining after the removal of the head, feet, skin (pelt), tail and visceral organs including kidneys. The dressed carcasses were split into retail cuts such as shoulder/forelegs, thigh/hindleg, rack and loin as described by Blasco et al. (1993). The dressed carcass and the retail cuts were weighed and expressed as percentage of slaughter weight.

Table 2. Proximate composition of experimental diets (on dry matter basis).

Nutrient (%)	Level of CBRCM in the diets (%)						
	0	10	20	30	40	SEM	CBRCM
Dry matter	92.11	91.23	92.01	92.31	92.30	1.03	91.63
Crude protein	19.20	19.01	18.94	18.63	18.24	0.48	36.40
Crude fibre	18.34[b]	19.34[ab]	20.12[a]	20.31[a]	20.43[a]	0.54*	20.43
Ether extract	4.50[a]	3.50[b]	3.40[b]	3.82[b]	3.66[b]	0.19*	4.01
Total Ash	2.00[c]	3.01[b]	3.08[b]	3.07[b]	3.55[a]	0.03*	4.90
Nitrogen-free extract	48.07	46.37	46.47	46.48	46.42	0.45	25.89
ME(Kcal/kg)	3,061.48	2953.57	2,909.51	2,861.02	2,42.96	-	2,819.33

SEM = Standard error of means; NS= Not significant (P>0.05), *= Means in the same row bearing different superscripts differ significantly (P<0.05), - = Not determine, ME = Metabolizable energy calculated according to the formula of Pauzenga (1985): ME = 37x% CP+81x%EE+35.5% x NFE.

Table 3. Mean apparent nutrients digestibility of rabbits fed graded levels of camel blood-rumen content mixture (CBRCM).

Nutrients (%)	Level of CBRCM in the diets (%)					
	0	10	20	30	40	SEM
Dry matter	66.68[a]	64.11[ab]	62.45[ab]	60.35[b]	58.56[b]	1.65*
Crude protein	78.32[a]	75.33[b]	73.03[c]	71.85[d]	70.53[e]	0.23*
Crude fibre	38.54[a]	37.43[ab]	36.28[b]	35.00[c]	34.83[c]	0.39*
Ether extract	70.56[a]	67.73[b]	65.87[c]	65.17[c]	53.60[d]	0.39*
Total ash	63.77[a]	56.07[ab]	57.72[b]	56.31[ab]	54.17[c]	1.21*
Nitrogen-free extract	80.16	86.27	84.46	75.89	75.98	3.83

SEM = Standard error of means; NS= Not significant (P>0.05), * = Means in the same row bearing different superscripts differ significantly (P<0.05).

$$\text{Dressing percentage} = \frac{\text{Dressed carcass Wt. (g)}}{\text{Slaughter weight (g)}} \times 100$$

Statistical analysis

All the data collected were subjected to analysis of variance (ANOVA) using the randomized complete block design (Steel and Torrie, 1980). Means were separated where applicable using the Duncan's multiple range test (Duncan, 1955).

RESULTS AND DISCUSSION

Proximate composition of the experimental diets

The proximate composition of the experimental diets and CBRCM are presented in Table 2. The dry matter (DM) and crude protein (CP) were similar (P>0.05) among the diets; these levels are adequate for growing rabbits as reported by Olumeyan et al. (1995), Mohammed et al. (2005a) and Dairo et al. (2005). Crude fibre (CF) in diets containing 20, 30 and 40% camel blood-rumen mixture (CBRCM) were higher compared to 0% CBRCM diet, but did not differ significantly (P>0.05) from 10% CBRCM diet. The crude fibre content increased linearly with increasing levels of camel blood-rumen content mixtures (CBRCM) in the diets. This was attributed to the fibrous

nature of the CBRCM compared to groundnut cake and maize in the diets. The values obtained (18.24 to 19.20%) were within ranges (15 to 20%) reported to be adequate for weanling rabbits (Cheeke et al., 1982). The control diet (0% CBRCM) had higher level of ether extract (EE) than the CBRCM-based diets. However, the values in all the diets were within the recommended level (≥ 3%) for growing rabbits in the tropical countries as reported by Cheeke (1987). Diet containing 40% CBRCM had higher value while control diet had the lowest total ash (TA) content compared to CBRCM-based diets. Nitrogen–free extract (NFE) did not differ significantly (P>0.05) among the various diets. The metabolizable energy levels of the diets decreased linearly as the level of CBRCM increased in the diets. This was attributed to the higher energy content of groundnut cake and maize compared to CBRCM and the higher fibre level of the CBRCM. The values obtained in the study were similar to the values earlier reported by Mohammed et al. (2005a) and Dairo et al. (2005).

Nutrient digestibility

The apparent nutrient digestibility of rabbits fed graded levels of CBRCM is presented in Table 3. The digestibility of DM, CP, CF, EE and TA were significantly affected (P<0.05) by the test material in the diets. Rabbits fed

Table 4. Effect of varying levels of camel blood-rumen content mixture (CBRCM) on the body components and organs of rabbits expressed as percentage of slaughter weight.

| Parameter | Level of CBRCM in the diets (%) | | | | | SEM |
	0	10	20	30	40	
No. of rabbits slaughtered	3	3	3	3	3	-
Slaughter weight (g)	1320.8	1277.0	1348.3	1200.0	1173.2	71.08
Dressed weight (g)	696.50	637.75	628.75	646.50	635.25	46.57
Dressing percentage (%)	54.33	53.35	55.28	52.82	52.26	5.64
As % of slaughter weight						
Shoulder/forelegs	15.44[a]	15.30[ab]	15.46[a]	13.79[ab]	13.24[b]	0.68*
Rack	5.73	6.64	5.46	5.14	5.46	0.58
Loin	10.66[a]	8.51[b]	10.05[ab]	10.47[a]	9.11[ab]	0.57*
Thighs/Hind legs	22.15	20.01	20.70	19.40	19.41	0.44
Head	9.15	9.14	9.18	9.37	9.63	0.50
Tail	0.37	0.33	0.37	0.33	0.36	0.04
Skin	8.49	8.80	8.30	9.26	7.43	0.59
Feet	2.50	2.58	2.62	2.66	2.41	0.15
Heart	0.26	0.27	0.26	0.25	0.25	0.03
Liver	2.63[b]	2.77[a]	2.75[a]	2.36[b]	2.65[b]	0.07*
Lungs	0.66	0.79	0.72	0.64	0.64	0.05
Kidneys	0.64	0.66	0.63	0.60	0.65	0.03
Caecum	5.49	5.67	5.26	5.04	5.50	046
Large intestine	2.22	2.53	2.48	2.65	2.58	0.61
Small intestine	3.79	3.56	3.80	3.74	3.69	0.40
Stomach	4.45	4.68	4.73	5.16	5.18	0.42
Body length (cm)	28.40	27.92	28.11	27.01	29.00	0.45

SEM = Standard error of means; NS= Not significant (P>0.05), * = Means in the same row bearing different superscripts differ significantly (P< 0.05).

control (0% CBRCM) had significantly (P<0.05) higher DM digestibility than those fed 30 and 40% CBRCM diets. There were no significant differences amongst the rabbits fed control, 10 and 20% CBRCM for DM digestibility. Apparent CP digestibility of the rabbits fed control (0%) was higher than those of rabbits on the CBRCM-based diets. The poorest CP digestibility was recorded in the rabbits fed 40% CBRCM. The values (70.53 to 78.52%) obtained here were higher than values (44.58 to 57.71 and 20.50 to 34.90) reported by Mohammed et al. (2005b) and Adeniji (2008).

Rabbits fed control (0% CBRCM) utilized the CF in their diets better (P<0.05) than those on the 20, 30 and 40% CBRCM diets. There were no significant differences amongst rabbits fed control and 10% CBRCM. The poorest CF digestibility was recorded in rabbits fed 30 and 40% CBRCM diets. Both nitrogen and CF utilization decreased linearly with increasing CBRCM in the diets. This may be attributed to poor utilization of the nutrient as a result of increasing fibre content of the diets. This is in agreement with the report of Adeniji (2008) that linked the low digestibility values for bovine rumen content to its high fibre content.

The rabbits fed control diet showed better EE digestibility than those fed test material in their diets. The poorest EE was recorded by the rabbits fed 40% CBRCM diet. EE digestibility tended to decrease with increasing levels of CBRCM in the diets. This may be attributed to fibrous nature of the CBRCM as earlier reported in Table 2. The decrease in the digestibility of EE with increasing levels of CBRCM in the diets, agree with the work of Igwebuike et al. (1998) who reported that increase in CF levels of the diets depressed EE digestibility. Rabbits fed 0% CBRCM showed better (P<0.05) TA absorption compared to those fed 20 and 40% CBRCM diets. Rabbits fed control did not differ from 10 and 30% CBRCM diets. The poorest TA absorption was recorded in rabbits fed 40% CBRCM diet. The NFE digestibility was not significantly affected by the test material in the diets and thus suggesting efficient utilization of soluble carbohydrates in all the diets. This was in line with the findings of Onifade and Tewe (1993) who reported high digestibility of readily available carbohydrates by rabbits in a feeding trial involving.

Carcass parameter

The carcass parameters presented in Table 4 showed that slaughter weight, dressed weight, dressing percentage, rack, thighs, head, tail, skin, feet, heart, lungs, kidneys, caecum, large intestine, small intestine, stomach and body length were not significantly (P>0.05) different among the treatment groups. This is an indication that the organ developments of the growing rabbits were not compromised by the various levels of CBRCM included in their diets. The shoulder, loin and liver were significantly (P<0.05) different amongst the treatment groups. Rabbits fed control (0% CBRCM) diet had significantly (P<0.05) heavier shoulder than those fed 40% CBRCM diet. There were no significant (P>0.05) differences in shoulder amongst the rabbits fed the control, 10, 20 and 30% CBRCM diets. The weight of the shoulder is a reflection of the heavier slaughter weights of rabbits on these four diets. Rabbits fed control and 30% CBRCM diets had significantly (P<0.05) heavier weight of loin than those on 10% CBRCM diet. All other treatments were similar to the control (0% CBRCM). The liver weight were significant (P<0.05) higher in 10 and 20% CBRCM compared to other groups.

The carcass measurements of rabbits in this study compared favourably with the values reported by Mohammed (2010) who slaughter rabbits of similar weight and ages in the same environment. These also tallies with findings of Mohammed et al. (2005b) who fed goat rumen content to rabbits of similar ages.

Conclusion

From the result of this study, it can be concluded that inclusion of camel blood-rumen content mixture up to 40% level in the diets of growing rabbit will not adversely affect carcass measurement and digestibility of the growing rabbits. However, further studies are needed to evaluate blood parameters and histopathology for health status of rabbits which was not covered in this study.

REFERENCES

Adeniji AA (2008). Replacement value of maize with enzyme supplement decomposed bovine rumen content in the diet of weaner rabbit. J. Anim. Vet. Adv. 3:104-108.

Adeniji AA, Balogun OO (2001). Evaluation of blood–rumen content mixture in the diets of starter chicks. Nig. J. Anim. Prod. 28(2):153–157.

Ajayi AF, Farinu GO, Ojebiya OO, Olayeni TB (2007). Performance of male rabbits fed diets containing graded levels of blood-wild sunflower leaf meal mixture. World J. Agric. Sci. 3(2):250-255.

AOAC (2000). Official Methods of Analysis of Official Analytical Chemists. 17th Edition. Ed. Horwitz, W., Washington, D.C., Association of Official Analytical Chemists. P. 156.

Blasco A, Ouhayoun J, Masoero G (1993). Harmonization of criteria and terminology in rabbit meat research. World Rabbit Sci. 1(1):3–10.

Cheeke PR (1987). Rabbit Feeding and Nutrition. Academic Press Inc. Orlando, Florida, U. S. A .p. 376.

Cheeke PR, Patton NM, Templeton GS (1982). Rabbit Production. The Interstate Printers and Publisher Inc Danvilte, Ilinios, U. S. A.

Dairo FA, Aina OO, Asafa AR (2005). Performance evaluation of growing rabbits fed varying levels of rumen content and blood- rumen mixture. Nig. J. Anim. Prod. 32(1):67 -72.

Duncan DB (1955). Multiple range and multiple F. tests. Biometrics 11:1–42.

FAO (1991). Guideline for Slaughtering Meat Cutting and Further Processing FAO Animal Production and Health Paper No. 91 Food and Agriculture Organisation, Rome Italy. p. 176.

FAO (1993). Production Year book of Food and Agricultural Organization of the United Nations, Rome, Italy 1:17.

Igwebuike JU, Abba I, Msheliza NKA (1998). Effect of feeding graded levels of sorghum waste on the nutrient and mineral utilization of growing rabbits. Res. J. Sci. 4(1,2):46-56.

Ijaiya AT, Awonusi EA (2002). Effect of replacing maize with yam peel meal on the growth performance of weaner rabbits, Proceeding 7thAnnual Conference Animal. Science Association of Nigeria (ASAN) Sept. 16- 19th University of Abeokuta, Nigeria, pp. 161-163.

Joseph KJ, Awosanya B, Adebua AA (1994). The effects of pre-slaughter withholding of feed and water from rabbits on their carcass yield and meat quality. Nig. J. Anim. Prod. 21:164–169.

Mann I (1960). Meat Handling in under developed Countries. FAO Agricultural Development paper No. 70. Food and Agriculture Organisation, Rome, Italy.

Mohammed G, Igwebuike JU, Kwari ID (2005a). Performance of growing rabbits fed graded levels of goat rumen content. Global J. Pure Appl. Sci. 11(1):39 – 43.

Mohammed G, Igwebuike JU, Kwari ID (2005b). Nutrient digestibility, haematological indices and carcass measurements of rabbits fed graded levels of goat rumen content. Global J. Agric. Sci. 4(1):1-6.

Mohammed G, Igwebuike JU, Medugu CI, Kwari ID, Ahamed U (2010). Effect of feeding graded levels of Tigernut (Cyperus esculentus) residue on haematological parameters and carcass characteristic of growing rabbits. Proceeding 15th Conference Animal Science Association of Nigeria. (ASAN) 13 – 15th Sept, 2010. University of Uyo, Nigeria, pp. 227–229.

Olumeyan B, Afolayan SB, Bawa GS (1995). Effect of graded levels of dried rumen ingesta on the performance of growing rabbits fed concentrate diets. Paper presented at the 20th Annual Conference of Nigeria Society Animal Production 26 – 30th March, 1995. Federal University of Technology, Minna, Nigeria.

Onifade AA, Tewe OO (1993). Alternative tropical energy feed resources in rabbits diets: growth performance, digestibility and blood composition. World Rabbit Sci. 1:17-24

Steel RGD, Torrie JA (1980). Principles and Procedures of Statistics. A Biometrical Approach, 2nd Ed. Mc Graw. Hill Book Co. Inc. NewYork. p. 633.

30

Clinico- sonographic evaluation based surgical management of urolithiasis in young calves

Mohmmad Aarif Khan[1], D. M. Makhdoomi[2], Mohsin A Gazi[2], G. N. Sheikh[2] and S. H. Dar[2]

[1]Sher-E-Kashmir University of Agricultural Sciences and Technology of Kashmir, India.
[2]Sher-E-Kashmir University of Agricultural Sciences and Technology of Shuhama, India.

The objectives of study were to screen the calves as stone and non-stone formers, to manage the calves with obstructive urolithiasis by diversified surgical approaches based on clinical and sonographic assessment. The study was conducted in bovine calves (n=27), manifesting clinical urolithiasis. All these animals and 52 calves of age group 3 to 18 months presented with any form of ailments were included in screening as normal and stone former. Anamnesis, clinical and physical examinations, ultrasonography, peri operative and post operative observations and, complications were recorded. On the basis of clinical symptoms, ultrasonographic findings of urinary system, duration of obstruction and position of calculi, were divided into five groups of 6 animals each except group T having 3 animals and most suited surgical procedure, that is, in dwelling Normograde catheterization, tube cystostomy, cystostomy and uretotomy, trocarization and Percutaneous catherizati on and abdominocentesis were awarded. The highest occurrence was found in the age group of under one year (60%) followed by 40% above one year. Cross bred calves were most affected (70%). Duration of illness in calves varied from 24 to 120 h. The rectal temperature increased slightly at different postoperative intervals in all the groups. The heart rate and respiration rate showed a gradual decrease at different postoperative intervals and became normal towards end of study. In 90% cases small multiple irregular and smooth concretions were retrieved. The calves suffering from obstructive urolithiasis had the history of feeding on diets containing wheat bran, commercial cattle feed, rice bran and rice straw few days before manifesting symptoms. Trocharization followed by peritoneal catheterization helped to prolong the life of the severely diseased animals and made them more suitable for major surgery subsequently.

Key words: Urolithiasis, abdominocentesis, trocarization, ultrasonography.

INTRODUCTION

Male bullocks continue to be exclusive source of agricultural power in Kashmir valley as mechanization in agriculture using tillers is not possible due to terrain of land. Urolithiasis a disease of bullocks is now exceptionally a disease of male calves in Kashmir. It is invariable accompaniment of series of events starting with obstruction to free flow of urine, increased intra cystic pressure, vesico-urethral reflex, hydronephrosis and alterations in urea nitrogen, creatinine, calcium, phosphorous and other metabolites. Urolithiasis results in considerable economic losses to the owner in terms of death of the animal, lower weight gain, medicenal and surgical treament costs, loss of precious germplasm. The calculi formation results from a combination of physiological, nutritional and managemental factors. The factors that favour development of obstruction include anatomical long convoluted urethra-sigmoid flexures, urethral process in small ruminants, surgical factors early castration and exogenous factors, estrogens as growth promoting implants. Excessive or imbalance intake of minerals has been reported as one of the factors for formation of urinary calculi (Larson, 1996; Lonsdale et al., 1968). The incidence among male calves is enormously as high as 12% (Makhdoomi and Sheikh, 2008).

Diagnosis of Urolithiasis and rupture of the urinary bladder or urethra in cattle, sheep and goats is based on the clinical symptoms, physical examination, adbomeninocentesis, radiography and sonograph. Treatment comprises use of antibiotics, muscle relaxants (Gasthuys et al., 1993), urine acidifiers, litholytic drugs (Joshi et al., 1988) and dialysis (Radostitis et al., 2000). Complete obstructive urolithiasis demands surgical treatment (Radostitis et al., 2000). The surgical interventions include penile catheterization (Winter et al., 1987), removal of urethra or cystic calculi (Gera and Nigam, 1980), urethrostomy and bladder fistulation (Lund vall, 1974), intrapelvic cystic catheterization and penile amputation (Jenning, 1984; Winter et al., 1987. The present paper documents the clinico-sonographic evaluation based surgical managemnt urolithiasis in calves.

MATERIALS AND METHODS

The study was conducted at University Veterinary Clinical Services Complex, for a period of nine (9) months in bovine calves (n=27), aged 3 to 18 months manifesting clinical urolithiasis. All the 27 animals with urolithiasis and 52 calves of age group 3 to 18 months presented with any form of ailments were included in screening as normal and stone former by method described by Teotia, (1975).The systematic and complete pre-operartive evaluation and magement of the cases is designated here under.

Anamnesis

A record of previous and present history was taken. The previous history included questions eliciting information about previous treatment, nutritional history related to type of feed, change of feed. Present history included present ailment and symptoms. The clinical examination done, included status of the eyeballs, visible membrane with emphasis on urinary bladder examination regarding pattern of urination, bladder distension, ruptured and intact bladder.

Clinical observations

The clinical parameters recorded at day zero, 8, 16 and at the removal of catheter and included heart rate, respiratory rate, rectal temperature (°F), skin fold persistence test time (seconds) and extent of dehydration (Radostitis et al., 2000).

Physical examination

The urine was collected aseptically either by aseptically by percutaneous /cystocentesis in intact bladder cases and by abomenicentosis in cases with ruptured urinary bladder, using spinal needle. The urine was collected in sterile test tubes and centrifuged at 3000 rpm for 10 min. One drop of the sediment was discarded and the sediment one drop was taken on a slide for examination of casts and crystals.

Grouping

On the basis of clinical symptoms, ultrasonographic, findings of urinary system, duration of obstruction and position of calculi and kidney function tests, groups A, B, C, D and T consisting of 6 animals each except group T having 3 animals, as follows.

Group A (Alert clinically): They included animals with urinary obstruction of 24 h duration. They were clinically alert with viable reflexes, plasma urea nitrogen up to 50 mg/dl and creatinine 2 to 3 mg/dl. They had intact bladder and the calculi were lodged in urinary bladder.

Group B (Below danger line): They included animals with urinary obstruction of 48 h duration, were clinically alert with sluggish reflexes, plasma urea nitrogen up to 50 to 100 mg/dl and serum creatinine 3 to 4 mg/dl respectively, with intact bladder (distended) and calculi were lodged in the neck of urinary bladder.

Group C (Critical): They included animals with urinary obstruction of 96 h duration. They were clinically dull with poor reflexes with plasma urea nitrogen above 100 mg/dl but up to 150 mg/dl and serum creatinine 4 to 5 mg/dl. The animals had ruptured bladder and the calculi were lodged in the neck of urinary bladder and sigmoid flexure.

Group D (Danger line): They included animals with urinary obstruction of 120 h duration. They were clinically recumbent and areflexic with plasma urea nitrogen above 150 mg/dl up to -200 mg/dl and serum creatinine 5 to 6 mg/dl respectively. The animals had ruptured bladder and calculi lodgement was in the neck of urinary bladder, ischial urethra and sigmoid flexure.

Group T (Terminal): They included animals with urinary obstruction beyond 120 h duration. They were clinically recumbent and grossly are flexic with plasma urea nitrogen above 250 mg/dl and serum creatinine 7 mg/dl. The animals had ruptured bladder and calculi lodgement was in the neck of urinary bladder, ischial urethra, sigmoid flexure and penile urethra.

Ultrasonographic examination

Ultrasonography examination for the urinary conducts was done using a real time, B mode diagnostic ultrasound scanner (Sonaliza-32, Larson and Turbo) equipped with linear array 5 MHz linear probe. The animals were subjected to ultrasonography detailed as: The animals of Group A & B were subjected to ultrasound scanning in ventrodorsal approach with animal in supine recumbency, Group C animals were scanned in ventrodorsal approach in standing position, Group D animals in ventrodorsal, left lateral and trans-rectal approach in standing position however the animal of group T in terminal stage and grossly areflexic and hence were not subjected to sonography.

Surgical interventions

Preoperatively the dehydrated animals were given dextrose saline as per the dehydration status (Radostitis et al., 2000). The status of hydration was checked by status of eyes and skin fold test. The animals were allowed to stabilize and prepared for surgery at the earliest as per standard procedures.

Group A animals were subjected to normograde indwelling catheterization using polyvinyl chloride urinary catheter. The group B animals were divided into two subgroups of animals each

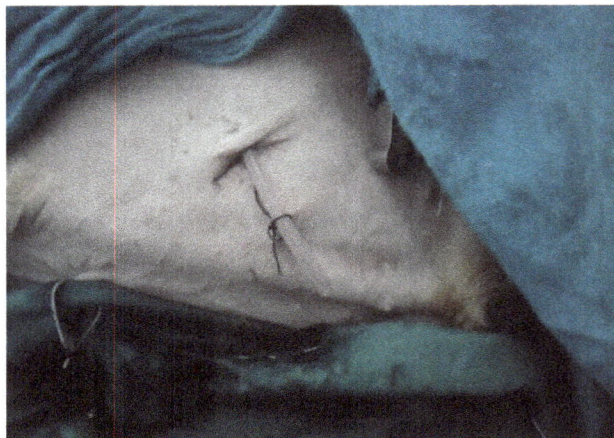

Figure 1. Tube cystotomy paramedian approach.

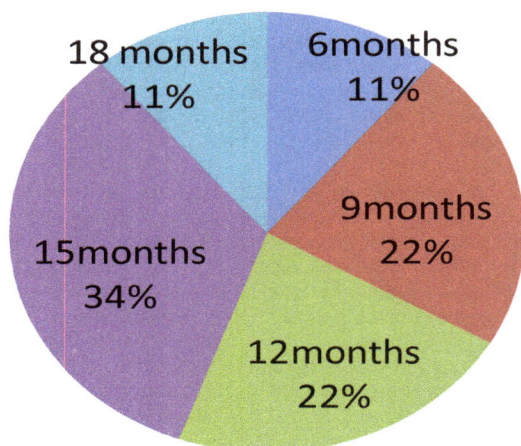

Figure 2. Age wise break up of cases.

subjected to tube cystotomy using Foleys catheter by two different approaches. In one subgroup para-median approach (Figure 1) and in another subgroup left paralumbar fossa was adopted. The group C animals were subjected to cystotomy and urethrotomy. The cystotomy was done using left supra pubic approach. The operation was done by standard procedure and the wound was sutured and sealed. Post-operatively animals received injections of Ampicillin-cloxacillin[1] and Meloxicam[2] at 12 and 0.5 mg/kg b.wt .respectively intravenously for 7 days.   roup D   animals were subjected   to   rocharization and percutaneous   peritoneal catheterization. The group T was recumbent and comatose, abdominocentesis was attempted.

Intraoperative and post operative observations

The observations were made by assessing the conditions of urinary bladder, location of calculi, technical ease, difficulty encountered and re-invasive surgery needed if any. The animals were examined

[1] and [2] AC-Vet Forte, 3g vial, and Melonex, 30 ml vial.Intas Pharmaceuticals Ltd, 2nd floor Ashram road, Ahmadabad- 380009, INDIA.

clinically 24 h and at 8 and 16 days and at the time of removal of catheter for vital signs like rectal temperature, heart rate, respiration rate, uremic breath and capillary refill time. The wound sites were examined for complications if any and urine dribbling free flow from the natural orifices. The catheter was checked for leakage and patency. Minimal invasive surgery as demanded by particular animal was done and the patient discharged under a post-operative advisory note.

Statistical analysis

The data so procured were classified and subjected to statistical analysis. The inferences were drawn using analysis of variance (ANOVA) and Duncan's Multiple Range test (Snedecor and Cochran, 1976).

RESULTS

Screening of the calves

A total of 27 aging 3 to 18 months (Figure 2) calves presented to the Teaching Veterinary Clinical Service Complex of the University for the Treatment of obstructive urolithiasis conditions were screened as positive stone formers (100%). A total number of 52(A) calves were presented for various treatments other than urolithiasis. Within 3 to 4 months, from (A), 35 cases were reported to the clinics and out of them 19 were positive for urolithiasis. The percentage of positive cases from "A" was 54.28% and 8 calves were from these animals treated, included in total number of 27 cases, which form the material of this study (Table 1). Thus screening can be a test for early detection of urolithiasis to the level of 67% accuracy. The microscopic examination of the urine collected at days 0, 8, 16 and at the time of removal of catheter for observation of casts and crystals revealed that heavy score of crystals was recorded in the animals of all groups on the day zero without any significant difference.

Postoperatively   the crystal   showed   progressive declining trend irrespective of the severity of the disease.

Anamnesis

The history record revealed that occurrence of the obstructive urolithiasis was age related. The age of the calves varied from 3 to 18 months. A study of age break up of cases revealed that 60% calves were in the age group up to 12 months and 40% above 12 months. Of all the cases presented with urolithiasis were Cross bred Jersey calves 70%, while local non-descript calves constituted 10% and only 20% cases were Frisian crosses. All these calves were uncastrated males except one calf which was female. The different feeds/fodders given to the calves of the study included wheat bran, rice bran, rice straw, commercial cattle feed, oats, oil cakes, soybean straw alone or in combination. Wheat bran alone

Table 1. Showing screening of calves as stone and non stone formers.

The animals without any clinical manifestation of clinical Urolithiasis but presented with other diseases	52
Number of animals that show clinical manifestation of Urolithiasis	24
Animals that were positive (stone former)	24
Percentage of positive animals that show clinical Urolithiasis	10%
Animals presented at earlier (normal), manifested clinical Urolithiasis	35
Percentage of positive cases	67.30%

was given to (70%) of the cases. Wheat bran was provided in combination with rice straw to 30% of the cases. A common factor recorded was that the calves were feed concentrates at too early age before their rumen started functioning.

Previous treatment given to calves revealed that, about 50% cases received Frusemide injections either from owner himself/herself or by the Vet or Paravet and another 50% received sedatives, analgesics and tablet Cystone[1].

Clinical observation

Groups A and B were clinically alert, group C were dull and depressed whereas groups D and T were recumbent. Results depicting Mean ±SE of vital signs like rectal temperature, heart rate, respiration rate, and capillary refill time were recorded and are presented in Table 2.

The rectal temperature at day zero was slightly higher in the animals which were clinically alert rectal temperature (102.86±0.16°F) and below danger line (102.60±0.33°F). However, the rectal temperature started declining to normal with the progression of the disease and it decreased significantly (p<0.05) from groups A and B to groups D and T.

At 0 day, tachycardia was recorded in all the groups, which varied from 115.83 to 129.00 beats per minute. With the progression of recovery from the disease, the heart rate in different groups at days 8th, 16th and on the day of removal of catheter varied between 78.66±1.50 and 93.66±1.00; 83.16±0.83to 94.00±0.16 and 82.00±0.50 to 83.33±1.00. Results from Table 2 revealed that at the time of catheter removal, the heart rate of the claves came within physiological limits irrespective of severity of the disease.

In all the cases in general there was significant (p<0.05) decreasing trend of heart rate post-operatively till day 8th, 16th and up to the end of study where heart rate touched near normal (Table 2).The capillary refill time in different groups varied between 2.33±0.16 and 6.33±0.33). It showed an increasing trend in the animals with different levels of severity of disease. On 8th post-treatment, it decreased from 2±0.00 to 2.83±0.16 s. The

capillary refill time fell within range in all the groups except in group D. On day 16th, this group also showed reduction in capillary refill time which was comparable to rest of the groups. There was significant (p<0.05) reduction in the capillary refill time from day 8th toward normal.

Ultrasonographic findings

Results of sonography in the groups under study revealed that there was intact bladder in group A and B as evident by the round hypo echoic image on bladder (Figure 3). It was however distended in group B. Sonography revealed a mild seepage of urine into the pelvic cavity, uroperitoneum, as evident by the hypo echoic image with floating of intestinal loop. The urethra was intact with mild urethritis. The concretions and calculi were located in the lumen and towards the neck of urinary bladder as evident by the multiple unevenly spread tiny hyper echoic patterns (Table 3).

In group C and D showed hyper echoic wall of the bladder. The tear was on the dorsal side in group C and at the neck in group D. There was moderate to severe seepage of urine into the pelvic cavity, uroperitoneum, as evident by the hypo echoic image. The urethra was ruptured in both the groups with severe urethritis and subcutaneous urine accumulation. The concretions and calculi were located in sigmoid flexure, ischial urethra, pelvic urethra and the penile urethra. Whole of the tract was occupied with the concretions as evident by the multiple unevenly spread tiny hyper echoic patterns (Table 3).

Perioperative results

Perioperative results were compared with the sonographic findings.

Group A and B

The urinary bladder was intact and shows sserosal haemorrhages. There were concretions the lumen of the bladder. The distension was however more in group B animals. In group C animals bladder was ruptured on its dorsal aspect, tear measuring about 3 inches in length.

---

[1] *Himalyan drug ,manufactured by Indian Herbs Limitted,Saharanpur,India.

Table 2. Mean±SE Rectal temperature (F°), heart rate (beats/minute), respiratory rate (per minute) and capillary refill time (in seconds) at different intervals in calves with obstructive Urolithiasis.

Groups	Parameter	0 day	8 day	16 day	*At the time of removal of catheter
Alert clinically (Group A)	Temperature (°F)	102.86±0.16[aA]	101.33±0.58[aB]	100.33±0.33[aB]	100.96±0.13[aB]
	Heart rate	115.83±2.83[aB]	88.00±1.33[aA]	85.33±0.33[aA]	83.33±0.33[aA]
	Respiration rate	43.16±0.83[aC]	14.66±1.50[aB]	11.33±0.66[abA]	12.16±0.50[aA]
	Capillary refill time	2.33±0.16[aC]	2.00±0.00[aB]	1.00±0.00[aA]	1.00±0.00[aA]
Below danger line (Group B)	Temperature (°F)	102.66±0.33[aA]	101.33±0.41[aAB]	100±0.66[aB]	101.5±0.25[aAB]
	Heart rate	117.66±2.66[abC]	91.16±0.66[aB]	87.33±2.00[aaB]	83.66±0.66[aA]
	Respiration rate	43.83±1.00[aB]	13.16±1.50[aA]	10.50±0.16[aA]	12.66±0.66[aA]
	Capillary refill time	2.83±0.33[abB]	2.66±0.33[abB]	1.00±0.00[aA]	1.00±0.00[aA]
Critical (Group C)	Temperature (°F)	100.91±0.41[bA]	101.25±0.41[aA]	100.83±0.33[aA]	100.75±0.25[aA]
	Heart rate	120.66±2.66[abB]	94.00 ±0.16[aA]	78.66±1.50[bA]	82.00±0.50[aA]
	Respiration rate	45.66±1.00[aB]	13.83±1.50[aA]	12.33±0.83[abA]	12.83±0.66[aA]
	Capillary refill time	3.50±0.50[bC]	2.00±0.00[aB]	1.00±0.00[aA]	1.00±0.00[aA]
Danger line (Group D)	Temperature (°F)	94.83±0.66[cB]	91.33±0.66[bA]	100.06±0.03[aC]	100.16±0.16[aC]
	Heart rate	124.16±4.33[abC]	93.66±1.00[aB]	83.16±0.83[aA]	83.33±1.00[aA]
	Respiration rate	56.00±1.16[bB]	15.16±1.33[aA]	13.16±0.83[bA]	12.50±0.16[aA]
	Capillary refill time	4.50±0.16[cD]	2.83±0.16[cC]	2.00±0.00[aB]	1.00±0.00[aA]
Terminal (Group T)	Temperature (°F)	88.33±1.66[d]	-	-	-
	Heart rate	129±4.96[b]	-	-	-
	Respiration rate	57.00±0.66[b]	-	-	-
	Capillary refill time	6.33±0.33[d]	-	-	-

*Above thirty days. Means with different superscripts differ significantly (p<0.01). Small letters show comparison between groups. Capital letters show comparison between treatments.

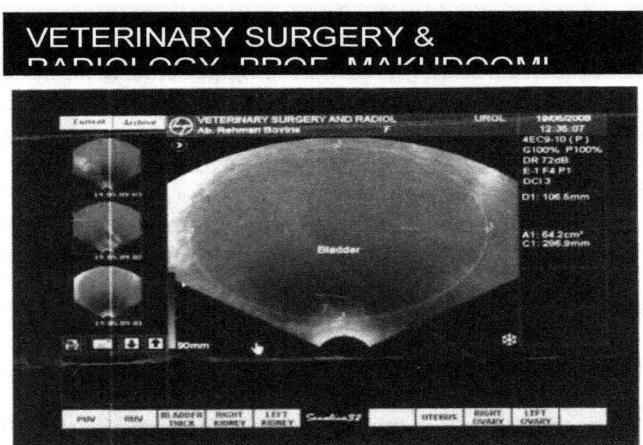

Figure 3. Sonogram showing distended intact bladder at 48 h obstruction.

Extensive haemorrhages on the serosal surface of the bladder were seen. There were concretions in the neck of the bladder, pelvic urethra and ischial urethra and group D animals showed ruptured bladder on the neck, tear measuring up to 4.5 inchs in length. There were concretions in the neck of the bladder, pelvic urethra and ischial urethra, sigmoid flexure and penile urethra.

Surgical interventions

The institution of surgical treatment in different groups was found more suited to clinical status, levels of clinical stress, duration of obstruction and ruptured or intact bladder. Earlier the time of diagnosis and presentation of the case for treatment earlier was the recovery. The animals with more derangement of clinical status received surgical approach with less intervention and duration The time of initiation of dribbling of urine in the animals of group A was at the time of accomplishment of the surgery and the average time of initiation of urination through natural orifice was 16 days (19 days in case of Para lumbar approaches) treated by tube cystotomy. In groups C and D the time of imitation of urination was 22 and 26 respectively. The free flow of urine through the external urethral orifice could be due to interplay of many factors. Reduction in inflammation and urethral spasm by administration of anti-inflammatory drugs, drying up of

Table 3. Status of urinary conducts and position of calculi in groups A, B, C and D as revealed by ultrasonography.

Sonographic findings	
Disease conditions	**Groups**
Cystitis	A,B,C and D
Intact bladder	A and B
Ruptured bladder	C and D
Intact urethra	A and B
Ruptured urethra	C and D
Urethritis	
Mild	A and B
Moderate	A and B
Severe	C and D
Location of the calculi/concretions	
Lumen of the bladder	A
Neck of the bladder	B
Pelvic urethra	C (20%) and D (24%)
Ischial urethra	C (10%) and D (10%).
Sigmoid flexure	C (60%) and D (56%).
Penile urethra/ Glans	C (10%) and D (10%).

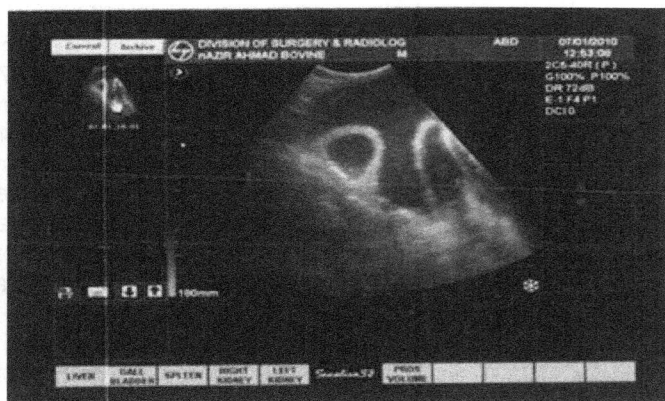

Figure 4. Sonogram showing ruptured bladder at 96 h obstruction.

calculi by diversion of urine through the tube cystotomy catheter, dissolution of urethral calculi by acidic urine caused by oral administration of ammonium chloride and of sodium chloride along with drinking water, pulverisation of calculi by litholytic effect of cystone tablets at 1 tablet twice daily, and occlusion of tube cystotomy catheter helped in achieving urethral patency by flushing the urethra of all debris and calculus material.

Post-operative complications

Different post-operative complications recorded in different groups included. Two cases of the group A showed catheter blockage with fibrin clots and clotted blood, which were relieved by flushing with normal saline. Catheter dislodgement was observed in 3 cases each in group A and D. Second surgical intervention was needed in 6 of cases, and minimal invasive surgery was performed because of loss of the PVC catheter. Second laparotomy was performed and the Foley's catheter was fixed through Para median approach. Multiple nicks on the ventral abdomen were made in the cases of secondary urethral rupture to drain accumulated urine, to prevent urine scald and the gangrene. These cases recovered without any further complications. The mean time for removal of different types of catheters in different groups was almost same and depended upon the establishment of normal urine flow, following removal of urinary obstruction, fixation of catheter and suturing of the ruptured bladder (Figure 4). It was however instant in animals of group A and B.

DISCUSSION

Screening of the calves

Large numbers of crystals in dilute urine, persistent crystalluria and large crystals have greater significance in relation to stone formation and the crystal number is of greater significance than crystal shape and size (Khan and Hackett, 1987). Screening proved a good index for prediction of urolithiasis cases to 54% accuracy. The percentage could be more as during the study, only 35 cases out of 52 recorded cases reported to the clinical complex, fate of the rest cases goes unrecorded. Teotia (1975) reported the test 60% accurate.

Anamnesis

The duration of obstruction in calves varied from 24 to 120 h, our 50% (three cases in each group) cases were presented within 24 to 96 h of illness. The delay in the presentation of cases for the surgical intervention could be due to the time taken by the field veterinary staff in diagnosing and then treating the cases with medical options.

Equal number of calves had ruptured (groups C and D) and intact urinary bladders (groups A and B) at the time of presentation. The high occurrence of ruptured urinary bladder could be due to the delay caused in the diagnosis of the disease and consequent delayed presentation of the animal to the hospital, administration of diuretics to increase the urine output (Adam, 1995). The high occurrence of ruptured bladder cases has also been reported by earlier researcher (Amarpal et al., 2004). Prasad et al. (1978) also reported rupture of bladder in 37.5% of obstructive urolithiasis cases in cattle. During the study, urethra was found ruptured in 50% cases. Rupture of urethra is more common with irregularly shaped stones, which cause pressure necrosis of the urethral wall (Radostitis et al., 2000).

Incidence of urolithiasis was highest (60%) in age group of about 12 months and lowest (40%) in the calves of above one year. The observations recorded by Larson (1996) reported that castration causes reduction in urethral diameter, predisposes castrates for the lodgement of calculi does not seem to be the cause in our study.

Seventy percent calves were jersey crosses, 20 percent Friesian crosses and rest 1 percent non-descript. The highest percentage of urolithiasis in Jersey cross bred calves cannot be attributed to breed, as our population had highest number of jersey cross animals. The massive jersey cross breeding drive started in Kashmir during the years, have replaced our livestock with jersey population. Amarpal et al. (2004) reported incidence of urolithiasis in mixed breeds of cattle, buffalo and goats.

Previous treatment provided

Hundred percent cases were treated in the field before being presented to us with analgesic and anti-inflammatory, antibiotics, B complex preparations, urinary antiseptics, urinary alkalizers, antispasmodics and sedatives various. Half of the population under study had received injections of Lasix (Frusemide Hcl), a loop diuretic. Lasix increases urine formation and flow by its loop diuretic action and might clear obstruction in a few

cases with partial obstruction (Adam, 1995). Excessive production of urine may cause the rupture of the urinary bladder due to development of retrograde intra cystic pressure in bladder resulting rupture (Radostitis et al., 2000). This was one of the factors responsible for a high percentage of ruptured urinary bladder cases in this study.

Clinical examination

The rectal temperature in intact urinary bladder cases was slightly higher than the cases with ruptured urinary bladder. Broadly temperature was within normal range in groups A, B and C. However, it was significantly ($p<0.05$) low in group D and T due to shock state as evidenced by clinical parameters for these groups. This substantiates the findings of Kulkarni et al. (1985); Sockett and Knight, 1986) and Radostitis et al. (2000). The results are not however in agreement with Jadon et al. (1987) and Singh and Sahu (1995).

Tachycardia recorded in the present study is in total agreement with the reports by (Jadon et al., 1987; Joshi et al., 1989; Monoghan and Boy, 1990; Tsuchiya and Sato, 1991; Singh and Sahu, 2005; Hooper, 1998; Smith, 1989). Increased heart rate could be attributed to the reflex response of baro-receptors and chemo-receptors, sympathetic stimulation or para-sympathetic inhibition of SA node (Sobti et al.,1986), progressive hyperkalaemia (Sharma et al., 2005; Bhokre and Deshpande, 1987), dehydration, biochemical alterations, inter-compartmental fluid shifts and myocardial asthenia (Kelly, 1984), accumulation of toxic metabolic waste products (Lavania et al., 1973), pain and progressive systemic disturbances (Monoghan and Boy, 1990). Inappetance and prolonged duration of illness and myocardial asthenia resulting from hyponatraemia and hyperkalaemia in ruminants could also be the possible.

Increased respiratory rate recorded corroborates with Hooper (1998) and Smith (2002). Increased respiratory rate could be attributed to toxaemia as a result of retention of excretory metabolites due to pain caused by urethral calculi, abdominal crisis, electrolyte alterations like hypocalcaemia, hypomagnesaemia and hypovolumic shock (Wilson and Lofstedt, 1990) during obstructive urolithiasis However, Radostits et al. (2000) attributed the increased respiratory rate in uraemic animals to dehydration, myocardial asthenia, with hyponatraemia and hyperkalaemia as main causes. The delayed capillary refill time (2 to 4 s), an index of cardiovascular dynamics, could be due to dehydration and haemoconcentration, which is in total agreement with the statements given by Chew and Bateman (1999) and Radostits et al. (2000).

Ultrasonography

Ultrasonography anon invasive method for diagnosis

has undoubtly improved success rate of urolithiasis because of being earliest mode of diagnosis (Makhdoomi and Sheikh, 2008). It helps in localization of urethral calculi, detection of dilated urethra, cystitis, urethritis and rupture of the urethra or the urinary bladder (Braun, 1992; Cartee et al., 1980). Ultrasonography is done without sedation (Braun, 1993: Magda, 2006; Makhdoomi and Sheikh, 2008). During the present study, transducers with different frequencies viz (3.5 MHz) for urinary bladder, 6.5 MHz for urinary bladder and pelvic urethra. Trans rectal scanning with trans-rectal transducers having frequency of (7.5 MHz) and percutaneous scanning of penile urethra with necessary adjustments of machine gains were used. In animals of group A, B,C and D cystitis revealed highly thickened and single layered hyperechoic cystic wall, with or without hyperechoic material within the lumen and/or attached to the mucosal layer but without any acoustic shadow. These hyper echoic materials without any acoustic shadows were blood clots and were differentiated ultrasonographically from the intramural growth and cystoliths, by subjecting the animals to moderate shaky movements to allow movement of the gravity dependent debris and clots. All the cases were confirmed intera-operatively by thickened cystic wall, hemorrhagic inflammation of cystic mucosa and tissue debris and massive blood clots in the cystic lumen and on mesentery.

Uroperitoneum is evident as anechoic fluid accumulation in the abdomen, within which the organs appear to be floating (Braun et al., 2006). In 18 cases of this study uroperitoneum was predominant and was correctly diagnosed on ultrasonographic examination as confirmed on laparo cystotomy. During this study acoustic shadow due to a single calculus or mass of calculi in a single unit was seen in three cases only from groups C and D, in rest cases ultrasonographic multiple small hyperechoic structures of varying size swirling in the anechoic fluid (urine) without any acoustic shadows were seen.

Peri-operatively, urinary bladder was found intact in all the cases of groups A and B (100%). This could be attributed to their early presentation with median duration of illness of 24 h. Haemorrhagic serosal surface could be due to the rupture of vessels and capillaries at the time of bladder rupture, while chocolaty colour of bladder surface observed could be because of decreased perfusion due to overstretching of the bladder before rupture. Adhesions of urinary bladder with omentum and peritoneum could be due to the fact that upon rupture, eroded serosal surface could come into contact with adjacent omentum and peritoneum. Secondly in both such cases there was history of field diagnostic abdominocentesis/cystocentesis, which could have damaged both the serosal surface of the bladder and peritoneum/omentum and brought them together for healing. Bovine peritoneum lacks the plasminogen activators, which convert plasminogen to plasmin, a specific fibrinolytic inhibitor, which blocks the lysis of fibrin

(Trent and Bailey, 1986). A favourable environment for adhesion formation is thus provided after initial trauma of the peritoneum in cattle. The highest quantity of urine could be obtained from the cases with the prolonged duration of illness.In group C and D prolonged duration of illness with (96 to 120 h) seemed to be the predisposing factor for the rupture of urinary bladder. More ruptures on ventral side of the urinary bladder contradict the observations of Monoghan and Boy (1990) and Sockett and Knight, (1986). Ruptures were distributed throughout the bladder wall, that is, apex, body and neck, thereby suggesting that weak points anywhere in the bladder wall could rupture by the intraluminal urine pressure. Variable degree of roughness, haemorrhagic necrosis and leathery appearance of bladder surface in ruptured urinary bladder cases could be due to rupture of vessels and capillaries at the time of bladder rupturing. Bladder rupture seemed to have no effect on the uroliths retrieval sites within the bladder, as uroliths could be retrieved equally from cases of intact and ruptured urinary bladders.

Post-operatively, mild leakage of urine at 24 post-operative hours in tube cystotomy groups (Paralumber fossa approaches) was due to seepage of urine along the sides of catheter from uroperitoneum. Leakage from urethrotomy site as complication of this procedure has been documented (Sharma et al., 2005) which has been attributed to pressure necrosis of urethra caused by calculi which makes urethra more prone to leakage (Radostits et al., 2000). Weaver and Schulte (1962) and Edwards and Trott (1973) were of the view that urethral stents like catheters may sometime have deleterious effects as they over distended urethral lumen at the injured site and thus leading to leakage of urine. However, tube cystotomy via paralumber fossa was easy, safe and simple procedure. The leakage was the main constraint. There was blockade of catheter in a few animals due to urinary sludge, blood clots, sandy material left in urinary bladder, and mucosal shreds.

REFERENCES

Adam S (1995). Veterinary Pharmacology and Therapeutics. J.B. Lippincott and Co., pp. 529-530.

Amarpal, Kinjavdekar P, Aithal HP, Pawde AM, Tarunbir S, Pratap K, Singh T (2004). Incidence of urolithiasis: a retrospective study of five years. Indian J. Ani. Sci. 74(2):175-177.

Bhokre AP, Deshpande KS (1987). Experimental studies on peritonitis in buffalo calves. Indian Veter. J. 64:137-141.

Braun U, Schefer U, Fohn J (1992). Urinary Tract Ultrasonography in Normal Rams and in Rams with Obstructive Urolithiasis. Canadian Veter. J.. 33:654-659.

Cartee RE, Selcer BA, Patton CS (1980). Ultrasonographic diagnosis of renal disease in small animals. J. Am. Veter. Med. Assoc. 176(5): 426-430.

Chew DJ, Bateman SW (1999). Fluid therapy for dogs and cats. In: Saunders Manual of Small Animal Practice, by Birchard, S. J. and Sherding, R. G. 2nd ed. pp. 64-78.

Edwards L, Trott PA (1973). Catheter induced urethral inflammation. J. Urol. 110:678-681.

Gasthuys F, Steenhaut M, De Moor A, Sercu K (1993). Surgical treatment of urethral obstruction due to urolithiasis in male cattle.A review of 85 cases. Veter. Record. 20:522-526.

Gera KL, Nigam JM (1979). Urolithiasis in bovines (a report of 193 clinical cases). Indian veter. J. 56(5):417-423.

Hooper NR (1998). Management of urinary obstruction in small ruminants. Proceeding, Western Veter. Confer. pp. 99-106.

Jadon NS, Joshi HC, Singh B, Kumar A (1987). Urological and biochemical changes in urethral obstruction in buffaloes.Indian J. Veter. Med. 9: 29-30.

Jenning PB (1984). Practice of Large Animal Surgery, W.B. Saunders, Philadelphia. P. 1072.

Joshi HC, Zangana IK, Saleem AN (1989). Haematobiochemical and electrocardiographic changes in uraemia in sheep. Indian J. veter. Med.. 9:69-73.

Kelly WR (1984). Veterinary Clinical Diagnosis .edn 3 , Typeret by Mc Millian India ltd ., Bangalore Printed and Bound in Great Britain by William Clowes Limited , Beccles and London P. 254.

Khan SR, Hackett RL (1987). Crystal-matrix relationships in experimentally induced urinary calcium oxalate monohydrate crystals, an ultra structure study. Calcification Tissue Internal. 41(3):157-163.

Kulkarni BB, Chandna IS, Peshin PK, Singh J, Singh AP (1985). Experimental evaluation of treatment of uraemia due to ruptured urinary bladder in calves-I. Conventional Treatment. Indian J. Veter. Surgery 6 (1):19-24.

Larson BL (1996). Identifying, treating and preventing bovine urolithiasis. Veter. Med. 91:366-377.

Lavania JP, Misra SS, Angelo SJ (1973). Repair of rupture of bladder in a calf - a clinical case. Indian Veter. J. 50:477-481.

Lonsdale K, Osborne CA, Lulich JP (1968). Epistasis as a growth factor in urinary calculi and gall stones. Nature 217(213):56-58.

Lundvall RL (1974). Textbook of Large Animal Surgery, Williams and Wilkins, Baltimore P. 450.

Magda A (2006). Diagnosis of obstructive urolithiasis in cattle and buffal o by using ultrasonography.Online J. Veter. Res. 10:26-30.

Makhdoomi DM, Shiekh GN (2008). Diagnosis of obstructive urolithiasis i calves by ultrasonograpy J. Remount Veter. Corps. 47(1):27-38.

Monoghan ML, Boy MG (1990).Diseases of the Renal System – Ruminant Renal System. In: Large Animal Internal Medicine. SMITH. B. P.; Mosby, USA, pp. 895 -899.

Prasad B, Sharma SN, Singh J, Kohli RN (1978). Surgical repair and management of bladder rupture in bullocks. Indian Veter. J. 55(11):905-911.

Radostitis OM, Blood DC, Gray GC, and Hinchcliff KW (2008). Veterinary Medicine: A text book of the disease of cattle, sheep, pig, goat and horse. BailliereTindall, London, P. 1877.

Radostitis OM, Blood DC, Gray GC, Hinchcliff KW (2000). Veterinary Medicine:A text book of the disease of Cattle, sheep, pig, goat and horse. Bailliere Tindall, London, P. 1877.

Sharma AK, Mogha IV, Singh GR, Amarpal, Aithal HP (2005). Clinico physiological and haematobiochemical changes in urolithiasis and its management in bovine. Indian J. An. Sci. 75(10):1131-1134.

Singh H, Sahu S (1995). Peritoneal lavage as an adjunct therapeutic measure for uraemia in bovines. Indian Veter. J. 72(11):1174-1176

Smith LH (1989). The medical aspects of urolithiasis.An overview. J. Urol.141:707-710.

Smith KR (2002).Indoor air pollution in developing countries: recommendations for research. Indoor Air.12:198-207.

Snedecor GW, Cochran WG (1976). Statistical Methods.Iowa State University Press. pp. 20-28:58-59.

Sobti VK, Jalaludin AM, Kumar VR, Kohli RN (1985). Electrocardiographic changes following experimental rupture of urinary bladder in calves. Indian J. Veter. Surgery.7:36-41.

Sockett DS, Knight AP (1986). Metabolic changes associated with obstructive urolithiasis in cattle. Comp. Cont. Educ. Pract. 6:311-315.

Teotia G and (1975). Screening of calves as normal and stone formers. Veter. Record. 128:234-238.

Trent MA, Bailey JV (1986). Bovine peritoneum: Fibrinolytic activity and adhesion formation. Am. J. Veter. Res. 47(6):53-659.

Tsuchiya R, Sato M (1991). Correlation between meat inspection findings and the BUN levels in urolithiasis in fattening cattle. J. Japanese Veter. Med. Assoc. 44(6):632-636.

Weaver RG, Schulte JW (1962). Experimental and clinical studies on urethral regeneration. Surgery. Gynaecology Obstetrics 115:729-736.

Wilson WD, Lofstedt J (1990). Alterations in respiratory functions. In: Large Animal Internal Medicine, by Smith, B.P. C. V. S. Mosby Co. pp. 47-99.

Winter RB, Hawkins LL, Holterman DE, Jones SG (1987). Catheterization- an effective method of treating bovine urethral calculi. Veter. Med. 82:1261-1266.

31

The effect of *Origanum onites* L., *Rosmarinus officinalis* L. and *Schinus molle* L. on *in vitro* digestibility in lamb

Sibel SOYCAN-ÖNENÇ

Department of Animal Science, Namık Kemal University, Faculty of Agriculture, Tekirdağ-Turkey.

The objective of this study was to evaluate the effects of different levels of aromatic plants (AP);"*Origanum onites L.* (ORE), *Rosmarinus officinalis L.* (ROS) and *Schinus molle L.* (SHN)" on *in vitro* gas production (GP), organic matter digestibility (OMD) and net energy lactation (NEL) using an *in vitro* gas-production method. Two rumen-fistulated sheep were used in the experiment. The sheep were fed with 60% of alfalfa hay and 40% of concentrate feed twice daily. Five different levels of ORE, ROS, and SHN were added to the concentrate (CON) to produce 200 mg KM (C - without , AP; 1 - 0.02 mg AP + 198 mg CON; 2 - 0.04 mg AP + 196 mg CON; 3 - 0.06 mg AP + 194 mg CON; 4 - 0.08 mg AP + 192 mg CON; 5 - 0.10 mg AP + 190 mg CON). The volume of gas produced was recorded at 2, 4, 8, 12, and 24 h after incubation. The results showed that, GP, OMD and NEL contents decreased significantly as the level of ORE, ROS, and SHN added to CON increased. It was concluded that, ORE, ROS, and SHN are potential supplements that alter rumen fermentation. To obtain exact results, the findings obtained under *in vitro* conditions should be supported by *in vivo* studies.

Key words: Ruminant, aromatic plants, feed digestibility, gas production, organic matter digestibility.

INTRODUCTION

In Europe, the utilisation of such plant extracts as essential oils (EOs) in livestock production has expanded following the ban of the use of antibiotic growth promoters, including iontophores, in livestock nutrition (OJEU; 2003; Regulation (EC) 1831/2003). EOs are naturally occurring volatile components that can be extracted from plants by different distillation methods, particularly steam distillation (Greathead, 2003). Different parts of the plants can be used to obtain EOs, including the flowers, leaves, seeds, roots, stems, and bark. For many centuries, EOs have been used for their essence, flavour, antiseptic and/or preservative properties (Chaves et al., 2008).

The antimicrobial properties of EOs have been related to a number of small terpenoid and phenolic compounds (Helander et al., 1998). EOs that is particularly rich in phenolic compounds has been shown to possess the high levels of antimicrobial activity against both Gram-positive and -negative bacteria (Fraser et al., 2007).

Recent studies have shown that, at low concentrations, some EOs, their active components or mixtures of EOs have the potential to modify rumen N metabolism by reducing the degradation of proteins and ammonia production in the rumen (McIntosh et al., 2003; Molero et al., 2004; Newbold et al., 2004; Busguet et al., 2005 a,b; Chaves et al., 2008). Busguet et al. (2006) showed that, at 3.000 mg/L, capsicum oil, carvacrol, carvone, cinnamaldehyde, cinnamon oil, clove bud oil, eugenol, fenugreek and oregano oils resulted in a 30 to 50% reduction in the ammonia N concentration.

Additionally, these authors reported that, the careful selection and combination of these extracts may allow

Table 1. The nutrient composition of CON and AP.

Samples	DM (%)	CA (%)	CP (%)	EE (%)	CF (%)	NFE (%)	ME (kcal/kg)
CON	89.38	4.8	15.15	4.39	6.27	58.03	2725
ORE	88.77	7.69	9.49	3.60	18.20	49.79	2078
ROS	91.01	5.17	6.12	10.99	26.25	42.48	2153
SHN	88.62	3.49	6.34	7.59	16.69	54.51	2397

DM:Dry matter, CA:crude ash, CP:crude protein, EE:ether extract ,CF:crude fiber, NFE:nitrogen free extract, ME:metabolic energy.

Table 2. Essential oil ratios of aromatic plants.

Aromatic plant	Essential oil ratio (% DM)
ORE	1.35
ROS	0.60
SHN	0.23

the manipulation of rumen microbial fermentation.

The objective of the present study was to evaluate the effects of increasing levels of aromatic plants (AP) on *in vitro* gas production (GP), organic matter digestibility (OMD), and net energy lactation ((NEL) using an *in vitro* gas-production method.

MATERIALS AND METHODS

Experimental material and procedures

Two rumen-fistulated sheep (Tahirova breed, East Friesian 75% × Kivircik25%) were used in the test. Sheep were fed with 60% of alfalfa hay and 40% concentrate feed twice daily as described by Steingass and Menke (1986). The feed material consisted of fattening concentrate (CON), and the AP consisted of *Origanum onites* L. (ORE), *Rosmarinus officinalis* L. (ROS) and *Schinus molle* L. (SHN). Five different levels of ORE, ROS, and SHN (all of them called aromatic plant; AP) were added to the concentrate (CON) to produce 200 mg KM (C - without , AP; 1 - 0.02 mg AP + 198 mg CON; 2 - 0.04 mg AP + 196 mg CON; 3 - 0.06 mg AP+194 mg CON; 4 - 0.08 mg AP+192 mg CON; 5 - 0.10 mg AP+190 mg CON). The results of the crude nutrient analysis of CON and each AP are presented in Table 1.

The rates of ORE, ROS, and SHN EOs used in the study are presented in Table 2, and the EOs components are presented in Table 3.

Chemical analyses

The concentrate feed and AP were grounded through a 1 mm screen in preparation for the chemical analysis. The dry matter (DM), crude protein (CP), ether extract (EE), crude ash (CA) and crude fibre (CF) were analysed according to Verband Deutscher Landwirtschaftlicher Untersuchungs-und Forschungsanstalten, VDLUFA (Naumann and Bassler, 1993). The metabolisable energy (ME) was calculated based on the chemical composition (Anonymous, 1991).

The rumen fluid was collected before the morning feeding from two ruminally fistulated lambs. The estimates of GP were obtained using the method of Menke and Steingass (1988). A buffer solution (macro and microminerals) was prepared on the day prior to the analysis and incubated in a waterbath at 39°C under a continuous CO_2 stream (DLG, 1981). Incubations were terminated after 24 h for the OMD and NEL estimations of the concentrate and AP mixtures. The volumes of gas produced were recorded at 2, 4, 8, 12, and 24 h after inoculation, and the (GP) results were applied to calculate OMD and NEL using the following equations.

OMD (%) = 0.889 × GP + 0.448 × CP^* + 0.651 × CA^* + 14.88 in % DM. (Menke ve Huss, 1987).

NEL (MJ/kg DM) = 3.95 + 0.3305 × GP-0.0023 × GP^2 + 0.0535 × CP + 0.0132 × EE^2-0.0336 × CF-0.1073 × CA (Aiple, 1993).

GP: 24-h cumulative GP in DM.

The EOs from 10 g of dry plant materials were extracted by hydro-distillation for 3 h using a Clavenger-type apparatus, according to the European Pharmacopoeia (1975), with three replications. The GC analyses were performed at the Central Laboratory of Aegean University using a Carlo Erba Fractovap Series 2350 gas chromatograph equipped with a flame ionisation detector. A glass column (3 m long, 3.18 mm internal diameter) packed with 3% OV-1 50 chromosorb 80/100-mesh was used. The carrier gas was N_2 at a flow rate of 25 ml/min. Each GC run lasted for 20 min. The oven temperature was isothermal at 110°C, and the injector and detector temperatures were 225 and 250°C, respectively.

Statistical analysis

The data obtained were evaluated using the GLM procedure of SPSS V10 software. Duncan's test was employed for the comparison of the differences between the group averages (Efe et al., 2000).

RESULTS

As shown in Figure 1A and B, the highest and lowest GP

Table 3. The chemical composition of essential oils.

	ORE		ROS		SHN
Compounds	Percent composition (%)	Compounds	Percent composition (%)	Compounds	Percent composition (%)
Carvacrol	69.10	Borneol	26.16	α-Phellandrene	27.60
Thymol	10.70	Camphor	23.54	β- Phellandrene	21.95
P-Cymene	4.00	1,8-cineole	14.34	Myrcen	8.50
Borneol+ α-Terpineol	3.00	α-Terpineol	11.99	(+)-Spathulenol	6.30
γ-Terpinene	2.50	Limonene	11.86	α-Eudesmol	5.45
		Bornyl acetate	7.56	P-Cymene	5.20
		α-Pinene	2.48	β-Eudesmol	3.40
				α-Pinene	3.27
Unknown	10.70		2.00		18.33
Total compounds	100		100		100

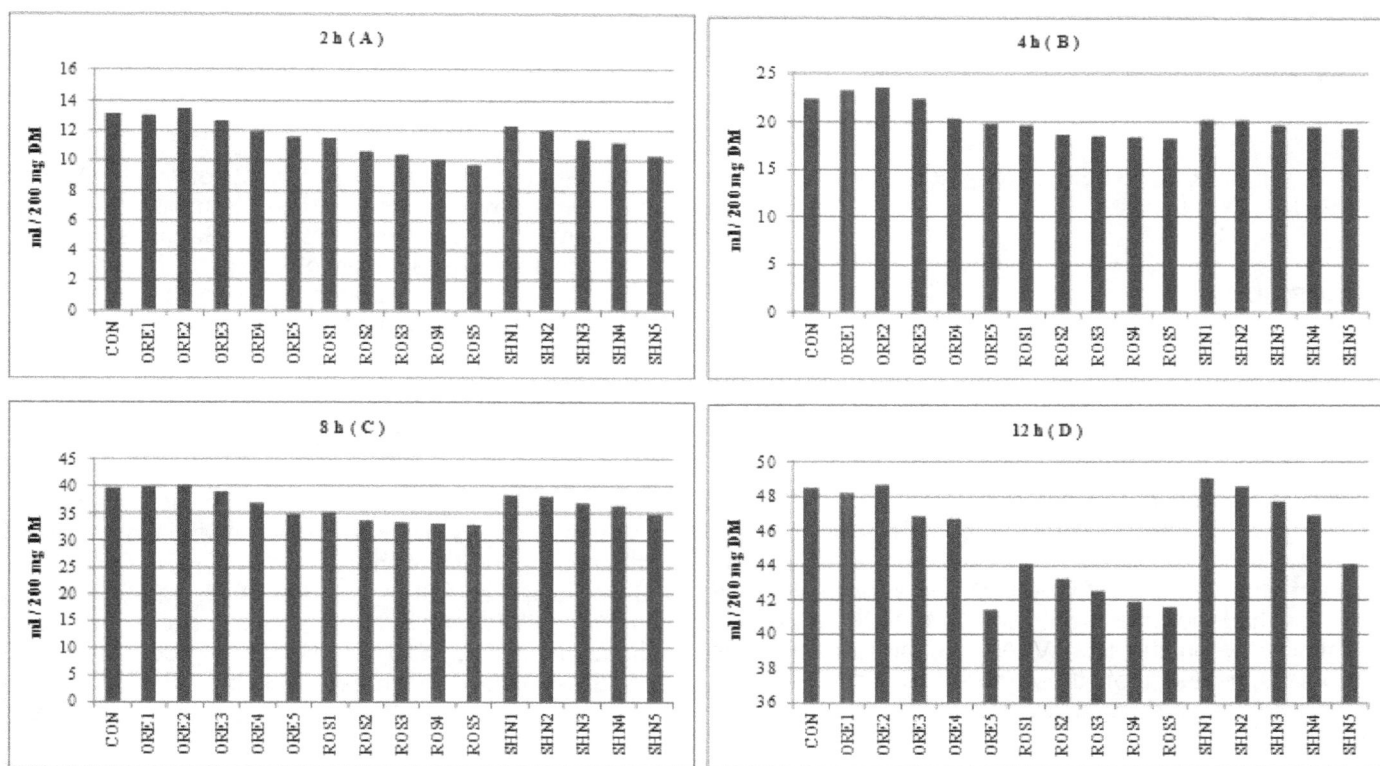

Figure 1. Incubation periods of GP, 2 h (A), 4 h (B), 8 h(C), and 12 (D) h.

values were found after 2 and 4 h of incubation, respectively, for ORE_2 and ROS_5. The GP values determined after 8 h of incubation (Figure 1C) showed a significant difference between the treatments except for ORE_1, ORE_2, ORE_3, and SHN_1 ($P < 0.01$). After 12 h, the highest GP was found in SHN_1 (49.06 ± 0.28), whereas

the lowest value was determined in ORE_5 (41.45 ± 0.38) (Figure 1D). The values for GP, OMD, and NEL determined under *in vitro* conditions are presented in Table 4. The highest total GP was found in ORE_2 (60.28 ± 0.37), and the lowest was found in ORE_5 (52.63 ±0.35). The OMD and NEL values decreased significantly

Table 4. GP, OMD, and NEL contents of CON and APs.

Sample	24-h GP (ml/200 mg DM)	OMD (%)	NEL (MJ/kg DM)
CON	$60.13^{ab} \pm 0.52$	$79.47^{ab} \pm 0.47$	$8.01^{a} \pm 0.03$
ORE$_1$	$59.66^{ab} \pm 1.26$	$79.06^{ab} \pm 1.12$	$7.97^{ab} \pm 0.08$
ORE$_2$	$60.28^{a} \pm 0.37$	$79.61^{a} \pm 0.33$	$8.02^{a} \pm 0.02$
ORE$_3$	$59.21^{abc} \pm 0.20$	$78.66^{abc} \pm 0.18$	$7.96^{ab} \pm 0.01$
ORE$_4$	$58.08^{bcdef} \pm 0.35$	$77.66^{bcdef} \pm 0.32$	$7.89^{abcd} \pm 0.02$
ORE$_5$	$52.63^{g} \pm 0.35$	$72.81^{g} \pm 0.31$	$7.48^{f} \pm 0.03$
ROS$_1$	$57.52^{cdef} \pm 0.20$	$77.15^{cdef} \pm 0.17$	$7.85^{bcde} \pm 0.01$
ROS$_2$	$57.05^{def} \pm 0.27$	$76.74^{def} \pm 0.24$	$7.82^{cde} \pm 0.01$
ROS$_3$	$56.82^{ef} \pm 0.31$	$76.53^{ef} \pm 0.28$	$7.81^{cde} \pm 0.02$
ROS$_4$	$56.08^{f} \pm 0.37$	$75.88^{f} \pm 0.33$	$7.75^{e} \pm 0.03$
ROS$_5$	$53.51^{g} \pm 0.24$	$73.59^{g} \pm 0.22$	$7.55^{f} \pm 0.02$
SHN$_1$	$60.25^{a} \pm 0.22$	$79.58^{a} \pm 0.20$	$8.02^{a} \pm 0.01$
SHN$_2$	$59.44^{abc} \pm 0.31$	$78.86^{abc} \pm 0.28$	$7.97^{ab} \pm 0.02$
SHN$_3$	$59.00^{abcd} \pm 0.90$	$78.47^{abcd} \pm 0.80$	$7.94^{abc} \pm 0.06$
SHN$_4$	$58.71^{abcde} \pm 0.35$	$78.21^{abcde} \pm 0.31$	$7.93^{abcd} \pm 0.02$
SHN$_5$	$56.72^{ef} \pm 0.26$	$76.45^{ef} \pm 0.23$	$7.80^{de} \pm 0.02$
P	0.001	0.001	0.001

[abc] Means with different letters in the same column are statistically significant ($p < 0.01$), CON: concentrate, ORE; *O. onites*, ROS; *R. officinalis*, SHN; *S. molle*, 1,2, 3, 4, 5; aromatic plant level in concentrate (200 mg).

($P < 0.01$) in ROS$_1$, ROS$_2$, ROS$_3$, ROS$_4$, ROS$_5$, ORE$_5$, and SHN$_5$.

DISCUSSION

The highest GP values were found in the ORE$_2$ and SHN$_1$ groups at 60.28 ± 0.37 and 60.25 ± 0.25 ml/200 mg KM, respectively. Although the increase was not significant, the groups supplemented AP had slightly higher GP values than the control group. Similarly, no significant difference was found for OMD and NEL ($P < 0.01$), indicating that supplementing CON with ORE and SHN does not positively affect OMD in the rumen. Busquet et al. (2005a,b) found that, cinnamaldehyde and garlic oil had no effect on DM, OM, NDF, and ADF digestibility or on the total VFA concentration, and the authors suggested that, these additives did not modify the overall diet fermentability. Castillejos et al. (2005) determined that 1.5 mg/l BEO (Crina ruminants) supplemented to high concentrate and coarse feed rations (that is, 100 forage and 900 concentrate versus 600 forage and 400 concentrate) did not affect DM, OM, NDF, ADF, and CP digestion, though BEO did increase the total VFA concentration (122.8 mM versus 116.2 mM). In another study, Newbold et al. (2004) observed a reduction in the *in situ* DM degradation of soya-bean meal after 8 and 16 h of incubation when 110 mg/d EO$_s$ (a mixture of thymol, guaiacol, and limonene) was added to the diet of sheep. However, the mixture had no effect on the DM degradability of rapeseed meal and hay.

In the present study, the lowest total GP, OMD, and NEL values were found in ORE$_5$. Examining the effects of carvacrol on rumen fermentation, Garcia et al. (2007) reported that, the addition of carvacrol reduced *in vitro* DM, CP, and neutral-detergent fibre (NDF) digestion. The effects induced by 250 mg/l carvacrol on DM digestion after 72 h of incubation were comparable to those of monensin, whereas a greater reduction was obtained when carvacrol was supplemented at a concentration of 500 mg/l. The researchers explain that, the reduced CP potential degradability by the supplementation was mainly caused by a reduction of the slowly degradable fraction. Indeed, the GP-reducing effect that occurred in the early periods of incubation in ORE$_3$, ORE$_4$, ORE$_5$, SHN$_3$, SHN$_4$ and SHN$_5$ (Figure 1A, B, and C) was found in the ORE$_5$ and SHN$_5$ groups at the end of the incubation. ORE$_5$, SHN$_5$ and all ROS groups affected the rapidly and slowly degraded fractions.

Castillejos et al. (2006) reported that, the effect of thymol on *in vitro* DM, OM, NDF, and ADF digestion varies according to the level used, determining that DM, OM, NDF, and ADG digestion did not change with the addition of 5 mg/l thymol yet decreased with the use of 500 mg/l thymol. In another study, Benchaar et al. (2007) found that, although 400 mg/l carvacrol and 200 mg/l thymol reduced GP and NDF digestibility in an *in vitro* 24 h batch culture environment, 200 mg/l oregano and thyme oil caused a reduction of NDF digestibility while not affecting GP digestibility.

In agreement with the findings of Garcia et al. (2007), Castillejos et al. (2006) and Benchaar et al. (2007), in the present study, it was determined that, GP, OMD, and NEL were not affected in the groups supplemented with low amounts of ORE (ORE$_1$, ORE$_2$, ORE$_3$, and ORE$_4$), whereas these parameters are significantly ($P < 0.01$) reduced at high levels of (ORE$_5$) addition.

Helander et al. (1998) reported that, the capacity of carvacrol and thymol to degrade the outer membrane of Gram-negative bacteria and observed the release of membrane lipopolysaccharides and increased permeability of the plasma membrane. Therefore, the small molecular weight of these compounds may allow them to be active in Gram-positive and Gram-negative bacteria.

The ORE EOs used in this research consists of 69.10% carvacrol and 10.70% thymol as the principal components. The increasing carvacrol and thymol concentrations in the ORE-supplemented groups caused decreases in GP, OMD, and NEL by increasing the antimicrobial effect. According to Helander (1998) and Covan (1999), this is a result of the antimicrobial effect of carvacrol and thymol on rumen microorganisms.

In all groups in which ROS was added to CON, GP decreased significantly ($P < 0.01$) from the 2nd h of incubation, and OMD and NEL decreased accordingly. Smith-Palmer et al. (1998) report that, the bacteriostatic concentrations of ROS EOs against S. aureus and L. monocytogenes range from 0.02 to 0.04%, bacteriocidal concentrations are less or equal to 1% and that concentrations over 1% can be used for inhibiting Gram-negative bacteria. In another study, Santoyo et al. (2005) reported that, ROS EOs shows antimicrobial activity against Gram-positive bacteria (Staphylococcus aureus and Bacillus subtilis), Gram-negative bacteria (Escherichia coli and Pseudomonas aeruginosa), a yeast (Candida albicans) and a fungus (Aspergillus niger) and that the effects increase as the camphor, borneol, and verbenone contents in the composition of the EOs increase.

Pattnaik et al. (1997) stated that, 1,8-cineole and camphor exhibit strong antimicrobial effects. However, Hayouni et al. (2008) reported that, the antimicrobial activity of S. officinalis EOs is related to the 1,8-cineole, α / β-thujone and borneol content in the oil.

It is believed that the reducing effect of ROS addition on GP, OMD, and NEL becomes more evident with increasing rates of ROS and is related to the presence of 26.16% borneol and 23.54% camphor in the composition of the ROS essential oil. Similarly, Santoyo et al. (2005), Pattnaik et al. (1997), and Hayouni et al. (2008) reported that the antimicrobial effect is related to the borneol and camphor concentrations in ROS essential oil.

It was observed in this study that, in the SHN-including groups of CON, GP, OMD, and NEL decreased proportionally with the amount of ROS addition. Having determined the α-phellandrene and germacrene D contents of SHN EOs as 6.94, 6.54, 3.53, and 20.77%, respectively, Deveci et al. (2010) reported that these compounds show potential in terms of antimicrobial and

repellent activity. With regard to sensitivity to SHN EOs, Hayouni et al. (2008) determined that Enterococcus faecalis ATCC 2912, E. coli ATCC 25922 and E. coli (clinical strain 1) are very sensitive, followed by Pseudomonas aeruginosa ATCC 9027, Staphylococcus aureus ATCC 6539, E. coli (clinical strain 2) and Salmonella anatum. The researchers reported that the SHN EOs they used contained 35.86 % α-phellandrene and 29.30 % β- phellandrene as the principal components and that these compounds are related to the antimicrobial activity of the oil. Similar to the EO used by Hayouni et al. (2008), the SHN EOs used in this study contained 27.60 % α-phellandrene and 21.95 % β-phellandrene as the principal components. Thus, the decrease of GP, OMD, and NEL in SHN$_5$ may be related to the increase in the α-phellandrene and β- phellandrene contents, in accordance with the increase in the amount of SHN added to CON.

It is reported that, there is a high level of correlation ($r = 0.82$) between the amount of gas created with the in vitro incubation of feed with rumen liquid for 24 h and the level of digestion of organic substances (Menke et al., 1979). The increasing amount of EOs and antimicrobial effect was correlated with the increase in the levels of ORE, ROS, and SHN used in this study, and lower GP, OMD, and NEL were determined in an inversely proportional manner with this increase. In a previous study by Soycan-Onenc (2008) it was determined, that varying levels of ORE, ROS and SHN addition to barley decreased total GP, OMD, and ME and that, this result was associated with a decrease or inhibition of the activities of amylolytic bacteria in the rumen. In addition, Oh et al. (1967, 1968) observed that, high levels of different plant EOs decreased the production of gas and total VFA in the in vitro fermentations of mixed ruminal microorganisms, suggesting that, high levels resulted in a general inhibition of rumen microbial fermentation.

Conclusions

In this study, it was found that, GP, OMD, and NEL decreased significantly as the level of ORE, ROS, and SHN added to CON increased. It was concluded that, ORE, ROS, and SHN are potential supplements to alter rumen fermentation. To obtain exact results, the findings obtained under in vitro conditions should be supported by in vivo studies.

REFERENCES

Aiple KP (1993). Vergleichende untersuchungen mit pansensaft und kot als inokulum im Hohenheimer Futterwerttest, Dissertation.

Anonymous (1991). Animal feeds-determination ofmetabolizable energy (chemical method). Turkish Standards Institute (TSE). Publ. 9610:1-3.

Benchaar C, Chaves AV, Fraser GR, Wang Y, Beauchemin KA, McAllister TA (2007). Effects of essential oils and their components on in vitro rumen microbial fermentation. Can. J. Anim. Sci. 87(3):413-419.

Busquet M, Calsamiglia S, Ferret A, Cardozo PW, Kamel C (2005a). Screening for effects of plant extracts and active compouds of plants on dairy cattle rumen microbial fermentation in a continuous culture system. Anim. Feed Sci. Techn. 124:597-613.

Busquet M, Calsamiglia S, Ferret A, Cardozo PW, Kamel C (2005b). Effects of cinnamaldehyde and garlic oil on rumen microbial fermentation in a dual flow continuous culture. J. Dairy Sci. 88:2508-2516.

Castillejos L, Calsamiglia S, Ferret A, Losa R (2005). Effects of a spesific blend of essential oil compounds and the type of diet on rumen microbial fermentation and nutrient flow from a continuous culture system. Anim. Feed Sci. Techn. 119:29-41.

Castillejos L, Calsamiglia S, Ferret A (2006). Effect of essential oil active compounds on rumen microbial fermentation and nutrient flow in in vitro systems. J. Dairy Sci. 89:2649-2658.

Chaves AV, Stanford K, Dugan MER, Gibson LL, McAllister TA, Van Herk F, Benchaar C (2008). Effects of cinnaldehyde, garlic and juniper berry essential oils on rumen fermentation, blood metabolites, growth performance, and carcass characteristics of growing lambs. Livestock Sci. 117:215-224.

Covan MM (1999). Plant products as antimicrobial agents. Clin. Microbiol. Rev. 12:564-582.

Deveci O, Sukan A, Tuzun N, Kocabaş EEH (2010). Cemical composition, repellent and antimicrobial activity of Schinus molle L. J. Med. Plants Res. 4 (21):2211-2216.

DLG (1981). Methode zur schätzung des NEL-gehaltes im milchleistungsfutter, DLG-Forschungsbericht. Nr.538014, Frankfurt.

Efe E, Bek Y, Şahin M (2000). Statistical methods with solutions in SPSS. Kahramanmaraş Sütçü İmam University Pub. Textbook Pub 73:9.

European Pharmacopoeia (1975). Maissonneuve SA, Sainte-Ruffine, Vol. 3:68-71.

Fraser GR, Chaves AV, Wang Y, McAllister TA, Beauchemin KA, Benchaar C (2007). Assessment of the effects of cinnamon leaf oil on rumen microbial fermentation using two continuous culture systems. J. Dairy Sci. 90:2315-2328.

Garcia V, Catala-Gregori P, Madrid J, Hernandez F, Megias MD (2007). Potential of carvacrol to modify in vitro rumen fermentation as compared with monensin. Animal. 1:675-680.

Greathead H (2003). Plants and plant extracts for improving animal productivity. Proc. Nutr. Soc. 62:279-290.

Hayouni EA, Chraiefl, Abedrabba M, Bouiix M, Leveau J, Mohammed H, Hamdi M (2008). Tunisian Salvia officinalis L. and Schinus molle L. essential oils: their chemical compositions and their preservative effects against salmonella inoculated in minced beef meat. Int. J. Food Mic. 125:242-251.

Helander IM, Alakomi HL, Latva-Kala K, Mattila-Sandholm T, Pol I, Smid EJ, Gorris LGM, Von Wright A (1998). Characterization of the action of selected essential oil components on gram-negative bacteria. J. Agric. Food Chem. 46:3590-3595.

McIntosh FM, Williams P, Losa R, Wallace RJ, Beever DA, Newbold CJ (2003). Effects of essential oils on ruminal microorganisms and their protein metabolism. Appl. Environ. Microbiol. 69:5011-5014.

Menke KH, Raab L, Salewski A, Steingass H, Fritz D, Scheider W (1979). The estimation of digestibility and metabolizable energy content of ruminant feedingstuffs from the gas production when they are incubated with rumen liquor in vitro. J. Agric. Sci. 93:217-222.

Menke KH, Huss W (1987). Tierernährung und Futtermittelkunde.

Menke KH, Steingass H (1988). Estimation of the energetic feed value obtained from chemical analysis and in vitro gas production using rumen fluid. Anim. Res. Dev. 28:7-55.

Molero R, Ibars M, Calsamiglia S, Ferret A, Losa R (2004). Effect of a spesific blend of essential oil compounds on dry matter and crude protein degradability in heifers fed diets with different forage to concentrate rations. Anim. Feed Sci. Technol. 114:91-104.

Naumann C, Bassler R (1993). Methodenbuch, Band III. Die Chemische Untersuchung von Futtermitteln. VDLUFA-Verlag, Darmstadt, Germany.

Newbold CJ, McIntosh FM, Williams P, Losa R,Wallace RJ (2004). Effects of a spesific blend of essential oil compouns on rumen fermentation. Anim. Feed Sci. Techn. 114:105-112.

OJEU (2003). Regulation (EC) No 1831/2003 of the European Parliament and the Council of 22 September 2003 on Additives for Use in Animal Nutrition. Official J. European Union. Page L268/36 in OJEU of 18/10/2003.

Oh HK, Sakai T, Jones MB, Longhurst WM (1967). Effect of various essential oils isolated from douglas needles upon sheep and deer rumen microbial activity. Appl. Microbiol. 15:777-784.

Oh HK, Jones MB, Longhurst WM (1968). Comparision ofrumen microbial inhibition resulting from various essential oils isolated from relatively unpalatable plant species. Appl. Microbiol. 16:39-44.

Pattnaik S, Subramanyam VR, Bapaji M, Kole CR (1997). Antibacterial and antifungal activity of aromatic constituents of essential oils. Microbios. 89:39-46.

Santoyo S, Cavero S, Jaime L, Ibanez E, Senorans FJ, Reglero G (2005). Chemical composition and antimicrobial activity of Rosmarinus officinalis L. essential oil obtained via supercritical fluid extraction. J. Food Protect. 68(4):790-795.

Smith-Palmer A, Stewart J, Fyfe L (1998). Antimicrobial properties of plant essential oils and essences against five important food-borne pathogens. Letters in Applied Microbiol. 26:118-122.

Steingass H, Menke KH (1986). Schätzung des energetischen futterwerts aus der in vitro mit pansensaft bestimmten gasbildung und der chemischen analayse. I. Untersuchungen zur Methode. Übers. Tierernährung. 14:251-270.

Permissions

The contributors of this book come from diverse backgrounds, making this book a truly international effort. This book will bring forth new frontiers with its revolutionizing research information and detailed analysis of the nascent developments around the world.

We would like to thank all the contributing authors for lending their expertise to make the book truly unique. They have played a crucial role in the development of this book. Without their invaluable contributions this book wouldn't have been possible. They have made vital efforts to compile up to date information on the varied aspects of this subject to make this book a valuable addition to the collection of many professionals and students.

This book was conceptualized with the vision of imparting up-to-date information and advanced data in this field. To ensure the same, a matchless editorial board was set up. Every individual on the board went through rigorous rounds of assessment to prove their worth. After which they invested a large part of their time researching and compiling the most relevant data for our readers.

The editorial board has been involved in producing this book since its inception. They have spent rigorous hours researching and exploring the diverse topics which have resulted in the successful publishing of this book. They have passed on their knowledge of decades through this book. To expedite this challenging task, the publisher supported the team at every step. A small team of assistant editors was also appointed to further simplify the editing procedure and attain best results for the readers.

Apart from the editorial board, the designing team has also invested a significant amount of their time in understanding the subject and creating the most relevant covers. They scrutinized every image to scout for the most suitable representation of the subject and create an appropriate cover for the book.

The publishing team has been an ardent support to the editorial, designing and production team. Their endless efforts to recruit the best for this project, has resulted in the accomplishment of this book. They are a veteran in the field of academics and their pool of knowledge is as vast as their experience in printing. Their expertise and guidance has proved useful at every step. Their uncompromising quality standards have made this book an exceptional effort. Their encouragement from time to time has been an inspiration for everyone.

The publisher and the editorial board hope that this book will prove to be a valuable piece of knowledge for researchers, students, practitioners and scholars across the globe.

List of Contributors

Changmin Hu
Faculty of Animal Science and Veterinary Medicine, Huazhong Agricultural University, Wuhan 430070, China

Aizhen Guo
Faculty of Animal Science and Veterinary Medicine, Huazhong Agricultural University, Wuhan 430070, China

Simone de Melo Santana-Gomes
Universidade Estadual de Maringa, Pós Graduação em Agronomia, Avenida Colombo, no 5790 - CEP: 87020-900, Maringa, PR, Brazil

Claudia Regina Dias-Arieira
Universidade Estadual de Maringa, Pós Graduação em Agronomia, Avenida Colombo, no 5790 - CEP: 87020-900, Maringa, PR, Brazil

Miria Roldi
Departamento de Ciências Agronômicas, Universidade Estadual de Maringa, Avenida Colombo, no 5790 - CEP: 87020-900, Maringa, PR, Brazil

Tais Santo Dadazio
Departamento de Ciências Agronômicas, Universidade Estadual de Maringa, Avenida Colombo, no 5790 - CEP: 87020-900, Maringa, PR, Brazil

Patricia Meiriele Marini
Departamento de Ciências Agronômicas, Universidade Estadual de Maringa, Avenida Colombo, no 5790 - CEP: 87020-900, Maringa, PR, Brazil

Davi Antonio de Oliveira Barizão
Universidade Estadual de Maringa, Pós Graduação em Agronomia, Avenida Colombo, no 5790 - CEP: 87020-900, Maringa, PR, Brazil

F. Kabir
Faculty of Animal Science and Veterinary Medicine, Patuakhali Science and Technology University, Babugonj, Barisal, Bangladesh

J. J. Kisku
Faculty of Animal Science and Veterinary Medicine, Patuakhali Science and Technology University, Babugonj, Barisal, Bangladesh

Hajkhodadadi
Department of Animal Science, Faculty of Agriculture and Natural Resources, Tehran University, Karaj, Iran

M. Shivazad
Department of Animal Science, Faculty of Agriculture and Natural Resources, Tehran University, Karaj, Iran

H. Moravvej
Department of Animal Science, Faculty of Agriculture and Natural Resources, Tehran University, Karaj, Iran

A. Zare-shahneh
Department of Animal Science, Faculty of Agriculture and Natural Resources, Tehran University, Karaj, Iran

Fernando Sánchez
Universidad Autónoma de Nuevo León, Department of Agronomy, Laboratory of Animal Reproduction, Agricultural Sciences Campus UANL, Francisco Villa S/N, Ex Hacienda Canadá, P.O. Box 66050 General Escobedo, Nuevo León, México

Hugo Bernal
Universidad Autónoma de Nuevo León, Department of Agronomy, Laboratory of Animal Reproduction, Agricultural Sciences Campus UANL, Francisco Villa S/N, Ex Hacienda Canadá, P.O. Box 66050 General Escobedo, Nuevo León, México

Alejandro S. del Bosque
Universidad Autónoma de Nuevo León, Department of Agronomy, Laboratory of Animal Reproduction, Agricultural Sciences Campus UANL, Francisco Villa S/N, Ex Hacienda Canadá, P.O. Box 66050 General Escobedo, Nuevo León, México

Adán González
Universidad Autónoma de Nuevo León, Department of Agronomy, Laboratory of Animal Reproduction, Agricultural Sciences Campus UANL, Francisco Villa S/N, Ex Hacienda Canadá, P.O. Box 66050 General Escobedo, Nuevo León, México

Emilio Olivares
Universidad Autónoma de Nuevo León, Department of Agronomy, Laboratory of Animal Reproduction, Agricultural Sciences Campus UANL, Francisco Villa S/N, Ex Hacienda Canadá, P.O. Box 66050 General Escobedo, Nuevo León, México

Gerardo Padilla
School of Medicine, Agricultural Sciences Campus UANL, Francisco Villa S/N, Ex Hacienda Canadá, P.O. Box 66050 General Escobedo, Nuevo León, México

Rogelio A.Ledezma
Department of Veterinary Medicine, Laboratory of Reproduction, Agricultural Sciences Campus UANL, Francisco Villa S/N, Ex Hacienda Canadá, P.O. Box 66050 General Escobedo, Nuevo León, México

S. Kumar
Regional Research Station, Bathinda Punjab Agricultural University, Ludhiana, Punjab, India-141004, India

J. S. Kular
Department of Entomology, Punjab Agricultural University, Ludhiana, Punjab, India-141004, India

M. S. Mahal
Department of Entomology, Punjab Agricultural University, Ludhiana, Punjab, India-141004, India

A. K. Dhawan
Department of Entomology, Punjab Agricultural University, Ludhiana, Punjab, India-141004, India

Krzysztof Pawlak
Department of Poultry and Fur Animal Breeding and Animal Hygiene, University of Agriculture in Krakow, Aleja Adama Mickiewicza 21, Kraków, Poland

Małgorzta Dżugan
Department of Food Chemistry and Toxicology, University of Rzeszów, Aleja Tadeusza Rejtana 16, Rzeszów, Poland

Dorota Wojtysiak
Department of Animal Reproduction and Anatomy, University of Agriculture in Kraków, Aleja Adama Mickiewicza 21, Kraków, Poland

Marcin Lis
Department of Poultry and Fur Animal Breeding and Animal Hygiene, University of Agriculture in Krakow, Aleja Adama Mickiewicza 21, Kraków, Poland

Jerzy Niedziółka
Department of Poultry and Fur Animal Breeding and Animal Hygiene, University of Agriculture in Krakow, Aleja Adama Mickiewicza 21, Kraków, Poland

M. Y. Al-Saiady
ARASCO R & D Department, P. O. Box 53845, Riyadh 11593, Kingdom of Saudi Arabia

Mogawer
ARASCO R & D Department, P. O. Box 53845, Riyadh 11593, Kingdom of Saudi Arabia

S. E. Al-Mutairi
Camel Breeding, Range Protection and Improvement Center in Al-Jouf area, Kingdom of Saudi Arabia

M. Bengoumi
Camel Breeding, Range Protection and Improvement Center in Al-Jouf area, Kingdom of Saudi Arabia

A. Musaad
Camel Breeding, Range Protection and Improvement Center in Al-Jouf area, Kingdom of Saudi Arabia

A. Gar-Elnaby
Animal Production Department, College of Food and Agricultural Sciences, P. O. Box 2460, King Saud University, Riyadh 11451, Kingdom of Saudi Arabia

B. Faye
Camel Breeding, Range Protection and Improvement Center in Al-Jouf area, Kingdom of Saudi Arabia

A. A. Kamel
Beekeeping Research Department, Plant Protection Research Institute (PPRI), Agricultural Research Center, Giza, Egypt

A. A. Moustafa
Department of Biochemistry, Faculty of Agriculture, Cairo University, Egypt

E. A. Nafea
Beekeeping Research Department, Plant Protection Research Institute (PPRI), Agricultural Research Center, , Giza, Egypt

E. N. Nwachukwu
Department of Animal Breeding and Physiology, Okpara University of Agriculture, Umudike, Umuahia, Abia State, Nigeria

K. U. Amaefule
Department of Animal Nutrition and Forage Sciences, Okpara University of Agriculture, Umudike, Umuahia, Abia State, Nigeria

F. O. Ahamefule
Department of Animal Production and Management, Okpara University of Agriculture, Umudike, Umuahia, Abia State, Nigeria

S. C. Akomas
College of Veterinary Medicine, Okpara University of Agriculture, Umudike, Umuahia, Abia State, Nigeria

T. U. Nwabueze
Department of Food Processing and Analysis, Okpara University of Agriculture, Umudike, Umuahia, Abia State, Nigeria

U. A. U Onyebinama
College of Agricultural Economics and Rural Sociology,
Michael Okpara University of Agriculture, Umudike,
Umuahia, Abia State, Nigeria

O. O. Ekumankama
College of Agricultural Economics and Rural Sociology,
Michael Okpara University of Agriculture, Umudike,
Umuahia, Abia State, Nigeria

Ivo Ngundu WOOGENG
Department of Zoology and Animal Physiology,
University of Buea, P. O. Box 63 Buea, Cameroon
Department of Genetics, University of the Free State, P.
O. Box 339, Bloemfontein, 9300, South Africa

J. Paul GROBLER
Department of Genetics, University of the Free State, P.
O. Box 339, Bloemfontein, 9300, South Africa

Kingsely Agbor ETCHU
Institute of Agricultural Research for Development
(IRAD) Ekona, P. M. B. 25 Buea, Cameroon

Kenneth Jacob N. NDAMUKONG
Department of Zoology and Animal Physiology,
University of Buea, P. O. Box 63 Buea, Cameroon

Oluwatosin Jesuyon
Department of Animal Science, University of Ibadan, Oyo
State, Nigeria

Adebowale E. Salako
Department of Animal Science, University of Ibadan, Oyo
State, Nigeria

Matiwos Habte
Departement of Animal and Range Sciences, Dilla
University, P.O. Box 419, Dilla Ethiopia

Negassi Ameha
School of Animal and Range Sciences, Haramaya
University, P. O. Box 138, Dire Dawa, Ethiopia

Solomon Demeke
Departement of Animal and Range Sciences, Jimma
University College of Agriculture, P.O. Box 307, Jimma
Ethiopia

K. DEGHNOUCHE
Department of Agronomy, Mohamed Kheider University,
Biskra 07000, Algeria

M. TLIDJANE
Department of Veterinary, Hadj Lakhder University,
Batna 05000, Algeria

T. MEZIANE
Department of Veterinary, Hadj Lakhder University,
Batna 05000, Algeria

A. TOUABTI
Biochemistry Laboratory, CHU Setif 19000, Algeria

D. Luseba
Department of Animal Sciences, Tshwane University of
Technology, Private Bag X680, Pretoria 0001, Republic of
South Africa

Amanzhol Kusaynovich Dnekeshev
Faculty of Veterinary Medicine and Biotechnology,
West-Kazakhstan Agrarian Technical University Named
after Zhangir Khan, West Kazakhstan Region, Uralsk,
Kazakhstan

Abzal Kenesovich Kereyev
Faculty of Veterinary Medicine and Biotechnology,
West-Kazakhstan Agrarian Technical University Named
after Zhangir Khan, West Kazakhstan Region, Uralsk,
Kazakhstan

Erika Cosendey Toledo de Mello Peixoto
Universidade Estadual do Norte do Paraná (UENP/
Bandeirantes), BR-369, Km 54, Villa Maria, Caixa Postal
261, CEP 86360-000, Bandeirantes, Paraná, Brasil

Andressa de Andrade
Zootecnista, Rua Acapulco, n 293, Bairro Vila Industrial,
CDEP – 85904-120, Toledo, Paraná, Brazil

Fillipi Valadares
Zootecnista, Rua Acapulco, n 293, Bairro Vila Industrial,
CDEP – 85904-120, Toledo, Paraná, Brazil

Luciana Pereira da Silva
Faculdade de Ciências e Letras de Assis, Universidade
Estadual Paulista "Júlio de Mesquita Filho",Laboratório
de Fisiologia Vegetal e Fitoterápicos, Avenida Dom
Antônio, 2100, CEP: 19806-900, Assis-SP, Brasil

Regildo Márcio Gonçalves da Silva
Faculdade de Ciências e Letras de Assis, Universidade
Estadual Paulista "Júlio de Mesquita Filho", Laboratório
de Fisiologia Vegetal e Fitoterápicos, Avenida Dom
Antônio, 2100, CEP: 19806-900, Assis-SP, Brasil

S. Basavarajappa
Apidology Laboratory, Department of Studies in Zoology,
University of Mysore, Mysore-570 006, India

K. S. Raghunandan
Apidology Laboratory, Department of Studies in Zoology,
University of Mysore, Mysore-570 006, India

Ogundu Uduak E
Department of Animal Science and Technology, Federal University of Technology, Owerri, Imo State, Nigeria

V. M. O. Okoro
Department of Animal Science and Technology, Federal University of Technology, Owerri, Imo State, Nigeria

G. U. Okeke
Department of Animal Science and Technology, Federal University of Technology, Owerri, Imo State, Nigeria

N. Durugo
Department of Animal Science and Technology, Federal University of Technology, Owerri, Imo State, Nigeria

G. A. C. Mbaebie
Department of Animal Science and Technology, Federal University of Technology, Owerri, Imo State, Nigeria

C. I. Ezebuike
Department of Animal Science and Technology, Federal University of Technology, Owerri, Imo State, Nigeria

Anselm Ego ONYIMONYI
Department of Animal Science, University of Nigeria, Nsukka, Enugu State, Nigeria

Ndubuisi Samuel MACHEBE
Department of Animal Science, University of Nigeria, Nsukka, Enugu State, Nigeria

Jervas UGWUOKE
Department of Animal Science, University of Nigeria, Nsukka, Enugu State, Nigeria

Endale Mekuria
College of Veterinary Medicine, Haramaya University, P. O. Box 05 Hawassa, Ethiopia

Shihun Shimelis
College of Veterinary Medicine, Haramaya University, P. O. Box 05 Hawassa, Ethiopia

Jemere Bekele
School of Veterinary Medicine, Hawassa University, P. O. Box 05 Hawassa, Ethiopia

Desie Sheferaw
School of Veterinary Medicine, Hawassa University, P. O. Box 05 Hawassa, Ethiopia

D. K. Giri
Department of Veterinary Pathology, College of Veterinary Science and Animal Husbandry, Anjora, Durg, Chhattisgarh, India

R. C. Ghosh
Department of Veterinary Pathology, College of Veterinary Science and Animal Husbandry, Anjora, Durg, Chhattisgarh, India

M. Mondal
Department of Veterinary Pathology, College of Veterinary Science and Animal Husbandry, Anjora, Durg, Chhattisgarh, India

Govina Dewangan
Department of Veterinary Pathology, College of Veterinary Science and Animal Husbandry, Anjora, Durg, Chhattisgarh, India

Deepak Kumar Kashyap
Department of Veterinary Pathology, College of Veterinary Science and Animal Husbandry, Anjora, Durg, Chhattisgarh, India

A. H. Abu
Department of Veterinary Physiology, Pharmacology and Biochemistry, College of Veterinary Medicine, University of Agriculture, P.M.B. 2373, Makurdi, Benue State, Nigeria

L. I. Mhomga
Department of Animal Health and Production, College of Veterinary Medicine, University of Agriculture, Makurdi, Benue State, Nigeria

E. I. Akogwu
College of Veterinary Medicine, University of Agriculture, Makurdi, Benue State, Nigeria

Manishkumar R. Dabhi
SRA, I.F.T.C., Directorate of Extension Education, Anand Agricultural University, Anand-388 110 (Gujarat), India

Dhirubhai M. Korat
Anand Agricultural University, Anand-388 110 (Gujarat), India

Piyushbhai R. Vaishnav
Department of Agricultural Statistics, BACA, AAU, Anand-388 110 (Gujarat), India

Mammo Mengesha
Ethiopian Institute of Agricultural Research, Debre Zeit Agricultural Research Center; P. O. Box 32, Debre Zeit, Ethiopia

Abdi Husein
Wollega University, School of Veterinary Medicine, P. O. Box 395, Nekemte, Ethiopia
Jijiga Regional Veterinary Laboratory, Jijiga, Ethiopia

Berihu Haftu
Wollega University, School of Veterinary Medicine, P. O. Box 395, Nekemte, Ethiopia

Addisalem Hunde
Wollega University, School of Veterinary Medicine, P. O. Box 395, Nekemte, Ethiopia

Asamenew Tesfaye
Wollega University, School of Veterinary Medicine, P. O. Box 395, Nekemte, Ethiopia

Kefyalew Alemayehu
Department of Animal Production and Technology, Faculty of Agriculture and Environmental Sciences, Bahir Dar, University P. O. Box 21 45, Bahir Dar, Ethiopia

B. C. Majekodunmi
Physiology Unit, Department of Animal Science, University of Ibadan, Ibadan, Oyo State, Nigeria

O. A. Sokunbi
Physiology Unit, Department of Animal Science, University of Ibadan, Ibadan, Oyo State, Nigeria

O. A. Ogunwole
Physiology Unit, Department of Animal Science, University of Ibadan, Ibadan, Oyo State, Nigeria

O. A. Adebiyi
Physiology Unit, Department of Animal Science, University of Ibadan, Ibadan, Oyo State, Nigeria

G. Mohammed
Department of Animal Science, University of Maiduguri, Maiduguri, Nigeria

S. B. Adamu
Department of Animal Science, University of Maiduguri, Maiduguri, Nigeria

J. U. Igwebuike
Department of Animal Science, University of Maiduguri, Maiduguri, Nigeria

N. K. Alade
Department of Animal Science, University of Maiduguri, Maiduguri, Nigeria

L. G. Asheikh
Department of Animal Science, University of Maiduguri, Maiduguri, Nigeria

Mohmmad Aarif Khan
Sher-E-Kashmir University of Agricultural Sciences and Technology of Kashmir, India

D. M. Makhdoomi
Sher-E-Kashmir University of Agricultural Sciences and Technology of Shuhama, India

Mohsin A Gazi
Sher-E-Kashmir University of Agricultural Sciences and Technology of Shuhama, India

G. N. Sheikh
Sher-E-Kashmir University of Agricultural Sciences and Technology of Shuhama, India

S. H. Dar
Sher-E-Kashmir University of Agricultural Sciences and Technology of Shuhama, India

Sibel SOYCAN-ÖNENÇ
Department of Animal Science, Namık Kemal University, Faculty of Agriculture, Tekirdağ-Turkey